Selected Titles in This Series

79 **Joseph A. Cima and William T. Ross,** The backward shift on the Hardy space, 2000
78 **Boris A. Kupershmidt,** KP or mKP: Noncommutative mathematics of Lagrangian, Hamiltonian, and integrable systems, 2000
77 **Fumio Hiai and Dénes Petz,** The semicircle law, free random variables and entropy, 2000
76 **Frederick P. Gardiner and Nikola Lakic,** Quasiconformal Teichmüller theory, 2000
75 **Greg Hjorth,** Classification and orbit equivalence relations, 2000
74 **Daniel W. Stroock,** An introduction to the analysis of paths on a Riemannian manifold, 2000
73 **John Locker,** Spectral theory of non-self-adjoint two-point differential operators, 2000
72 **Gerald Teschl,** Jacobi operators and completely integrable nonlinear lattices, 1999
71 **Lajos Pukánszky,** Characters of connected Lie groups, 1999
70 **Carmen Chicone and Yuri Latushkin,** Evolution semigroups in dynamical systems and differential equations, 1999
69 **C. T. C. Wall (A. A. Ranicki, Editor),** Surgery on compact manifolds, second edition, 1999
68 **David A. Cox and Sheldon Katz,** Mirror symmetry and algebraic geometry, 1999
67 **A. Borel and N. Wallach,** Continuous cohomology, discrete subgroups, and representations of reductive groups, second edition, 2000
66 **Yu. Ilyashenko and Weigu Li,** Nonlocal bifurcations, 1999
65 **Carl Faith,** Rings and things and a fine array of twentieth century associative algebra, 1999
64 **Rene A. Carmona and Boris Rozovskii, Editors,** Stochastic partial differential equations: Six perspectives, 1999
63 **Mark Hovey,** Model categories, 1999
62 **Vladimir I. Bogachev,** Gaussian measures, 1998
61 **W. Norrie Everitt and Lawrence Markus,** Boundary value problems and symplectic algebra for ordinary differential and quasi-differential operators, 1999
60 **Iain Raeburn and Dana P. Williams,** Morita equivalence and continuous-trace C^*-algebras, 1998
59 **Paul Howard and Jean E. Rubin,** Consequences of the axiom of choice, 1998
58 **Pavel I. Etingof, Igor B. Frenkel, and Alexander A. Kirillov, Jr.,** Lectures on representation theory and Knizhnik-Zamolodchikov equations, 1998
57 **Marc Levine,** Mixed motives, 1998
56 **Leonid I. Korogodski and Yan S. Soibelman,** Algebras of functions on quantum groups: Part I, 1998
55 **J. Scott Carter and Masahico Saito,** Knotted surfaces and their diagrams, 1998
54 **Casper Goffman, Togo Nishiura, and Daniel Waterman,** Homeomorphisms in analysis, 1997
53 **Andreas Kriegl and Peter W. Michor,** The convenient setting of global analysis, 1997
52 **V. A. Kozlov, V. G. Maz'ya, and J. Rossmann,** Elliptic boundary value problems in domains with point singularities, 1997
51 **Jan Malý and William P. Ziemer,** Fine regularity of solutions of elliptic partial differential equations, 1997
50 **Jon Aaronson,** An introduction to infinite ergodic theory, 1997
49 **R. E. Showalter,** Monotone operators in Banach space and nonlinear partial differential equations, 1997
48 **Paul-Jean Cahen and Jean-Luc Chabert,** Integer-valued polynomials, 1997

(*Continued in the back of this publication*)

The Backward Shift on the Hardy Space

Mathematical
Surveys
and
Monographs

Volume 79

The Backward Shift on the Hardy Space

Joseph A. Cima
William T. Ross

American Mathematical Society

Editorial Board

Georgia Benkart
Peter Landweber

Michael Loss
Tudor Ratiu, Chair

2000 *Mathematics Subject Classification.* Primary 47B38;
Secondary 46E10, 46E15.

ABSTRACT. This book is a thorough treatment of the classification of the backward shift invariant subspaces of the well-known Hardy spaces H^p. For $1 < p < \infty$, the characterization was done by Douglas, Shapiro, and Shields. The case $0 < p \leq 1$ was done by A. B. Aleksandrov in a paper which was not translated into English and as a result is not readily available in the West. This book puts all of these results, along with the necessary background material, under one roof.

Library of Congress Cataloging-in-Publication Data

Cima, Joseph A. , 1933–
    The backward shift on the Hardy space / Joseph A. Cima, William T. Ross.
       p. cm. (Mathematical surveys and monographs, ISSN 0076-5376; v. 79)
    Includes bibliographical references and index.
    ISBN 0-8218-2083-4 (alk. paper)
    1. Hardy spaces. I. Ross, William T., 1964– . II. Title. III. Series: Mathematical surveys and monographs; no. 79.
QA331.C53 2000
515′.94–dc21
                                                                                                     00-028032
                                                                                                         CIP

    Copying and reprinting. Individual readers of this publication, and nonprofit libraries acting for them, are permitted to make fair use of the material, such as to copy a chapter for use in teaching or research. Permission is granted to quote brief passages from this publication in reviews, provided the customary acknowledgment of the source is given.
    Republication, systematic copying, or multiple reproduction of any material in this publication is permitted only under license from the American Mathematical Society. Requests for such permission should be addressed to the Assistant to the Publisher, American Mathematical Society, P. O. Box 6248, Providence, Rhode Island 02940-6248. Requests can also be made by e-mail to reprint-permission@ams.org.

        © 2000 by the American Mathematical Society. All rights reserved.
            The American Mathematical Society retains all rights
           except those granted to the United States Government.
                Printed in the United States of America.

      ∞ The paper used in this book is acid-free and falls within the guidelines
          established to ensure permanence and durability.
        Visit the AMS home page at URL: http://www.ams.org/

        10 9 8 7 6 5 4 3 2 1   05 04 03 02 01 00

Contents

Preface   ix

Numbering and notation   xi

Chapter 1.   Overview   1

Chapter 2.   Classical boundary value results   9
    2.1.   Limits   9
    2.2.   Pseudocontinuations   13

Chapter 3.   The Hardy space of the disk   17
    3.1.   Introduction   17
    3.2. H^p and boundary values   17
    3.3.   Fourier analysis and H^p theory   21
    3.4.   The Cauchy transform   23
    3.5.   Duality   28
    3.6.   The Nevanlinna class   39

Chapter 4.   The Hardy spaces of the upper-half plane   45
    4.1.   Motivation   45
    4.2.   Basic definitions   47
    4.3.   Poisson and conjugate Poisson integrals   49
    4.4.   Maximal functions   52
    4.5.   The Hilbert transform   54
    4.6.   Some examples   55
    4.7.   The harmonic Hardy space   60
    4.8.   Distributions   61
    4.9.   The atomic decomposition   72
    4.10.   Distributions and $\mathcal{H}^p$ 75
    4.11.   The space $\mathcal{H}^p(\mathbb{C}\backslash\mathbb{R})$ 76

Chapter 5.   The backward shift on H^p for $p \in [1, \infty)$ 81
    5.1.   The case $p > 1$ 81
    5.2.   The first and most straightforward proof   82
    5.3.   The second proof - using Fatou's jump theorem   85
    5.4.   Application: Bergman spaces   87
    5.5.   Application: spectral properties   94
    5.6.   The third proof - using the Nevanlinna theory   97
    5.7.   Application: $VMOA$, $BMOA$, and $L^1/\overline{H^1_0}$ 99
    5.8.   The case $p = 1$ 101
    5.9.   Cyclic vectors   105

5.10.	Duality	109
5.11.	The commutant	109
5.12.	Compactness of the inclusion operator	111

Chapter 6. The backward shift on H^p for $p \in (0,1)$ — 115
- 6.1. Introduction — 115
- 6.2. The parameters — 120
- 6.3. A reduction — 133
- 6.4. Rational approximation — 136
- 6.5. Spectral properties — 185
- 6.6. Cyclic vectors — 186
- 6.7. Duality — 187
- 6.8. The commutant — 188

Bibliography — 191

Index — 195

Preface

Shift operators on Hilbert spaces of analytic functions play an important role in the study of bounded linear operators on Hilbert spaces since they often serve as "models" for various classes of linear operators. For example, "parts" of direct sums of the backward shift operator on the classical Hardy space H^2 model certain types of contraction operators and potentially have connections to understanding the invariant subspaces of a general linear operator.

In this book, we do not want to give a general treatment of the backward shift on H^2 and its connections to problems in operator theory. This has been done quite thoroughly by Nikolskiĭ in his book [65]. Instead, we wish to work in the Banach (and F-space) setting of H^p ($0 < p < \infty$) where we will focus primarily on characterizing the backward shift invariant subspaces of H^p. When $p \in (1, \infty)$, this characterization problem was solved by R. Douglas, H. S. Shapiro, and A. Shields in a well known paper [29] which employed the concept of a 'pseudocontinuation' developed earlier by Shapiro [84]. When $p \in (0,1)$, the characterization problem is more difficult, due to some topological differences between the two settings $p \in [1, \infty)$ and $p \in (0,1)$, and was solved in a paper of A. B. Aleksandrov [3] which was never translated from its original Russian and hence is not readily available in the West. The Aleksandrov paper is also quite complicated and makes use of the distribution theory and Coifman's atomic decomposition for the Hardy spaces of the upper half plane, a topic we feel is not always at the fingertips of those schooled, as we were, in classical function theory and operator theory. It is for these reasons that we gather up these results, along with the necessary background material, and put them all under one roof.

In developing the necessary background results, we do not wish to reproduce the material in the books of Duren [31] or Garnett [39] (for a general treatment of Hardy spaces) or Stein [95] (for a detailed treatment of harmonic analysis and real variable H^p theory). Instead, we will only review this material and refer the interested reader to the appropriate places in these texts for the proofs. The reader is expected to have a reasonable background in functional analysis and function theory (including the basics of H^p theory), but might want to have Rudin's functional analysis book [78], Duren's H^p book [31], and Stein's harmonic analysis book [95] at the ready while reading this book. We will try to develop the more specialized topics as we need them.

The authors wish to thank several people who helped us along the way. First, we thank A. B. Aleksandrov, who, through many e-mails, helped us understand the more difficult parts of his papers. Secondly, we thank Alec Matheson and Don Sarason, who read a draft of this book and provided us with useful suggestions and corrections. Thirdly, we thank Olga Troyanskaya, who translated the Aleksandrov paper [3] from the original Russian. Finally, the second author wishes to thank the mathematics department of the University of North Carolina, Chapel Hill, for the comfortable setting for the semester in which he finally got to work with the first author face to face (and not over the Internet) where they assembled the final version of this book.

JAC AND WTR

Numbering and notation

1. *List of symbols:* The list of symbols is incorporated as part of the index and can be found at the end of the book.
2. *Definitions:* When defining functions, sets, operators, etc., we will often use the notation $A := xxx$. By this we mean A 'is defined to be' xxx.
3. *Estimates:* We use the notation $A \asymp B$ to mean there are (positive) constants c_1 and c_2 such that
$$c_1 A \leq B \leq c_2 A.$$
As is traditional in analysis, the constants c_1 and c_2 can change from one line to the next.
4. *Closures vs. conjugates:* For a set $A \subset \mathbb{C}$, we use $\overline{A}$ to denote the complex conjugates of the points in A. For a set U in some topological vector space, we use U^- to denote the closure of U.
5. *Manifold vs. subspace:* If U (as above) is closed under the vector space operations, we will say that U is a 'linear manifold'. A 'subspace' will be a *closed* linear manifold.
6. *Numbering:* Numbering is done by chapter and section, and *all* equations, theorems, propositions, and such are numbered consecutively.
7. *Errors:* Though we have made every attempt to avoid any errors, we realize that we are probably not perfect. We will maintain a list of corrections (mathematical, attributions, etc.) which the reader can find off Ross' web page at

www.richmond.edu/~wross

Please feel free to contact us with your comments.

Joseph A. Cima
Department of Mathematics
University of North Carolina, Chapel Hill
Chapel Hill, North Carolina 27599
cima@math.unc.edu

William T. Ross
Department of Mathematics and Computer Science
University of Richmond
Richmond, Virginia 23173
wross@richmond.edu

CHAPTER 1

Overview

In this monograph, we will discuss the invariant subspaces of the backward shift operator on the Hardy space. Here, for $p \in (0, \infty)$, the Hardy space H^p is the set of analytic functions f on the open unit disk $\mathbb{D} = \{|z| < 1\}$ for which the quantity

$$(1.0.1) \qquad \|f\|_p := \sup_{0 < r < 1} \left\{ \int_0^{2\pi} |f(re^{i\theta})|^p \, \frac{d\theta}{2\pi} \right\}^{1/p}$$

is finite. The backward shift operator B on H^p is the continuous linear operator defined for a function $f = \sum_{n=0}^{\infty} a_n z^n \in H^p$ by

$$Bf := \frac{f - f(0)}{z} = a_1 + a_2 z + a_3 z^2 + \cdots .$$

By "invariant subspace" for the operator B, we mean a closed linear manifold (i.e., a subspace) $\mathcal{M}$ of H^p for which $B\mathcal{M} \subset \mathcal{M}$. Although there are various aspects of the backward shift on the Hardy spaces that are certainly worthy of attention, this book will focus on only one of these topics:

the characterization of the backward shift invariant subspaces.

For $p \in (1, \infty)$ the B-invariant subspaces of H^p were described in a well known paper of R. Douglas, H. S. Shapiro, and A. Shields [29]; while for $p \in (0, 1]$, they were described by A. B. Aleksandrov [3] in a remarkable paper which was never translated into English and as a result, is not readily available in the West. The main purpose of this book is to give a thorough treatment of all this material, replete with the appropriate background material and full details of the proofs.

Although the backward shift operator is interesting to study in its own right, it has important connections to the general study of bounded linear operators on Hilbert spaces. Work of Rota [76], de Branges and Rovnyak [26], and Foiaş [36] show that the restriction of a direct sum of backward shifts on H^2 to an invariant subspace is often the "model" for important classes of bounded linear operators. For example, Rota showed that every strict contraction (a bounded operator T from a Hilbert space to itself with operator norm $\|T\|$ strictly less than one) is similar to a direct sum of backward shift operators on H^2 restricted to an invariant subspace (of the direct sum). One often uses the phrase T is similar to "part of a direct sum of backward shifts". de Branges and Rovnyak, and Foiaş brought this model theory to fruition and proved that if T satisfies the two conditions

$$\|T\| \leq 1 \text{ and } \|T^n x\| \to 0 \text{ as } n \to \infty \; \forall x,$$

then T is unitarily equivalent to part of a direct sum of backward shifts. We refer the reader to [65] [75] [98] for a more detailed treatment of model theory and its connections to backward shifts.

Though the invariant subspaces of B on H^2 are certainly not as complicated as the invariant subspaces of a direct sum of backward shifts, they do form interesting classes of functions whose description involves the boundary behavior of analytic functions near (and across) the unit circle $\mathbb{T} = \partial \mathbb{D}$. To describe these results, we recall some basic facts about H^p functions and refer the reader to Chapter 3 where they will be discussed in greater detail.

It can be shown from eq.(1.0.1) that for $p \in [1, \infty)$, the quantity $\|f\|_p$ defines a norm on H^p which makes it a Banach space; while for $p \in (0,1)$, the metric

$$\rho(f,g) := \|f - g\|_p^p$$

makes H^p an F-space. It is also known that functions belonging to H^p are reasonably well-behaved near $\mathbb{T}$ and in fact, for any $f \in H^p$, the radial limit

$$f(\zeta) := \lim_{r \to 1^-} f(r\zeta)$$

exists for almost every $\zeta \in \mathbb{T}$. One of the most important properties of this boundary function is that $f(\zeta)$ belongs to $L^p = L^p(\mathbb{T}, dm)$ (where $dm := d\theta/2\pi$ is normalized Lebesgue measure on the unit circle $\mathbb{T}$) and the quantity in eq.(1.0.1) is equal to the L^p-norm of the boundary function; that is,

$$\|f\|_p := \sup_{0 < r < 1} \left\{ \int_\mathbb{T} |f(r\zeta)|^p \, dm(\zeta) \right\}^{1/p} = \left\{ \int_\mathbb{T} |f(\zeta)|^p \, dm(\zeta) \right\}^{1/p}.$$

The linear mapping which takes a function $f \in H^p$ to its boundary function $f(\zeta)$ in L^p is an isometric mapping from H^p onto a closed subspace of L^p which we shall call $H^p(\mathbb{T})$. For $p \in [1, \infty)$, one can identify $H^p(\mathbb{T})$ by examining the Fourier coefficients

$$\hat{f}(n) := \int_\mathbb{T} f(\zeta) \overline{\zeta}^n dm(\zeta), \; n \in \mathbb{Z},$$

of f. Indeed, a theorem of F. and M. Riesz says that for $f \in L^p$,

$$f \in H^p(\mathbb{T}) \Leftrightarrow \hat{f}(n) = 0, \; \forall \, n < 0.$$

Moreover, the power series coefficients of $f \in H^p$ are the same as the Fourier series coefficients of the boundary function $f(\zeta)$ in $H^p(\mathbb{T})$. The behavior of this boundary function is not only useful in describing the topology of H^p, as seen above, but as we shall see shortly, is also instrumental in describing the backward shift invariant subspaces of H^p.

The primary tool used in describing the B-invariant subspaces of H^p, at least for $p \in [1, \infty)$, is the duality theory of H^p spaces. For $p \in (1, \infty)$, the norm dual, $(H^p)^*$, of H^p can be identified with H^q, where $1/p + 1/q = 1$, via the pairing

(1.0.2) $$<f, g> := \int_\mathbb{T} f(\zeta) \overline{g}(\zeta) dm(\zeta), \; f \in H^p, g \in H^q.$$

Furthermore, if $S : H^q \to H^q$ is the forward shift operator $(Sg)(z) := zg(z)$, then it is easy to check that

$$<Bf, g> = <f, Sg>, \; \forall \, f \in H^p, g \in H^q$$

and so for a subspace $\mathcal{M} \subset H^p$,

$$B\mathcal{M} \subset \mathcal{M} \Leftrightarrow S\mathcal{M}^\perp \subset \mathcal{M}^\perp,$$

where

$$\mathcal{M}^\perp := \{ g \in (H^p)^* : <f, g> = 0 \; \forall \, f \in \mathcal{M} \}$$

is the annihilator of $\mathcal{M}$. By a celebrated result of Beurling [**14**], every (non-trivial) forward shift invariant subspace of H^q is of the form IH^q, where I is a bounded analytic function on $\mathbb{D}$ whose boundary values are unimodular almost everywhere on $\mathbb{T}$ ($|I(\zeta)| = 1$ almost everywhere on $\mathbb{T}$). Such a function is called an "inner function". By the Hahn-Banach separation theorem,

$$\mathcal{M} = (\mathcal{M}^\perp)^\perp = (IH^q)^\perp = \{\, f \in H^p : \,<f, Ih> = 0 \,\, \forall\, h \in H^q\,\}.$$

Using the F. and M. Riesz theorem and equating functions in $\mathcal{M}$ with their boundary functions on $\mathbb{T}$, one can show that

$$\mathcal{M} = H^p(\mathbb{T}) \cap I\overline{H_0^p(\mathbb{T})},$$

where $H_0^p = \{f \in H^p : f(0) = 0\}$ and $H_0^p(\mathbb{T})$ are the boundary functions for H_0^p. For notational convenience, we write $\mathcal{M} = H^p \cap I\overline{H_0^p}$ and understand this as a space of functions on $\mathbb{T}$.

Douglas, Shapiro, and Shields, were able to further describe the functions in $(IH^q)^\perp = H^p \cap I\overline{H_0^p}$ by means of a "continuation" across $\mathbb{T}$. To motivate this idea, notice that

$$f \in (IH^q)^\perp \Rightarrow\, <f, Iz^n> = 0 \,\, \forall\, n \in \mathbb{N} \cup \{0\}.$$

However,

$$<f, Iz^n> = \int_\mathbb{T} f(\zeta)\overline{I}(\zeta)\overline{\zeta}^n dm(\zeta) = \widehat{f\overline{I}}(n)$$

which is the non-negative Fourier coefficient of the L^p function $f\overline{I}$. This means that $f\overline{I}$ has Fourier expansion

(1.0.3) $$f\overline{I} \sim b_1\overline{\zeta} + b_2\overline{\zeta}^2 + \cdots.$$

If one defines the function g on the extended exterior disk $\mathbb{D}_e := \mathbb{C}_\infty \setminus \mathbb{D}^-$ by

$$g(z) := \frac{b_1}{z} + \frac{b_2}{z^2} + \cdots, \quad z \in \mathbb{D}_e,$$

one shows (see Chapter 5) that

(1.0.4) $$\sup_{r>1} \int_\mathbb{T} |g(r\zeta)|^p \, dm(\zeta)/r$$

is finite. Analytic functions g on $\mathbb{D}_e$ for which eq.(1.0.4) is finite are said to belong to $H^p(\mathbb{D}_e)$, the Hardy space of the extended exterior disk. One can also show, as suggested by eq.(1.0.3), that the boundary values of g are almost always the same as the boundary values of the meromorphic function f/I, that is to say

$$\lim_{r \to 1^+} g(r\zeta) = \lim_{r \to 1^-} \frac{f(r\zeta)}{I(r\zeta)}$$

almost everywhere. Thus if $f \in (IH^q)^\perp$, then the boundary values of the meromorphic function f/I are the same, almost everywhere, as the boundary values of some $H^p(\mathbb{D}_e)$ function g which vanishes at infinity. Using a term of H.S. Shapiro [**84**], we say that g is a pseudocontinuation of f/I. A more formal definition will be given in Chapter 2. Douglas, Shapiro, and Shields [**29**] were able to show the converse to the above statement which is the first main result of this monograph.

THEOREM 1.0.5 (Douglas, Shapiro, Shields [**29**]). *Let $p \in (1, \infty)$ and $\mathcal{M}$ be a non-trivial B-invariant subspace of H^p. Then there is an inner function I such that $\mathcal{M} = H^p \cap I\overline{H_0^p}$. Moreover, $f \in \mathcal{M}$ if and only if f/I has a "pseudocontinuation" to a function belonging to $H^p(\mathbb{D}_e)$ which vanishes at infinity.*

In fact, even more can be said about functions belonging to $(IH^q)^\perp$. If
$$\sigma(I) := \{\, \lambda \in \mathbb{D}^- : \liminf_{z \to \lambda} |I(z)| = 0 \,\}$$
denotes the "lim inf zero set" for I (also called the "spectrum" of I), then every $f \in (IH^q)^\perp$ has an analytic continuation to
$$\mathbb{C}_\infty \setminus \{\, 1/\overline{z} : z \in \sigma(I) \,\}.$$

For $p = 1$, the description of the B-invariant subspaces is the same as before, namely $\mathcal{M} = H^1 \cap I\overline{H_0^1}$, except that the proof needs to be different due to some technicalities which arise from a slightly different dual pairing. For instance, the norm dual of H^1 can be identified not with H^∞, but with the somewhat larger class of functions $BMOA$, the analytic functions of bounded mean oscillation (see Chapter 3 for a definition) via the pairing

$$(1.0.6) \qquad \lim_{r \to 1} \int_{\mathbb{T}} f(r\zeta)\overline{g}(\zeta)dm(\zeta), \quad f \in H^1, \, g \in BMOA.$$

The limit is needed here since $f\overline{g}$ is not always integrable. Unfortunately, this identification makes the Fourier analysis used in the $p > 1$ case unworkable. Despite these difficulties, Aleksandrov is able to identify the B-invariant subspaces of H^1 and we will give a thorough treatment of his proof (which is hinted at in [**3**]) in Chapter 5 of this monograph. His technique is quite delicate and involves representing the duality in eq.(1.0.6) in a more usable form by using the truncation operator on functions of bounded mean oscillation.

Also included in our discussion of the backward shift on H^p ($1 \le p < \infty$) will be a closer look at some of the functional analysis properties of a typical invariant subspace $H^p \cap I\overline{H_0^p}$ as well as the spectral properties of the operator $B | H^p \cap I\overline{H_0^p}$. It is well known that
$$\sigma(B) = \mathbb{D}^-, \quad \sigma_p(B) = \mathbb{D}.$$
The description of $\sigma(B|H^p \cap I\overline{H_0^p})$, due to Moeller [**63**], is the following.

THEOREM 1.0.7 (Moeller [**63**]). *Let $p \in [1, \infty)$, I be an inner function, and let $\mathcal{M} := H^p \cap I\overline{H_0^p}$. Then*
$$\sigma(B|\mathcal{M}) = \overline{\sigma(I)}.[1]$$
Moreover

1. $\sigma_p(B|\mathcal{M}) \cap \mathbb{D} = \overline{\sigma(I)} \cap \mathbb{D} = \{\lambda \in \mathbb{D} : (1 - \lambda z)^{-1} \in \mathcal{M}\}$.
2. $\zeta \notin \sigma(B|\mathcal{M}) \cap \mathbb{T}$ *if and only if there is an open neighborhood U of the point ζ such that every $f \in \mathcal{M}$ has an analytic continuation to U.*

For $p \in (1, \infty)$, the dual of H^p can be identified with H^q ($1/p + 1/q = 1$) via the pairing eq.(1.0.2); while for $p = 1$, the dual of H^1 can be identified with $BMOA$ via the pairing eq.(1.0.6). The following theorem identifies the dual of $H^p \cap I\overline{H_0^p}$.

---

[1]As mentioned in the section on "numbering and notation", $\overline{\sigma(I)} = \{\overline{\lambda} : \lambda \in \sigma(I)\}$ and *not* the closure of $\sigma(I)$. The closure of a set A is denoted by A^-.

THEOREM 1.0.8. *If $p \in (1, \infty)$ and I is an inner function, then the dual of $H^p \cap I\overline{H_0^p}$ can be identified with $H^q \cap I\overline{H_0^q}$ via the pairing*

$$\int f\bar{g}\, dm, \quad f \in H^p \cap I\overline{H_0^p},\ g \in H^q \cap I\overline{H_0^q}.$$

For $1 \leq p' < p < \infty$, the inclusion operator $i : H^p \cap I\overline{H_0^p} \to H^{p'} \cap I\overline{H_0^{p'}}$ is continuous. In fact, more is true.

THEOREM 1.0.9 (Aleksandrov [6]). *For $1 \leq p' < p < \infty$, the inclusion operator $i : H^p \cap I\overline{H_0^p} \to H^{p'} \cap I\overline{H_0^{p'}}$ is compact.*

Due to significant differences in both the functional analysis and the function theory when moving from $p \in [1, \infty)$ to $p \in (0, 1)$, the description of the B-invariant subspaces of H^p ($0 < p < 1$) is intensely more complicated as is the technique used to prove it. Some of the classical tools of functional analysis, for example, the Hahn-Banach separation theorem, fail in the $p \in (0, 1)$ setting. There are indeed proper B-invariant subspaces $\mathcal{M}$ of H^p for which $\mathcal{M}^\perp = (0)$. Here the dual of H^p can be identified with a space of Lipschitz or Zygmund functions, see Chapter 3. In light of a general result of Kalton [46], the failure of the Hahn-Banach separation property should not be surprising due to the fact that the metric topology on H^p is not locally convex. Important results from function theory, for example, the Cauchy integral theorem, an important function theoretic tool which allows us to recapture a function in H^p ($1 \leq p < \infty$) from its boundary values, also fail when $p \in (0, 1)$. In fact, some of the most important examples of functions in H^p ($0 < p < 1$), for example,

$$\frac{1}{(1 - \bar{\zeta}z)^k}, \quad \zeta \in \mathbb{T},\ k \in \mathbb{N} \cap [1, 1/p),$$

cannot be represented as Cauchy integrals against their boundary values. Closed linear spans of certain subsets of these functions also serve as examples of proper B-invariant subspaces $\mathcal{M}$ of H^p for which $\mathcal{M}^\perp$ is zero.

Aleksandrov is able to overcome these obstacles and successfully describe the B-invariant subspaces of H^p ($0 < p < 1$). His description is complicated and involves several parameters which we now take a moment to describe. The first parameter is the integer

$$n_p := \max\{n \in \mathbb{N} \cap [1, 1/p)\} = [1/p].$$

A routine computation shows that for all $\zeta \in \mathbb{T}$,

$$\frac{1}{(1 - \bar{\zeta}z)^j} \in H^p \quad \forall j = 1, \cdots, n_p.$$

Informally, n_p is the largest order pole an H^p function can have at a point $\zeta \in \mathbb{T}$. The second parameter will be an inner function I. The third parameter will be a closed set $F \subset \mathbb{T}$ which we will assume (we will see why later) contains $\sigma(I) \cap \mathbb{T}$. The final parameter will be a function $k : F \to \mathbb{N} \cap [1, n_p]$. Consider the space $\mathcal{E}^p(I, F, k)$, which will turn out to be a typical B-invariant subspace, consisting of H^p functions f such that

(i) $f \in H^p \cap I\overline{H_0^p}$
(ii) f has an analytic continuation to a neighborhood of $\mathbb{T} \setminus F$
(iii) for each $\zeta \in F_0 \setminus \sigma(I)$, where F_0 is the set of isolated points of F, f has a pole of order at most $k(\zeta)$.

It is not difficult to show, as was the case when $p \in [1, \infty)$, that $f \in H^p \cap \overline{IH_0^p}$ ($0 < p < 1$) if and only if f/I has a pseudocontinuation to a function belonging to $H^p(\mathbb{D}_e)$ which vanishes at infinity. However, in contrast to the $p \in [1, \infty)$ case, functions in $H^p \cap \overline{IH_0^p}$ need not have analytic continuations to $\mathbb{C}_\infty \setminus \{1/\bar{z} : z \in \sigma(I)\}$. In fact for any inner function I,

$$\frac{1}{(1-\bar{\zeta}z)^j} \in H^p \cap \overline{IH_0^p}, \ \forall \, j = 1, \cdots, n_p, \ \zeta \in \mathbb{T}.$$

This is why the second and third properties in the definition of $\mathcal{E}^p(I, F, k)$ are important. Aleksandrov's result is the following.

THEOREM 1.0.10 (A. B. Aleksandrov [3]). *For each $p \in (0, 1)$ and triple I, F, and k as above, the linear manifold $\mathcal{E}^p(I, F, k)$ is a (closed) B-invariant subspace of H^p. Moreover, every non-trivial B-invariant subspace of H^p is of this form.*

We will certainly take the time in Chapter 6 to explain the parameters in detail. For example, why does F need to contain $\sigma(I) \cap \mathbb{T}$? Why do we only specify the order of the pole at $F_0 \setminus \sigma(I)$? Why is $\mathcal{E}^p(I, F, k)$ closed?

Given a B-invariant subspace $\mathcal{M}$, one might wonder how to obtain the parameters I, F, and k. This is done in the following way: Given $\mathcal{M}$, the subspace $\mathcal{M} \cap H^2$ is a (possibly zero) B-invariant subspace of H^2 which, by the Douglas-Shapiro-Shields result, is of the form $H^2 \cap \overline{IH_0^2}$.[2] With some work, one shows that $\mathcal{M} \subset H^p \cap \overline{IH_0^p}$. To get the closed set F, we define

$$F := \left\{ \zeta \in \mathbb{T} : \frac{1}{1-\bar{\zeta}z} \in \mathcal{M} \right\}$$

and prove that F is indeed a closed subset of $\mathbb{T}$. It can be shown that if $\zeta \in \sigma(I) \cap \mathbb{T}$, then $\zeta \in F$, which is why we assumed from the onset that F contains $\sigma(I) \cap \mathbb{T}$. To obtain the function k, define $k : F \to \mathbb{N} \cap [1, n_p]$ by

$$k(\zeta) := \max \left\{ j \in \mathbb{N} : \frac{1}{(1-\bar{\zeta}z)^j} \in \mathcal{M} \right\}.$$

We will see that if ζ belongs to either $F \setminus F_0$ (the non-isolated points of F) or $\sigma(I) \cap \mathbb{T}$, then $k(\zeta) = n_p$, which is why we only specify the poles at $F_0 \setminus \sigma(I)$. Given $\mathcal{M}$, we now have our three parameters I, F, and k.

The first important result here is that the B-invariant linear manifold $\mathcal{E}^p(I, F, k)$ is actually closed in H^p and

$$\mathcal{M} \subset \mathcal{E}^p(I, F, k).$$

To obtain the reverse inclusion, Aleksandrov defines

$$e^p(I, F, k) := \left(H^2 \cap \overline{IH_0^2} \right) \bigvee \left\{ \frac{1}{(1-\bar{\zeta}z)^j} : \zeta \in F, 1 \leq j \leq k(\zeta) \right\}.$$

By the definitions of our three parameters, we have

$$e^p(I, F, k) \subset \mathcal{M} \subset \mathcal{E}^p(I, F, k).$$

The equality $e^p(I, F, k) = \mathcal{E}^p(I, F, k)$ is obtained by an ingenious rational approximation argument developed especially for this situation using the Coifman atomic decomposition [21] to control the order and location of the poles. A major portion of Chapter 6 will be dedicated to giving the full details and background material needed to understand this argument.

---

[2] If $\mathcal{M} \cap H^2 = (0)$, take the inner function I to be the constant function one.

As we did when $p \in [1, \infty)$, we will look at some of the functional analysis properties of the space $\mathcal{E}^p(I, F, k)$ as well as the operator theory properties of $B|\mathcal{E}^p(I, F, k)$. For example, one can compute the spectrum of $B|\mathcal{E}^p(I, F, k)$.

THEOREM 1.0.11. $\sigma(B|\mathcal{E}^p(I, F, k)) = \sigma_p(B|\mathcal{E}^p(I, F, k)) = \overline{\sigma(I) \cup F}$.

When $p \in [1, \infty)$, one is able to identity the dual of a B-invariant subspace of H^p as well as the commutant of B when restricted to an invariant subspace. When $p \in (0, 1)$, these two problems become more complicated and we will make some comments on the difficulties that arise.

The layout of this book is as follows. Chapter 2 (Classical boundary value results): Since the description of the backward shift invariant subspaces of the Hardy space depends heavily on the notion of a non-tangential limit and a pseudo-continuation, we will review, without proof, the fundamentals of these topics. We start with the definitions of a radial limit and a non-tangential limit for analytic functions on the unit disk and recall the classical existence and uniqueness results of Fatou [**33**], F. and M. Riesz [**72**], Plessner [**66**], Lusin, and Privalov [**59**] [**69**] [**68**]. We then proceed to define the term pseudocontinuation, as developed by H. S. Shapiro [**84**], and give several examples.

Chapter 3 (The Hardy space of the disk): We will assume the reader is somewhat familiar with the results from the theory of Hardy spaces and we will not attempt to rewrite the material from the books of Duren [**31**], Garnett [**39**], Hoffman [**44**], Koosis [**51**], and Zygmund [**106**]. We quickly review, without proof, the basic definitions of the Hardy spaces and catalog the main properties of these spaces for their use in later chapters. The results include: definition of the Hardy spaces, boundary values, factorization, Hardy spaces and Fourier analysis, the F. and M. Riesz theorem, Riesz projections and harmonic conjugation, duality, bounded mean oscillation, Lipschitz classes, Cauchy transforms, Kolmogorov's theorem, and Fatou's jump theorem.

Chapter 4 (The Hardy space of the upper-half plane): It is well known that functions in H^p ($1 \leq p < \infty$) can be recovered from their non-tangential boundary values by means of the Cauchy integral formula. Although this is no longer true for H^p ($0 < p < 1$), there is a deep theory of Fefferman, Stein, and Coifman which says that functions in the Hardy space correspond to a certain class of distributions on the circle and that the functions can be recovered from their corresponding distributions by means of a distributional form of the Cauchy integral formula via the atomic decomposition. We will discuss the atomic decomposition of a distribution arising from a Hardy space function since it will be a key ingredient in the rational approximation arguments mentioned earlier. This entire theory was originally formulated for spaces of the upper half-plane, hence the title of this chapter. This slight detour to the upper half plane is beneficial to us since the atomic decomposition theory for the Hardy spaces of the upper half plane is developed, with complete proofs, in Stein's harmonic analysis book [**95**]. In this chapter, we will develop and state the needed results and will refer the reader to the appropriate places in Stein's book for the proofs. These results include: definition and basic properties of the Hardy spaces of the upper-half plane, Poisson and conjugate Poisson integrals, maximal functions, the Hilbert transform, the harmonic Hardy space, distributions and the harmonic Hardy space, and Coifman's atomic decomposition.

Chapter 5 (The backward shift on H^p for $p \in [1, \infty)$): We will include three different proofs of the description of the B-invariant subspaces of H^p. We do this not only to demonstrate some of the techniques in this field, but more importantly, to show the reader how the techniques in two of these three proofs can be used to examine the B-invariant subspaces of other spaces of analytic functions. As mentioned above, every B-invariant subspace is of the form $(IH^q)^\perp$. All three proofs describe this annihilator but do it in different ways. The first is the simplest and most direct and uses the F. and M. Riesz theorem. The second one is more complicated but allows the reader to see an actual formula for the pseudocontinuation of the functions in $(IH^q)^\perp$, a formula which is not readily transparent from the first proof. A variation of the second proof can be used to examine the backward shift invariant subspaces of the Bergman spaces as was done by Aleman, Richter, and Ross [8] [9]. The second proof can also be used to describe $\sigma(B|H^p \cap \overline{IH_0^p})$. The third proof is due to Aleksandrov [5] and gives a unified approach to characterizing the B-invariant subspaces of other spaces of functions such as $VMOA$, $BMOA$, and $L^1/\overline{H_0^1}$ (Chapter 3 will have the formal definitions of these spaces). We end the chapter with some remarks about some of the functional analysis properties of $H^p \cap \overline{IH_0^p}$ as well as the operator theory properties of $B|H^p \cap \overline{IH_0^p}$.

Chapter 6 (The backward shift on H^p for $p \in (0,1)$): We finally arrive at the full proof of Aleksandrov's characterization of the B-invariant subspaces for H^p. After proving some basic facts about the subspaces $\mathcal{E}^p(I, F, k)$, we then proceed to his rational approximation scheme which has been laid out in several papers [2] [3] [4]. As with many of the results in H^p theory for $p \in (0,1)$ (see [32] for example) we need to consider two cases: $1/p \in \mathbb{N}$ and $1/p \notin \mathbb{N}$, the former usually the most difficult case to handle. Finally, we make some remarks about dual of $\mathcal{E}^p(I, F, k)$ as well as the commutants of B and $B|\mathcal{E}^p(I, F, k)$.

CHAPTER 2

Classical boundary value results

As seen in the Introduction, the description of the B-invariant subspaces of H^p depends heavily on understanding boundary values of analytic functions and the concept of a pseudocontinuation. We begin with a review of these topics and point out that a nice general treatment of the boundary behavior of analytic functions can be found in Collingwood and Lohwater's book [23].

2.1. Limits

If Ω is a domain in the extended complex plane $\mathbb{C}_\infty = \mathbb{C} \cup \{\infty\}$, we let $\mathfrak{H}(\Omega)$ and $\mathfrak{M}(\Omega)$ denote the family of holomorphic (resp. meromorphic) functions on Ω. Usually Ω will either be the unit disk $\mathbb{D} = \{z \in \mathbb{C} : |z| < 1\}$, the extended exterior disk $\mathbb{D}_e = \mathbb{C}_\infty \setminus \mathbb{D}^-$, or the upper half plane $\mathbb{C}_+ = \{z \in \mathbb{C} : \Im z > 0\}$.

DEFINITION 2.1.1. Let $f \in \mathfrak{H}(\mathbb{D})$ and $\zeta \in \mathbb{T} = \partial \mathbb{D}$ (the unit circle).

1. f has a (finite) *radial limit* equal to A at ζ if
$$f(r\zeta) \to A \text{ as } r \to 1^-.$$

2. For $\alpha > 1$, we let
$$\Gamma_\alpha(\zeta) = \{ z \in \mathbb{D} : |z - \zeta| < \alpha(1 - |z|) \}$$
be the *non-tangential approach region* at ζ with opening α. Near the point ζ, $\Gamma_\alpha(\zeta)$ is the triangular shaped region with vertex at ζ whose angle equals $2\arccos(1/\alpha)$.

3. f has a (finite) *non-tangential limit* equal to A at ζ if for *all* $\alpha > 1$
$$f(z) \to A, \text{ as } z \to \zeta, z \in \Gamma_\alpha(\zeta).$$

4. f has a (finite) *unrestricted limit* equal to A at ζ if
$$f(z) \to A \text{ as } z \to \zeta, \quad z \in \mathbb{D}.$$

Analogous definitions can be defined for meromorphic functions on $\mathbb{D}$. Although in the later chapters, we will only be concerned with non-tangential limits, we include, for the sake of completeness, some remarks about the relationships between these different types of limits as well as the pathological behavior which often arises.

2.1.0.1. *Comments about the definition.* Clearly if f has an unrestricted limit at ζ, then f has both a non-tangential and a radial limit at ζ. It is also clear that

if f has a non-tangential limit at ζ, then f has a radial limit at ζ. However, in general, any other implication is false. For example, the function
$$f(z) := \exp\left(-\frac{1+z}{1-z}\right)$$
has a radial limit of zero at the point $\zeta = 1$. However, it is easy to check that $|f(z)| = e^{-1}$ along the circle $|z - 1/2| = 1/2$ (which is tangent to the unit circle at $\zeta = 1$). Thus $f(z)$ does not approach zero as $z \to 1$ along $|z - 1/2| = 1/2$.

In the definition of non-tangential limit, the criterion that $f(z) \to A$ as $z \to \zeta$ in $\Gamma_\alpha(\zeta)$ for *all* $\alpha > 1$ is important. For example the function
$$f(z) := \exp\left(-\left\{\frac{1+z}{1-z}\right\}^2\right)$$
goes to zero as $z \to 1$ in the triangle with vertex at $\zeta = 1$ whose angle is less than $\pi/2$. However, $f(z)$ does not have a limit as $z \to 1$ along the sides of a non-tangential approach region, with vertex at $\zeta = 1$, when the opening angle is larger than $\pi/2$.

2.1.0.2. *Pathological behavior.* In general, the radial limits of analytic functions need not exist. For example, a theorem of Littlewood [55] says that if $\{a_n : n \in \mathbb{N}\}$ is a sequence of complex numbers with
$$\limsup_{n\to\infty} |a_n|^{1/n} = 1 \quad \text{and} \quad \sum_{n=1}^\infty |a_n|^2 = \infty,$$
then for almost every choice of signs $\varepsilon_n = \pm 1$, the function
$$f(z) := \sum_{n=1}^\infty \varepsilon_n a_n z^n$$
has a radial limit almost nowhere.

Radial limits of holomorphic functions may exist nowhere. MacLane [60] produced an example of an $f \in \mathfrak{H}(\mathbb{D})$ satisfying
$$\limsup_{r\to 1^-} |f(r\zeta)| = \infty \quad \text{and} \quad \liminf_{r\to 1^-} |f(r\zeta)| = 0$$
for *every* $\zeta \in \mathbb{T}$. In fact this f can be chosen to have no zeros in the disk and satisfy a certain growth condition.

Results of Bagemihl and Seidel [12] and Rudin [77] extend this pathological behavior of radial limits even further. For example, they show that given *any* continuous function g on $\mathbb{D}$ and *any* set E of first category in $\mathbb{T}$, there is an $f \in \mathfrak{H}(\mathbb{D})$ such that

(2.1.2) $\qquad f(r\zeta) - g(r\zeta) \to 0, \ r \to 1^- \ \forall \zeta \in E.$

Since there are sets of first category with full measure in $\mathbb{T}$, we see that the behavior of f on radii can be bad, quite often. There are further results along these lines which show that one can even make an analytic function f, with bad boundary behavior, have certain given growth conditions [31], p. 86, [45]. In the above result of Bagemihl and Seidel, the hypothesis of "first category" is important since the result is false without it. See Remark 2.1.10(2) below.

Even for bounded analytic functions, one cannot expect reasonable behavior. For example, the unrestricted limit need not exist anywhere as the following result of Lohwater and Piranian [57] demonstrates: Let γ be a simple closed Jordan curve which is internally tangent to $\mathbb{T}$ at $\zeta = 1$ and intersects $\mathbb{T}$ only at $\zeta = 1$. Then there is a bounded analytic function f on $\mathbb{D}$ such that for any angle θ, the function $f(e^{i\theta}z)$ does not have a limit as $z \to 1$ along γ. We remark that an "almost everywhere" version of this result was originally shown by Littlewood [**54**] and simplified by Zygmund [**105**].

2.1.0.3. *Positive results.* The first positive result about radial/non-tangential limits came with Fatou in his 1906 thesis.

THEOREM 2.1.3 (Fatou [**33**]). *If $f \in \mathfrak{H}(\mathbb{D})$ and bounded, then f has a (finite) non-tangential limit at each $\zeta \in \mathbb{T}$ except possibly for a set of Lebesgue measure zero.*

Actually, for bounded analytic functions, the existence of a radial limit implies the existence of a non-tangential limit as demonstrated by the following theorem of Lindelöf.

THEOREM 2.1.4 (Lindelöf [**53**]). *If $f \in \mathfrak{H}(\mathbb{D})$ and bounded with $f(z) \to A$ as $z \to \zeta$ along some arc $\gamma \subset \mathbb{D}$ which terminates at ζ, then for any $\alpha > 1$, $f(z) \to A$ uniformly as $z \to \zeta, z \in \Gamma_\alpha(\zeta)$.*

A significant strengthening and localization of Fatou's theorem was done by Lusin and Privalov [**59**] [**68**], Marcinkiewicz and Zygmund [**61**], and Spencer [**91**]. A proof of the complete result can be found in Zygmund [**106**], Chapter 14 (alternatively [**94**], Chapter 7).

THEOREM 2.1.5. *Let $f \in \mathfrak{H}(\mathbb{D})$ and $E \subset \mathbb{T}$. Then for almost every $\zeta \in E$, the following are equivalent.*

1. *f has a non-tangential limit at ζ.*
2. *f is non-tangentially bounded at ζ, i.e., $f(\Gamma_\alpha(\zeta))$ is a bounded set for some $\alpha > 1$.*
3. *The "Lusin area function"*
$$\zeta \to \Big\{ \int_{\Gamma_\alpha(\zeta)} |f'|^2 dx\, dy \Big\}^{1/2}$$
is finite for some $\alpha > 1$.

REMARK 2.1.6. In conditions (2) and (3), the "bounds" need not be uniform as ζ runs through E.

There is also the following dichotomy result of Plessner.

THEOREM 2.1.7 (Plessner [**66**]). *Let $g \in \mathfrak{M}(\mathbb{D})$. Then for almost every $\zeta \in \mathbb{T}$, one of the following conditions hold:*

1. *g has a finite non-tangential limit at ζ.*
2. *$g(\Gamma_\alpha(\zeta))$ is dense in $\mathbb{C}_\infty$ for every $\alpha > 1$.*

Thus for a given $g \in \mathfrak{M}(\mathbb{D})$, the unit circle can be partitioned into three disjoint Borel sets as
$$\mathbb{T} = N \cup L \cup P,$$
where N is a set of Lebesgue measure zero, L is a set of points where the non-tangential limit of g exists, and P (called the set of *Plessner points*) is a set of points $\zeta \in \mathbb{T}$ where $g(\Gamma_\alpha(\zeta))^- = \mathbb{C}_\infty$. For a bounded analytic function, note that $P = \emptyset$ which implies Fatou's theorem.

We also comment that the exceptional sets of measure zero in Fatou's theorem can not be made any smaller. Lusin and Privalov [59] show that given any set $E \subset \mathbb{T}$ with $|E| = 0$, there is a bounded $f \in \mathfrak{H}(\mathbb{D})$ whose radial limit exists nowhere on E. Lohwater and Piranian [57] sharpen this result and show that for a given set $E \subset \mathbb{T}$ with $|E| = 0$, there is a bounded $f \in \mathfrak{H}(\mathbb{D})$ (in fact f can be chosen to be an inner function) such that

$$\liminf_{r \to 1^-} |f(r\zeta)| = 0, \quad \limsup_{r \to 1^-} |f(r\zeta)| = 1 \; \forall \, \zeta \in E.$$

2.1.0.4. *Uniqueness.* An important question now is: To what extent do the radial/non-tangential boundary values "determine" an $f \in \mathfrak{H}(\mathbb{D})$? Fatou [33] showed that if $f \in \mathfrak{H}(\mathbb{D})$ is bounded and whose radial limit vanishes on an arc of the circle, then f is the zero function. The Riesz brothers in 1916 improved this result to sets of positive measure.

THEOREM 2.1.8 (F. and M. Riesz [72]). *If $f \in \mathfrak{H}(\mathbb{D})$ is bounded and*

$$\lim_{r \to 1^-} f(r\zeta) = 0$$

on a set $E \subset \mathbb{T}$ with $|E| > 0$, then f is the zero function.

The uniqueness result we will make the most use of is due to Lusin and Privalov.

THEOREM 2.1.9 (Lusin-Privalov [59] [69]). *If f belongs to $\mathfrak{M}(\mathbb{D})$ and has non-tangential limits which are equal to zero on a set of positive measure, then f is the zero function.*

REMARK 2.1.10. 1. This theorem says that if two meromorphic functions have the same (non-tangential) boundary values on a set of positive measure, they must indeed be the same function. We also make the important remark that the Lusin-Privalov theorem is false if the non-tangential limits are replaced by radial limits since, by the above mentioned result of Bagemihl and Seidel, there are non-zero functions whose radial limits are equal to zero almost everywhere. [1]

2. In the above remark, the phrase "almost everywhere" above cannot be replaced by "everywhere". For example, if there were an analytic function f for which $f(r\zeta) \to 0$ for *every* $\zeta \in \mathbb{T}$, then for each ζ, $f(r\zeta)$ would be a continuous function of r on $[0,1]$. Letting

$$E_n = \{ \, \zeta \in \mathbb{T} : |f(r\zeta)| \leq n \; \forall \, r \in [0,1] \, \}, \quad n \in \mathbb{N},$$

we see that E_n is closed and $\bigcup_n E_n = \mathbb{T}$. By the Baire category theorem, at least one of the E_n must contain an interval and so f must be bounded (by n) on a sector. But by Theorem 2.1.5, f has finite non-tangential limits almost everywhere on the boundary arc of this sector which must be zero (since the radial limits are zero). It follows now from the Lusin-Privalov uniqueness theorem that f must be the zero function [2].

---

[1] Choose E to be a set of full measure and of first category and g to be the zero function in eq.(2.1.2).

[2] We wish to thank Boris Korenblum for showing us this proof.

The non-tangential approach regions $\Gamma_\alpha(\zeta)$ can be defined in an analogous way for the exterior disk (reflect the interior non-tangential approach regions to $\mathbb{D}_e$ using the map $z \to 1/\overline{z}$) with analogous results for functions in $\mathfrak{H}(\mathbb{D}_e)$ and $\mathfrak{M}(\mathbb{D}_e)$, e.g., Plessner's theorem, Fatou's theorem, and the Lusin-Privalov theorem.

2.2. Pseudocontinuations

Recall from the introduction that if $f \in (IH^q)^\perp$, where I is an inner function, the L^p boundary function $f\overline{I}$ has the Fourier series expansion
$$f\overline{I} \sim b_1\overline{\zeta} + b_2\overline{\zeta}^2 + \cdots.$$
Moreover, the function $g \in \mathfrak{H}(\mathbb{D}_e)$ defined by
$$g(z) := \frac{b_1}{z} + \frac{b_2}{z^2} + \cdots$$
belongs to $H^p(\mathbb{D}_e)$ and hence has (exterior) radial boundary values $g(\zeta)$ almost everywhere. One can also show that
$$\lim_{r \to 1^+} g(r\zeta) = \lim_{r \to 1^-} \frac{f(r\zeta)}{I(r\zeta)}$$
almost everywhere. So in a way, we can think of g as a "continuation" of the meromorphic function f/I across $\mathbb{T}$.

In general we would like to define a "continuation" of $f \in \mathfrak{M}(\mathbb{D})$ to a $T_f \in \mathfrak{M}(\mathbb{D}_e)$ which is more general than an analytic continuation but still preserves some desirable properties. Such desirable properties for a "continuation" $T_f \in \mathfrak{M}(\mathbb{D}_e)$ of $f \in \mathfrak{M}(\mathbb{D})$ are the following:

1. If T_f exists and $T_f = 0$, then $f = 0$.
2. If T_f, T_g exist, then for all complex constants a and b, T_{af+bg} exists and is equal to $aT_f + bT_g$.
3. If T_f, T_g exist, then T_{fg} exists and is equal to $T_f T_g$.
4. If T_f exists and if f admits an analytic continuation across some point $\zeta \in \mathbb{T}$, then this continuation agrees with T_f.

We are now ready to define a concept of "continuation" which satisfies the above properties.

DEFINITION 2.2.1. We say that $T_f \in \mathfrak{M}(\mathbb{D}_e)$ is a *pseudocontinuation* of $f \in \mathfrak{M}(\mathbb{D})$ across a set $E \subset \mathbb{T}$ with $|E| > 0$ if the non-tangential boundary limits of T_f and f exist and are equal almost everywhere on E.

By the linearity and multiplicative properties of limits, pseudocontinuations satisfy properties (2) and (3). By the Lusin-Privalov uniqueness theorem, Theorem 2.1.9, pseudocontinuations, whenever they exist, satisfy the uniqueness property (1). That is to say, if $F, G \in \mathfrak{M}(\mathbb{D}_e)$ are both pseudocontinuations of $f \in \mathfrak{M}(\mathbb{D})$ across some $E \subset \mathbb{T}$ ($|E| > 0$), then $F \equiv G$. Observe that the Lusin-Privalov uniqueness theorem also implies the compatibility with analytic continuation property (4).

Although for this type of "continuation" the above four properties were somewhat easy to check, there are other types of continuation [3], often called *generalized analytic continuation* for which these properties are non trivial to check **[84] [86]**.

We also mention that the definition of pseudocontinuation in the introduction used only radial limits while the definition above uses non-tangential limits. In the

---
[3]e.g., "Gonchar continuation" (see **[84]** or **[40]**).

general case, this would be a big difference in the two definitions and would make the radial limit definition an undesirable continuation since it would not satisfy the uniqueness property (1) above. From our earlier remarks, it is indeed possible for a non-zero analytic function on $\mathbb{D}$ to have zero radial limits almost everywhere. However, for the class of functions in the introduction, the non-tangential limits (from inside and outside the disk) exist on a set of full measure in $\mathbb{T}$ and so testing the equality of two non-tangential limits is the same as testing the equality of two radial limits, almost everywhere.

EXAMPLE 2.2.2. Here are several examples of functions with (or without) pseudocontinuations.

1. Any polynomial has a pseudocontinuation across $\mathbb{T}$. The function $f(z) = e^z$, although it has an analytic continuation across $\mathbb{T}$, does not have a pseudocontinuation since it is not meromorphic in $\mathbb{D}_e$ (essential singularity at infinity).

2. Using the Lusin-Privalov uniqueness theorem, we note that if $f \in \mathfrak{H}(\mathbb{D})$ has an analytic continuation across all points of $\mathbb{T}$ except for an isolated branch point at say $z = 1$, e.g., $\log(1-z)$, then f does not have a pseudocontinuation across $\mathbb{T}$.

3. Assume $\{c_j : j \in \mathbb{N}\}$ is a sequence of complex numbers with $\sum_{j=1}^{\infty} |c_j| < \infty$ and $\{\zeta_j : j \in \mathbb{N}\}$ is a sequence in $\mathbb{T}$. The measure $d\mu = \sum c_j \delta_{\zeta_j}$, where δ_{ζ_j} is a unit point mass at ζ_j, is a finite Borel measure on $\mathbb{T}$ and one can show, using a theorem of Kolmogorov and Smirnov, Theorem 3.4.1, that the Cauchy transform

$$(C\mu)(z) = \int_{\mathbb{T}} \frac{d\mu(\zeta)}{\zeta - z} = \sum_j \frac{c_j}{\zeta_j - z}$$

is a holomorphic function on $\mathbb{C}_\infty \setminus \mathbb{T}$ which has (finite) non-tangential limits (from both inside and outside $\mathbb{D}$). Moreover, by Fatou's jump theorem, Theorem 3.4.8, the limits will agree almost everywhere. Hence $(C\mu)(z)$ has a pseudocontinuation across $\mathbb{T}$. If $\{\zeta_j : j \in \mathbb{N}\}^- = \mathbb{T}$, then the pseudocontinuation will not be an analytic continuation across any arc of $\mathbb{T}$.

4. If f is an inner function (a bounded analytic function on $\mathbb{D}$ with $|f(\zeta)| = 1$ almost everywhere), then f has a pseudocontinuation given by the formula

$$(T_f)(z) = \frac{1}{\overline{f(1/\bar{z})}}, \quad |z| > 1.$$

Again, by making the zeros of f accumulate at every point of $\mathbb{T}$ (which is possible), one shows that f will not have an analytic continuation across any arc of $\mathbb{T}$.

5. H. S. Shapiro [85] showed that the function

$$f(z) = \sum_{n=0}^{\infty} z^{2^n} 2^{-n}, \quad |z| < 1$$

(which is analytic on $\mathbb{D}$ and continuous on $\mathbb{D}^-$) does not have a pseudocontinuation across any arc of $\mathbb{T}$ (to any contiguous domain). In fact, there are more general results which say that lacunary series are, in general, "noncontinuable" (see [29] or more recent results in [1] and [7]). For example,

Aleksandrov [7], Lemma 6.3, uses lacunary series to create examples of functions which are not pseudocontinuable across any arc of $\mathbb{D}$. More specifically, let f be any analytic function on $\mathbb{D}$ whose radius of convergence is equal to one. Then, there is a $\psi \in C^\infty(\mathbb{T})$ such that the convolution

$$(f * \psi)(z) := \int_{\mathbb{T}} f(z\bar{\zeta})\psi(\zeta)\, dm(\zeta), \quad z \in \mathbb{D},$$

does not have a pseudocontinuation across any arc of $\mathbb{T}$ (to any contiguous domain) .

6. Meromorphic functions on $\mathbb{C}_\infty \setminus \mathbb{T}$ of the form

(2.2.3)
$$f(z) = \sum_j \frac{c_j}{z - z_j},$$

where $\{c_j : j \in \mathbb{N}\}$ and $\{z_j : j \in \mathbb{N}\}$ are complex numbers with

$$|z_j| > 1 \quad \text{and} \quad \lim_{j \to \infty} |z_j| = 1,$$

are often a rich source of pathological examples. Such series are often called *Borel series*. For example, J. Wolff [102] found a summable sequence $\{c_j : j \in \mathbb{N}\}$ such that $f = 0$ on $\mathbb{D}$ but not equivalently zero on $\mathbb{D}_e$.

Borel series can be used to create pathological examples of functions without pseudocontinuations. For example, in [9] the authors find a sequence $\{c_j : j \in \mathbb{N}\}$ so that the Borel series f in eq.(2.2.3) is continuous (even differentiable) on $\mathbb{D}^-$ but f does not have a pseudocontinuation across any set of positive measure in $\mathbb{T}$. One does this by first choosing a sequence $\{z_j : j \in \mathbb{N}\} \subset \mathbb{D}_e$ which has accumulation points (on $\mathbb{T}$) in every non-tangential approach region for almost every point of $\mathbb{T}$. Such a sequence is called a "dominating sequence" [16]. Then choose the constants $\{c_j : j \in \mathbb{N}\}$ so that the above Borel series converges uniformly on compact subsets of $\mathbb{C} \setminus \{z_j : j \in \mathbb{N}\}$ and defines a continuous function on $\mathbb{D}^-$. Using some estimates on harmonic measure, one can show that if f (on $\mathbb{D}$) has a pseudocontinuation to $\mathbb{D}_e$, then the pseudocontinuation must be given by the Borel series in eq.(2.2.3) on $\mathbb{D}_e$. But this Borel series on $\mathbb{D}_e$ cannot possibly have non-tangential limit values (from $\mathbb{D}_e$) since at almost every point of the circle, the poles accumulate in every non-tangential approach region.

CHAPTER 3

The Hardy space of the disk

3.1. Introduction

We will assume that the reader is somewhat familiar with the basics from the theory of Hardy spaces of the disk. We will, however, remind the reader of the fundamental results of the subject so we can refer to them in subsequent chapters.

The maximum modulus principle says that for $f \in \mathfrak{H}(\mathbb{D})$, the maximum modulus function

$$M_\infty(r, f) := \max\left\{ |f(r\zeta)| : \zeta \in \mathbb{T} \right\}, \quad r \in (0, 1),$$

is an increasing function of r. For fixed $p \in (0, \infty)$, G. H. Hardy [41] proved the same result for the integral means

$$M_p(r, f) := \left\{ \int_\mathbb{T} |f(r\zeta)|^p \, dm(\zeta) \right\}^{1/p}, \quad r \in (0, 1),$$

and from here, it is quite natural to consider the class of analytic functions f for which

$$\sup_{0<r<1} M_p(r, f) = \lim_{r \to 1^-} M_p(r, f)$$

is finite. The resulting class of analytic functions, called the *Hardy space* H^p, is the primary focus of this book. When $p = \infty$, we define H^∞ to be the class of analytic functions f for which

$$\sup_{0<r<1} M_\infty(r, f) < \infty;$$

namely, the bounded analytic functions on $\mathbb{D}$. As we shall see below, these classes have a remarkable structure which involves, via non-tangential boundary values, the L^p theory of functions on the circle.

For now, we note from the basic inequality

$$(a+b)^p \leq 2^p(a^p + b^p), \quad a \geq 0, \ b \geq 0,$$

that H^p is a (complex) linear space. Furthermore, a simple application of Hölder's inequality shows that for $p, q \in (0, \infty]$, $p < q$, the inequality

$$M_p(r, f) \leq M_q(r, f)$$

holds for all $r \in (0, 1)$ and so $H^q \subset H^p$.

3.2. H^p and boundary values

The first significant result in this area says that the non-tangential boundary values of H^p functions exist almost everywhere and are log-integrable.

THEOREM 3.2.1 (Non-tangential limits). *If $f \in H^p$ $(0 < p \leq \infty)$, then f can be written as the quotient of two bounded analytic functions on $\mathbb{D}$ and so (by Fatou's theorem and the Lusin-Privalov uniqueness theorem)*

$$\text{n.t.} \lim_{z \to \zeta} f(z) := f(\zeta)$$

exists for almost every $\zeta \in \mathbb{T}$. Moreover, if f is not the zero function, then

$$\int_{\mathbb{T}} \log |f(\zeta)|\, dm(\zeta) > -\infty.$$

REMARK 3.2.2. Meromorphic functions on $\mathbb{D}$ which can be written as the quotient of two bounded analytic functions form a special class of functions called the *functions of bounded type* and are denoted by $\mathfrak{N}(\mathbb{D})$. We will discuss this class in greater detail in Section 6 of this chapter.

This next result says the boundary function belongs to $L^p := L^p(\mathbb{T}, dm)$.

THEOREM 3.2.3. *If $f \in H^p$ $(0 < p \leq \infty)$, the (almost everywhere defined) non-tangential boundary function*

$$f(\zeta) = \text{n.t.} \lim_{z \to \zeta} f(z) \quad {}^1$$

belongs to L^p and

$$\lim_{r \to 1^-} M_p(r, f) = \Big\{ \int_{\mathbb{T}} |f(\zeta)|^p\, dm(\zeta) \Big\}^{1/p}.$$

Furthermore, for $p \in (0, \infty)$,

$$\lim_{r \to 1^-} \int_{\mathbb{T}} |f(r\zeta) - f(\zeta)|^p\, dm(\zeta) = 0.$$

This next result plays a critical role in H^p theory and in particular to the description of the backward shift invariant subspaces. It is known as the "factorization theorem" and is due to F. Riesz [71] and V. I. Smirnov [89].

THEOREM 3.2.4 (Factorization Theorem). 1. *Every non-zero function f in H^p $(0 < p \leq \infty)$ has a unique factorization of the form*

$$f = bsF$$

where b is Blaschke product

$$b(z) = z^m \prod_{n=1}^{\infty} \frac{|a_n|}{a_n} \frac{a_n - z}{1 - \overline{a_n} z}, \quad \{a_n : n \in \mathbb{N}\} \subset \mathbb{D} \setminus \{0\},$$

whose zeros, repeated according to multiplicity, satisfy

$$\sum_{n=1}^{\infty} (1 - |a_n|) < \infty,$$

s is a singular inner function

$$s(z) = \exp\left(-\int_{\mathbb{T}} \frac{\zeta + z}{\zeta - z} d\mu(\zeta)\right)$$

---

[1] From now on, whenever $\zeta \in \mathbb{T}$ and $f \in H^p$, we will understand that $f(\zeta)$ is the non-tangential boundary function defined at ζ (whenever it exists).

with positive singular measure (with respect to dm) μ, and F is an outer function

$$F(z) = e^{i\alpha} \exp\left(\int_{\mathbb{T}} \frac{\zeta+z}{\zeta-z} \log|f(\zeta)|\, dm(\zeta)\right), \quad \alpha > 0.$$

2. In the above factorization, the function $I_f := bs$ is called the *inner part* of f and has the property that $|I_f(z)| \leq 1$ on $\mathbb{D}$ and $|I_f(\zeta)| = 1$ almost everywhere. The function $O_f := F$, called the outer part of f, has no zeros in the disk and belongs to H^p.

3. Every product of the form

$$b(z)s(z)e^{i\alpha}\exp\left(\int_{\mathbb{T}} \frac{\zeta+z}{\zeta-z}\log|g(\zeta)|\, dm(\zeta)\right) = bsF,$$

where b is a Blaschke product, s is a singular inner function, and g is an L^p function with $\log|g| \in L^1$, belongs to the space H^p. Furthermore $|F(\zeta)| = |g(\zeta)|$ almost everywhere.

We make a few important remarks about the factorization theorem and some of its consequences.

REMARK 3.2.5.   1. *The inner function:*
   (a) In the definition of the singular inner function s, it is important that the singular measure μ is positive since otherwise, s might not belong to H^p.
   (b) Using the definition of convergence of a product, one can show that the Blaschke product converges uniformly on compact subsets of the plane which do not contain the points $1/\overline{a_n}$ and the accumulation points of $\{a_n : n \in \mathbb{N}\}$. At the points $1/\overline{a_n}$, the Blaschke product will have a pole. Moreover, it follows from the definition of b that

$$b(z) = 1/\overline{b}(1/\overline{z}), \quad 1/\overline{z} \notin b^{-1}(\{0\}).$$

   (c) The singular inner function s is analytic in the plane except at those points in $\mathbb{T}$ which lie in the support of the positive singular measure $\mathbb{T}$. In fact, this domain of analyticity is, in a sense, "maximal" since neither s (nor $|s|$) can be continuously extended from $\mathbb{D}$ to any point in the support of μ [**44**], p. 68 (alternatively [**39**], p. 76). Again, it follows from the definition of s that

$$s(z) = 1/\overline{s}(1/\overline{z}), \quad |z| > 1.$$

   (d) With the observations above, it seems natural to define the *spectrum* of an inner function $I = bs$, denoted by $\sigma(I)$, to be the points of $\mathbb{D}^-$ consisting of the closure of the zeros of the Blaschke product b along with the support of the singular measure μ. One can show (see [**39**], p. 76) that $\sigma(I)$ is also the "lim inf zero set" for I, specifically,

$$\sigma(I) = \left\{ z \in \mathbb{D}^- : \liminf_{\lambda \to z} |I(\lambda)| = 0 \right\}.$$

   From the above remarks, observe that I is analytic on the region

$$\mathbb{C}_\infty \setminus \left\{ 1/\overline{z} : z \in \sigma(I) \right\}.$$

(e) Recalling the definition of a pseudocontinuation, Definition 2.2.1, we notice that the inner part, I_f, of $f \in H^p$ always has a pseudocontinuation across $\mathbb{T}$ given by the formula
$$(T_{I_f})(z) = 1/\overline{I_f(1/\bar{z})}.$$
Thus f has a pseudocontinuation if and only if its outer part O_f has a pseudocontinuation.

2. *The outer function:*
   (a) There is the following useful characterization of outer functions (H^p functions f with $I_f = 1$). If $f \in H^p$ ($0 < p \leq \infty$) is not the zero function, then the following conditions are equivalent.
      (i) f is an outer function.
      (ii)
      $$\log|f(0)| = \int_{\mathbb{T}} \log|f(\zeta)|\, dm(\zeta)$$
      (iii) If $g \in H^p$ and if $|g(\zeta)| \leq |f(\zeta)|$ almost everywhere, then $|g(z)| \leq |f(z)|$ for all $z \in \mathbb{D}$.
   (b) The previous result is often useful in determining whether or not an H^p function is outer. For example, if $f \in H^p$ and $|f| \geq \delta > 0$ on $\mathbb{D}$, then f is an outer function. One sees this by noting that the hypothesis imply $\log|f(z)|$ is a harmonic function and so (ii) follows from the mean value property for harmonic functions. In fact, more is true: If $\Re f(z) > 0$ for all $z \in \mathbb{D}$, then applying the above to the functions $f + \varepsilon$ and using the monotone and dominated convergence theorems, one can show that condition (ii) holds. Hence f is an outer function.

Up to this point, H^p is a just a linear space. In order to introduce a topology on H^p which is compatible with both the L^p norm of the boundary values and the topology of $\mathfrak{H}(\mathbb{D})$, we need the following growth estimate which one obtains by using the Cauchy integral formula and the factorization theorem.

PROPOSITION 3.2.6 (Growth estimates). *If $f \in H^p$ ($0 < p < \infty$), then*
$$|f(z)| \leq 2^{1/p} \left\{ \sup_{0<r<1} M_p(r,f) \right\} (1-|z|)^{-1/p} \quad \forall z \in \mathbb{D}.$$

In order to define a metric topology on H^p we define, for $f \in H^p$,
$$\|f\| := \lim_{r \to 1^-} M_p(r,f) = \left\{ \int_{\mathbb{T}} |f(\zeta)|^p\, dm(\zeta) \right\}^{1/p}.$$
This last equality comes from Theorem 3.2.3. For $p \in (0,1)$, the basic inequality
$$(a+b)^p \leq a^p + b^p, \quad a,b \geq 0,$$
says that
$$d(f,g) := \|f - g\|^p$$
is a metric on H^p; while for $p \geq 1$, the quantity $\|f\|$ defines a norm on H^p. Using Proposition 3.2.6 and Theorem 3.2.3 along with Montel's theorem we obtain the following.

THEOREM 3.2.7. H^p is an F-space for $p \in (0,1)$ and a Banach space for $p \in [1, \infty)$.

REMARK 3.2.8. Here an *F-space* is a complete topological vector space whose topology is given by a translation invariant metric.

More will be said about the topology of H^p (duality, convexity, etc.) in a later section.

Before leaving this section, we remark that one can define the Hardy spaces of the exterior disk $\mathbb{D}_e$ and many of the same properties hold.

DEFINITION 3.2.9. For $p \in (0, \infty)$, define $H^p(\mathbb{D}_e)$ as the set of holomorphic functions f on $\mathbb{D}_e$ for which

$$\|f\|_p := \sup_{r>1} \left\{ \int_{\mathbb{T}} |f(r\zeta)|^p \, dm(\zeta)/r \right\}^{1/p} < \infty. \quad [2]$$

Functions in $H^p(\mathbb{D}_e)$ have (exterior) non-tangential boundary values which belong to $L^p(\mathbb{T}, dm)$ and moreover, the $H^p(\mathbb{D}_e)$ norm is the same as the L^p norm of the boundary values. We also mention the following relationship between $H^p(\mathbb{D})$ and $H^p(\mathbb{D}_e)$. If $f \in \mathfrak{H}(\mathbb{D})$, we can define the function $F \in \mathfrak{H}(\mathbb{D}_e)$ by

$$(3.2.10) \qquad F(z) := \overline{f(1/\overline{z})}, \quad z \in \mathbb{D}_e.$$

If f has a power series $f = \sum_{n \geq 0} a_n z^n$, then F has a Taylor series (about infinity) $F = \sum_{n \geq 0} \overline{a_n} z^{-n}$. If f has non-tangential limit $f(\zeta)$ at ζ, then so does F and $F(\zeta) = \overline{f(\zeta)}$. Furthermore, the sesquilinear mapping $f \to F$ maps $H^p(\mathbb{D})$ isometrically onto $H^p(\mathbb{D}_e)$.

3.3. Fourier analysis and H^p theory

From the above theory, we see that for $p \in (0, \infty)$, the map which takes an $f \in H^p$ to its L^p non-tangential boundary values is a linear isometry between H^p and a closed subspace of L^p which we will call $H^p(\mathbb{T})$. An obvious question is, when does an $f \in L^p$ belong to $H^p(\mathbb{T})$? When $p \in [1, \infty]$, there is a complete answer.

THEOREM 3.3.1 (F. and M. Riesz theorem [72]). 1. For $p \in [1, \infty]$ a function $f \in L^p$ belongs to $H^p(\mathbb{T})$ if and only if the Fourier coefficients

$$\hat{f}(n) := \int_{\mathbb{T}} f(\zeta) \overline{\zeta}^n \, dm(\zeta)$$

vanish for $n < 0$.

2. If $f \in H^p(\mathbb{T})$ has Fourier series $f \sim \sum_{n=0}^{\infty} \hat{f}(n) \zeta^n$, then the Taylor series of the corresponding H^p function is $f = \sum_{n=0}^{\infty} \hat{f}(n) z^n$. Conversely, if $f \in H^p$ has Taylor series $f = \sum_{n=0}^{\infty} a_n z^n$, then $f \in H^p(\mathbb{T})$ has Fourier series $f \sim \sum_{n=0}^{\infty} a_n \zeta^n$.

3. Let μ be a finite (complex) Borel measure on $\mathbb{T}$. The Fourier coefficients

$$\hat{\mu}(n) := \int_{\mathbb{T}} \overline{\zeta}^n \, d\mu(\zeta)$$

vanish for $n < 0$ if and only if $d\mu = f \, dm$ for some $f \in H^1$.

---

[2] The factor $1/r$ is here so that the constant functions can belong to $H^p(\mathbb{D}_e)$.

From this theorem, note that $f \in H^p$ $(1 \leq p \leq \infty)$ can be recovered from its $H^p(\mathbb{T})$ boundary values by the Cauchy integral formula

$$\int_{\mathbb{T}} \frac{f(\zeta)}{1-\bar{\zeta}z} \, dm(\zeta) = \sum_{n=0}^{\infty} \hat{f}(n) z^n = f(z). \tag{3.3.2}$$

REMARK 3.3.3. In a loose sense, H^1 is the largest of the H^p spaces for which the Cauchy integral formula is valid. Functions in H^p $(0 < p < 1)$ still have non-tangential boundary values almost everywhere. However, since f need not be integrable, we cannot recapture f back in the disk from its $H^p(\mathbb{T})$ boundary values via Cauchy's theorem. There is indeed a deep and subtle connection between functions in $H^p(\mathbb{T})$ and a special class of tempered distributions. We will explore this further in Chapter 4.

From the F. and M. Riesz theorem, a function $f \in L^p$ $(1 \leq p \leq \infty)$, belongs to $H^p(\mathbb{T})$ if and only if $\hat{f}(n) = 0$ for all $n < 0$. Some care must be taken with this result in order for it not to be misunderstood. If $f \in L^p$ has Fourier series

$$f \sim \sum_{n=-\infty}^{\infty} \hat{f}(n) \zeta^n,$$

one might be tempted to say that the analytic function $(Pf)(z)$, defined by

$$(Pf)(z) := \sum_{n=0}^{\infty} \hat{f}(n) z^n = \int_{\mathbb{T}} \frac{f(\zeta)}{1-\bar{\zeta}z} \, dm(\zeta), \quad z \in \mathbb{D},$$

belongs to H^p. [3] Unfortunately, this is not always the case as the following well-known examples show.

EXAMPLE 3.3.4. 1. $f \sim \sum_{n=1}^{\infty} \sin(n\theta)/n = \arg(1-e^{i\theta})$ is the Fourier series of a bounded function. However

$$(Pf)(z) = \frac{1}{2i} \sum_{n=1}^{\infty} \frac{z^n}{n} = \frac{1}{2i} \log(1-z)$$

does not belong to H^∞.

2. $f \sim \sum_{n=2}^{\infty} \cos(n\theta)/\log n$ is the Fourier series of an L^1 function [31], p. 64, but

$$(Pf)(z) = \frac{1}{2} \sum_{n=2}^{\infty} \frac{z^n}{\log n}$$

does not belong to H^1 (see Hardy's inequality, eq.(5.8.3)).

Thus, the process of "truncating off" the negative Fourier coefficients of on L^p function does not return one to $H^p(\mathbb{T})$, at least for $p = 1$ and $p = \infty$. For $p \in (1, \infty)$, however, we do have the following positive result due to M. Riesz.

THEOREM 3.3.5 (M. Riesz [73] [74]). *For $p \in (1, \infty)$, the operator P is a continuous projection from L^p to H^p.*

REMARK 3.3.6. 1. The operator P is called the *Riesz projection operator* and can be used to decompose L^p $(1 < p < \infty)$ in the following way: By the F. and M. Riesz theorem, observe that $H^p \cap \overline{H_0^p} = (0)$, where $H_0^p = \{f \in$

---

[3] Note that Pf does indeed define an analytic function on $\mathbb{D}$ since $|\hat{f}(n)| = O(1)$.

$H^p : f(0) = 0\}$. Moreover, $L^p = PL^p + (I_d - P)L^p$ which, by the Riesz theorem, implies

(3.3.7) $$L^p = H^p \oplus \overline{H_0^p}.$$

Notice that we are equating H^p with $H^p(\mathbb{T})$ in the above equation. By the continuity of P and $(I_d - P)$, we also conclude that for $f \in H^p$ and $g \in \overline{H_0^p}$,

(3.3.8) $$\|f\|^p + \|g\|^p \leq C_p \|f + g\|^p.$$

That is to say, the direct sum in eq.(3.3.7) is closed. For $p \in (0,1)$ it is also the case that $L^p = H^p + \overline{H_0^p}$ however the sum is far from direct. In fact, $H^p \cap \overline{H_0^p}$ is quite a rich class of functions, see Proposition 6.1.6.

2. Not only is the projection P not continuous on L^1, but by a result of D. J. Newman [64] (a nice treatment of this can be found in Hoffman's book [44]), there is no other bounded projection from L^1 to H^1. There also no continuous projection from L^∞ to H^∞ [44].

3. The Riesz operator can also be thought of in terms of harmonic conjugation. Namely, given a function $u \in L^p$, find a $v \in L^p$ such that $u + iv \in H^p(\mathbb{T})$. Notice that if $1 < p < \infty$, the function

$$\tilde{u} := \frac{1}{i}(2Pu - \hat{u}(0) - u)$$

belongs to L^p and $u + i\tilde{u} = 2Pu - \hat{u}(0) \in H^p(\mathbb{T})$ and the operator $u \to \tilde{u}$, called the *harmonic conjugation operator*, is continuous on L^p.

4. It turns out that $PL^1 \subset H^p$ for all $p < 1$ and PL^∞ is equal to the space of analytic functions of "bounded mean oscillation", see Theorem 3.4.1 and Theorem 3.5.11.

3.4. The Cauchy transform

3.4.1. Basic definitions. For $\mu \in M(\mathbb{T})$, the finite (complex) Borel measures on $\mathbb{T}$, define the *Cauchy transform* [4]

$$(C\mu)(z) := \int \frac{d\mu(\zeta)}{\zeta - z}, \quad z \in \mathbb{C}_\infty \setminus \mathbb{T}.$$

For a $f \in L^1(\mathbb{T})$ we let $(Cf)(z)$ denote the Cauchy transform of the measure fdm. Differentiating under the integral sign shows that $C\mu$ is analytic on $\mathbb{C}_\infty \setminus \mathbb{T}$ and furthermore, $(C\mu)(\infty) = 0$. This next result gives us more; namely, Cauchy transforms of measures belong to $\bigcap \{H^p : p < 1\}$.

THEOREM 3.4.1 (Kolmogorov [49]). *If $\mu \in M(\mathbb{T})$, then $C\mu \in H^p$ for all $p \in (0,1)$ and moreover,*

$$\|C\mu\| \leq c_p \|\mu\|,$$

where

$$\|\mu\| = \sup \Big\{ \sum_{j=1}^n |\mu(A_j)| : A_j, 1 \leq j \leq n, \text{ is a measurable partition of } \mathbb{T} \Big\}$$

is the total variation of μ. Furthermore, $c_p = O((1-p)^{-1/p})$.

---

[4] In several texts, the authors equate, in a natural way, $M(\mathbb{T})$ with the left continuous functions u on $[0, 2\pi]$ with $u(0) = 0$. In this setting, the Cauchy transform is a Cauchy-Stieltjes integral.

Though this theorem is standard fare in most treatments on H^p theory, we would like to give a short proof since we will be using a slight variation of this theorem in Chapter 5. We start with a lemma which can be found in [**39**], p. 114. The original proof of Kolmogorov's result is based on "weak-type" estimates of maximal functions, see Theorem 4.4.7.

LEMMA 3.4.2. *Let $F \in \mathfrak{H}(\mathbb{D})$ with $\Re F > 0$. Then for all $r \in (0,1)$ and $p \in (0,1)$,*
$$\int_{\mathbb{T}} |F(r\zeta)|^p \, dm(\zeta) \leq A_p |F(0)|^p.$$
Moreover, $A_p = O((1-p)^{-1})$.

PROOF. Since $\Re F > 0$, then $F = |F|e^{i\phi}$, where $-\pi/2 < \phi < \pi/2$. Also observe that since F has no zeros in the disk, F^p is also analytic on $\mathbb{D}$ and
$$F^p = |F|^p (\cos(p\phi) + i\sin(p\phi)).$$
For $p \in (0,1)$,
$$\Re(F^p) = |F|^p \cos(p\phi) \geq |F|^p \cos(p\pi/2).$$
From this we conclude
$$\int_{\mathbb{T}} |F(r\zeta)|^p \, dm(\zeta) \leq A_p \int_{\mathbb{T}} \Re(F^p(r\zeta)) \, dm(\zeta) = A_p \Re(F^p(0)).$$
The last equality follows from the fact that $\Re(F^p)$ is harmonic. The desired inequality follows from the observation $\Re(F^p(0)) \leq |F(0)|^p$. Finally, the constant A_p above is equal to $1/\cos(p\pi/2) = O((1-p)^{-1})$. $\square$

PROOF OF THEOREM 3.4.1. For the moment, let us consider a non-negative measure $\nu \in M(\mathbb{T})$ and the modified Cauchy transform
$$F_\nu(z) := \int \frac{\zeta}{\zeta - z} d\nu(\zeta).$$
Observe that for $z \in \mathbb{D}$,
$$\Re F_\nu(z) = \int \frac{1 - \Re(\bar{\zeta}z)}{|\zeta - z|^2} d\nu(\zeta) > 0$$
and so by the previous lemma, we have, for fixed $p \in (0,1)$,
$$\|F_\nu\|^p \leq A_p |F_\nu(0)|^p = A_p \Big| \int d\nu \Big|^p = A_p \|\nu\|^p.$$
Decomposing the measure $\bar{\zeta}d\mu$ by the Jordan decomposition
$$\bar{\zeta}d\mu = (d\nu_1 - d\nu_2) + i(d\nu_3 - d\nu_4),$$
where $\nu_j \geq 0$ we see that
$$C_\mu(z) = F_{\nu_1}(z) - F_{\nu_2}(z) + iF_{\nu_3}(z) - iF_{\nu_4}(z).$$
Now apply the previous lemma to each one of the summands and notice the inequality $\|\nu_j\| \leq A\|\bar{\zeta}d\mu\| \leq A\|\mu\|$. $\square$

REMARK 3.4.3. 1. The Cauchy transform is really just the Riesz projection operator, in the sense that $(Cf)(z) = P(\bar{\zeta}f)(z)$. Hence the above result says that although $PL^1 \not\subset H^1$, P is a continuous projection from L^1 (or even $M(\mathbb{T})$) to H^p for each fixed $p \in (0,1)$.

2. This result also says that $C\mu \in H^p(\mathbb{D}_e)$ with the same estimate as in Theorem 3.4.1. Indeed the function $f(z) := \overline{(C\mu)(1/\overline{z})} \in \mathfrak{H}(\mathbb{D})$ and a short computation shows $f(z) = -z(C\overline{\zeta\mu})(z)$ which belongs to $H^p(\mathbb{D})$ by Theorem 3.4.1. But by eq.(3.2.10), $(C\mu)(z)$, on $\mathbb{D}_e$, is equal to $\overline{f}(1/\overline{z}) \in H^p(\mathbb{D}_e)$ and has the same norm as f.

Since the Cauchy transform of a measure $\mu \in M(\mathbb{T})$ belongs to $H^p(\mathbb{D})$ (respectively $H^p(\mathbb{D}_e)$), the radial limits (from the inside and outside) exist almost everywhere. An important quantity to look at, almost everywhere, is the jump

$$(3.4.4) \qquad J(\zeta) = \lim_{r \to 1^-} \left\{ (C\mu)(r\zeta) - (C\mu)(\zeta/r) \right\}, \quad \zeta \in \mathbb{T}.$$

Notice that for $w, \zeta \in \mathbb{T}$ and $r \in (0,1)$,

$$(3.4.5) \qquad \frac{1}{w - r\zeta} - \frac{1}{w - \zeta/r} = \overline{w} P_{r\zeta}(w),$$

where

$$P_{r\zeta}(w) = (1 - r^2)/|w - r\zeta|^2$$

is the usual Poisson kernel function.

DEFINITION 3.4.6. For a point $e^{i\theta} \in \mathbb{T}$ and $s > 0$, let $I_s(e^{i\theta})$ be the arc of the circle subtended by the points $e^{i(\theta-s)}$ and $e^{i(\theta+s)}$. For $\mu \in M(\mathbb{T})$, the *symmetric derivative* of μ (with respect to normalized Lebesgue measure dm) at $e^{i\theta}$ is defined to be

$$\mu'(e^{i\theta}) := \lim_{s \to 0^+} \frac{\mu(I_s(e^{i\theta}))}{m(I_s(e^{i\theta}))},$$

whenever this limit exists.

From elementary measure theory, we know that $\mu'(e^{i\theta})$ exists m-almost everywhere and μ' is the Radon-Nikodym derivative of μ with respect to m [5].

THEOREM 3.4.7 (Fatou [33]). *Let* $\mu \in M(\mathbb{T})$. *If* $\mu'(\zeta_0)$ *exists at* $\zeta_0 \in \mathbb{T}$, *then*

$$n.t. \lim_{z \to \zeta} \int_{\mathbb{T}} P_z(\zeta) d\mu(\zeta) = \mu'(\zeta_0).$$

Combining this theorem with eq.(3.4.4) and eq.(3.4.5), one can conclude Fatou's "jump theorem".

THEOREM 3.4.8 (Fatou's "jump theorem"). *Let* $\mu \in M(\mathbb{T})$. *If* $\mu'(\zeta_0)$ *exists at* $\zeta_0 \in \mathbb{T}$, *then* $J(\zeta_0) = \overline{\zeta_0}\mu'(\zeta_0)$.

REMARK 3.4.9. 1. By the Cauchy integral formula, eq.(3.3.2), note that every H^1 function f is a Cauchy transform $f = (C\zeta f)$. However, not every Cauchy transform belongs to H^1; for example, $1/(1-z) = C(\delta_1)$.

2. By our discussion above, every Cauchy transform is an analytic function on $\mathbb{C}_\infty \setminus \mathbb{T}$, vanishing at infinity, which satisfies the three conditions

$$(3.4.10) \qquad f \in \bigcap_{0 < p < 1} H^p(\mathbb{C}_\infty \setminus \mathbb{T}).$$

---

[5] Note that we are equating $M(\mathbb{T})$ with the left continuous functions u of bounded variation on $[0, 2\pi]$ with $u(0) = 0$. The symmetric derivative of u corresponds to the Radon-Nikodym derivative of μ with respect to dm, see [43], p. 331. We bring up this point due to the differences in various authors' presentation of Fatou's theorem.

(3.4.11) $$\liminf_{p \to 1^-} \|f\|_{H^p(\mathbb{C}_\infty \setminus \mathbb{T})}(1-p) < \infty.$$

(3.4.12) $$(Jf)(\zeta) = \lim_{r \to 1^-} \{ f(r\zeta) - f(\zeta/r) \} \in L^1.$$

Here $f \in \mathfrak{H}(\mathbb{C}_\infty \setminus \mathbb{T})$ belongs to $H^p(\mathbb{C}_\infty \setminus \mathbb{T})$ if
$$\|f\|^p_{H^p(\mathbb{C}_\infty \setminus \mathbb{T})} = \|f|\mathbb{D}\|^p_{H^p(\mathbb{D})} + \|f|\mathbb{D}_e\|^p_{H^p(\mathbb{D}_e)} < \infty.$$

By a result of Aleksandrov [4], the above three conditions are actually the defining properties of Cauchy transforms. The result is the following: Let $f \in \mathfrak{H}(\mathbb{C}_\infty \setminus \mathbb{T})$ and vanishing at infinity. Then

(a) $f = C\mu$ for some measure $\mu \in M(\mathbb{T})$ if and only if eq.(3.4.10), eq.(3.4.11), and eq.(3.4.12) hold.

(b) $f = Cg$ for some $g \in L^1(\mathbb{T})$ if and only if eq.(3.4.10), eq.(3.4.12) hold, and
$$\liminf_{p \to 1^-} \|f\|_{H^p(\mathbb{C}_\infty \setminus \mathbb{T})}(1-p) = 0.$$

(c) $f = C\mu$ for some singular measure (with respect to dm) if and only if eq.(3.4.10), eq.(3.4.11) hold, and $Jf = 0$ almost everywhere.

3.4.2. The space of Cauchy transforms. Using the Cauchy transform, we can define the Riesz projection of a measure $\mu \in M(\mathbb{T})$ by
$$(P\mu) := (C\zeta\mu)(z), \quad z \in \mathbb{D}.$$

Notice that
$$(P\mu)(z) = \sum_{n=0}^{\infty} \hat{\mu}(n) z^n.$$

In this section, we will discuss a standard topology one can place on the space
$$K := \Big\{ (P\mu)(z) = \int_{\mathbb{T}} \frac{1}{1 - \bar{\zeta}z} d\mu(\zeta) : \mu \in M(\mathbb{T}) \Big\}$$
of Cauchy transforms of measures. The reason for representing a Cauchy transform as a Riesz projection will become clear in a moment.

For $\mu \in M(\mathbb{T})$, recall the *total variation* of μ to be
$$\|\mu\| := \sup \Big\{ \sum_{j=1}^{n} |\mu(A_j)| : A_j, 1 \leq j \leq n, \text{ is a measurable partition of } \mathbb{T} \Big\}.$$

Routine arguments show that the total variation norm behaves well with respect to Jordan and Lebesgue decompositions.

PROPOSITION 3.4.13. *If $\mu = (\mu_1^+ - \mu_1^-) + i(\mu_2^+ - \mu_2^-)$ ($\mu_j^\pm \geq 0$) is the Jordan decomposition of μ and $\mu = \mu_a + \mu_s$ ($\mu_a \ll m$ and $\mu_s \perp m$) is the Lebesgue decomposition with respect to m, then*

1. $\frac{1}{\sqrt{2}}(\|\mu_1^+\| + \|\mu_1^-\| + \|\mu_2^+\| + \|\mu_2^-\|) \leq \|\mu\| \leq \|\mu_1^+\| + \|\mu_1^-\| + \|\mu_2^+\| + \|\mu_2^-\|$
2. $\|\mu\| = \|\mu_a\| + \|\mu_s\|.$

We can regard $\overline{H_0^1}$ as a linear manifold of $M(\mathbb{T})$ by equating $\overline{f} \in \overline{H_0^1}$ with the measure $\overline{f}dm$. With this identification, $\overline{H_0^1}$ is closed in $M(\mathbb{T})$ (with the total variation norm).

3.4. THE CAUCHY TRANSFORM

Since $\overline{H_0^1}$ is a subspace (closed linear manifold) of $M(\mathbb{T})$, the quotient space $M(\mathbb{T})/\overline{H_0^1} = \{[\mu] = \mu + \overline{H_0^1} : \mu \in M(\mathbb{T})\}$ is a Banach space when endowed with the natural quotient norm

$$\|[\mu]\| = \inf\left\{\|\mu + \nu\| : \nu \in \overline{H_0^1}\right\}.$$

Furthermore, by the F. and M. Riesz theorem, the map

$$[\mu] \to (P\mu)(z) = \int_{\mathbb{T}} \frac{1}{1 - \bar{\zeta}z} d\mu(\zeta)$$

is a one-to-one map between $M(\mathbb{T})/\overline{H_0^1}$ and the space of Cauchy transforms K. From here, it is quite natural to norm K via this correspondence, that is

$$\|P\mu\|_K := \|[\mu]\|.$$

EXAMPLE 3.4.14. Consider δ_1, the unit point mass at $z = 1$. Then $(P\delta_1)(z) = (1-z)^{-1}$ and so, by Proposition 3.4.13(2),

$$\|(1-z)^{-1}\|_K = \inf\{\|\delta_1 + \nu\| : \nu \in \overline{H_0^1}\} = \|\delta_1\| + \inf\{\|\nu\| : \nu \in \overline{H_0^1}\} = 1.$$

In fact, for any singular measure μ, $\|P\mu\|_K = \|\mu\|$.

There is a natural decomposition of K which corresponds to the Lebesgue decomposition of $M(\mathbb{T})$. Let $M_a = M_a(\mathbb{T})$ and $M_s = M_s(\mathbb{T})$ be defined by

$$M_a := \{\mu \in M(\mathbb{T}) : \mu \ll m\} \quad (= L^1)$$

$$M_s := \{\mu \in M(\mathbb{T}) : \mu \perp m\}$$

and notice from Proposition 3.4.13 that M_a and M_s are closed subspaces of $M(\mathbb{T})$ with

$$M(\mathbb{T}) = M_a \oplus M_s.$$

Thus, if we define

$$K_a := PM_a \quad \text{and} \quad K_s := PM_s,$$

we have the following summary result.

PROPOSITION 3.4.15. 1. $K = K_a \oplus K_s$
2. K_a is isometrically isomorphic to $L^1/\overline{H_0^1}$.
3. K_s is isometrically isomorphic to M_s.
4. K_a is the closure in K of the analytic polynomials.
5. K_s is nonseparable.

PROOF. To prove (4) note that each $h \in M_a$ $(= L^1)$ is the limit in L^1 of a sequence of trigonometric polynomials $\{h_n : n \in \mathbb{N}\}$. Also note that $\|Ph - Ph_n\|_K \leq \|h - h_n\|_{L^1} \to 0$ and Ph_n is an analytic polynomial for each n.

To prove (5), we note that

$$\|\delta_{e^{it}} - \delta_{e^{is}}\| = 2$$

for each $0 \leq s < t \leq 2\pi$ and so

$$\left\|\frac{1}{1 - e^{-it}z} - \frac{1}{1 - e^{-is}z}\right\|_K = 2.$$

This proves that K_s is nonseparable. $\square$

3.5. Duality

3.5.1. Some general remarks. We begin this section on the dual spaces of H^p by reminding the reader of some elementary functional analysis. A thorough reference for the proofs of the results below is Rudin's functional analysis book [78]. Let X be a (complex) Banach space and let X^* denote the normed dual of X; namely, the linear maps $\ell : X \to \mathbb{C}$ such that there is a constant $c > 0$ with $|\ell(x)| \leq c\|x\|$ $\forall x \in X$. The infimum of all such constants c for which the above holds will be the norm of ℓ and will be denoted by $\|\ell\|$. A basic fact from functional analysis is

$$\begin{aligned} \|\ell\| &= \inf\left\{ c > 0 : |\ell(x)| \leq c\|x\| \; \forall x \in X \right\} \\ &= \sup\left\{ |\ell(x)| : \|x\| \leq 1 \right\} \end{aligned}$$

and that X^*, with the above norm, is also a Banach space. It is important to note that X^* can be endowed with another topology, the wk-* topology, which stems from the duality. A basis of open sets about the origin for the wk-* topology are the sets

$$\bigcap_{j=1}^{n} \left\{ \ell \in X^* : |\ell(x_j)| < \varepsilon \right\}, \quad x_1, \cdots, x_n \in X, \; \varepsilon > 0.$$

The space $(X^*, wk-*)$ is a LCTVS (locally convex topological vector space) and $(X^*, wk-*)^*$, the continuous linear functionals on X^* (in the wk-* topology), can be identified with X. [6]

DEFINITION 3.5.1. If $\mathcal{M} \subset X$ and $\mathcal{N} \subset X^*$ are linear manifolds, one can define the *annihilator*

$$\mathcal{M}^\perp := \left\{ \ell \in X^* : \ell(x) = 0 \; \forall x \in \mathcal{M} \right\}$$

and the *pre-annihilator*

$$^\perp \mathcal{N} := \left\{ x \in X : \ell(x) = 0 \; \forall \ell \in \mathcal{N} \right\}.$$

A routine exercise shows that $\mathcal{M}^\perp$ is a wk-* closed linear manifold of X^* and $^\perp\mathcal{N}$ is a norm closed linear manifold of X. A result which we will exploit many times in these notes is the Hahn-Banach separation theorem.

THEOREM 3.5.2 (Hahn-Banach separation theorem).
1. Let A be a norm closed linear manifold of X and $x \in X \setminus A$. Then, there is an $\ell \in A^\perp$ such that $\ell(x) = 1$.
2. Let B be a wk-* closed linear manifold of X^* and $\ell \in X^* \setminus B$. Then there is an $x \in {}^\perp B$ such that $\ell(x) = 1$.

For our purposes, the most useful form of the Hahn-Banach theorem is the following corollary.

COROLLARY 3.5.3. Let $\mathcal{M}$ be a linear manifold of X and $\mathcal{N}$ be a linear manifold of X^*.
1. $^\perp(\mathcal{M}^\perp)$ is the norm closure of $\mathcal{M}$.
2. $(^\perp\mathcal{N})^\perp$ is the wk-* closure of $\mathcal{N}$.

---

[6] For the record, a *topological vector space* (TVS) is a linear space with a topology such that the vector space operations are continuous with respect to the topology. A *locally convex topological vector space* (LCTVS) is a TVS whose topology has a base consisting of convex sets.

The Hahn-Banach theorem as stated in Theorem 3.5.2, for a Banach space X, is an application of a more general Hahn-Banach theorem for LCTVS (locally convex topological vector spaces): If Y is an LCTVS, E is a closed linear manifold of Y, and $x \in Y \setminus E$, then there is an $\ell \in E^\perp$ such that $\ell(x) = 1$. In statement (1) of Theorem 3.5.2 we let $Y = X$ and note that a Banach space is certainly locally convex. For statement (2) of Theorem 3.5.2 we let $Y = X^*$ and note that $(X^*, wk-*)$ is an LCTVS with $(X^*, wk-*)^* = X$.

If $\mathcal{M}$ is a closed linear manifold of a Banach space X, one can consider the quotient space $X/\mathcal{M}$ consisting of the cosets $[x] = x + \mathcal{M}, x \in X$. If we define the norm
$$\|[x]\| := \inf\{\|x - y\| : y \in \mathcal{M}\},$$
then $X/\mathcal{M}$ becomes a Banach space. If $\pi : X \to X/\mathcal{M}$ is the natural quotient map, then for $\ell \in (X/\mathcal{M})^*$ define $\phi_\ell \in X^*$ by
$$\phi_\ell := \ell \circ \pi.$$
Observe that $\phi_\ell \in \mathcal{M}^\perp$ and moreover the map $\ell \to \phi_\ell$ is an isometric isomorphism between $(X/\mathcal{M})^*$ and $\mathcal{M}^\perp$, i.e.,

(3.5.4) $$(X/\mathcal{M})^* \simeq \mathcal{M}^\perp.$$

In a similar way, if $\ell \in \mathcal{M}^*$, then by the Hahn-Banach extension theorem, there is an $L \in X^*$ with $L|\mathcal{M} = \ell$ and $\|L\| = \|\ell\|$. For $\ell \in \mathcal{M}^*$, define $\psi_\ell \in X^*/\mathcal{M}^\perp$ by
$$\psi_\ell := L + \mathcal{M}^\perp$$
and note that the map $\ell \to \psi_\ell$ is an isometric isomorphism between $\mathcal{M}^*$ and $X^*/\mathcal{M}^\perp$, i.e.,

(3.5.5) $$\mathcal{M}^* \simeq X^*/\mathcal{M}^\perp.$$

In these next several sections, we examine the duals of several spaces of analytic functions related to the Hardy spaces.

3.5.2. H^p ($1 < p < \infty$). From eq.(3.5.5), we know that for $p \in (1, \infty)$, $(H^p)^*$ is isometrically isomorphic to $L^q/(H^p)^\perp$ which (by the F. and M. Riesz theorem) is equal to $L^q/\overline{H^q_0}$. Note that $q = p/(p-1)$ is the conjugate index to p. The Riesz projection is instrumental in equating $L^q/\overline{H^q_0}$ with a space of analytic functions on the disk. A simple application of Hölder's inequality shows that the linear functional
$$\ell_g(f) := \int f\overline{g}\, dm$$
is continuous on H^p whenever g belongs to H^q. Conversely, by the Hahn-Banach extension theorem, any continuous linear functional ℓ on H^p can be extended to a continuous linear functional on L^p and so, by the Riesz representation theorem,
$$\ell(f) = \int f\overline{g}\, dm \quad f \in H^p$$
for some $g \in L^q$. For trigonometric polynomials p_1 and p_2, a simple computation reveals that
$$\int (Pp_1)\overline{p_2}\, dm = \int p_1 \overline{Pp_2}\, dm.$$

Using the continuity of the Riesz projection $P : L^p \to H^p$, Theorem 3.3.5, and the density of the trigonometric polynomials in L^p, we can replace p_1 and p_2 with arbitrary functions in L^p (respectively L^q). Putting this altogether, we have

$$\ell(f) = \int (Pf)\overline{g}\, dm = \int f\overline{Pg}\, dm = \ell_{Pg}(f)$$

and the first two parts of the following theorem:

THEOREM 3.5.6.   1. *For $p \in (1, \infty)$, $(H^p)^*$ is isometrically isomorphic to $L^q/\overline{H_0^q}$.*
2. *$\ell \in (H^p)^*$ if and only if there is a $g \in H^q$ such that*

$$\ell(f) = \ell_g(f) := \int_{\mathbb{T}} f(\zeta)\overline{g}(\zeta)\, dm(\zeta)$$

for all $f \in H^p$.
3. *The norm of the above linear functional is comparable to the H^q-norm of g.*

REMARK 3.5.7. In the above theorem, the mapping $g \to \ell_g$ is sesquilinear but not linear. If the pairing were written as

$$\ell_g(f) = \int f(\zeta) g(\overline{\zeta}) dm(\zeta),$$

then the mapping $g \to \ell_g$ would be an isomorphism between $(H^p)^*$ and H^q. With either pairing, it is traditional to say that the dual of H^p "is" H^q.

3.5.3. H^1 and bounded mean oscillation. Again using eq.(3.5.5), $(H^1)^*$ is isometrically isomorphic to $L^\infty/(H^1)^\perp = L^\infty/\overline{H_0^\infty}$. Moreover, by our previous analysis with the Hahn-Banach extension theorem, every $\ell \in (H^1)^*$ is of the form

$$\ell(f) = \int f\overline{g}\, dm \quad f \in H^1$$

for some $g \in L^\infty$. However, we are unable use the Riesz projection as before to assume that $g \in H^\infty$ (since $PL^\infty \not\subset H^\infty$). Instead, with a slightly different linear functional, g can be taken to belong to $BMOA$, the analytic functions of *bounded mean oscillation*. A good general reference for BMO and VMO are the notes of Sarason [81].

DEFINITION 3.5.8. A function $g \in L^1$ is of *bounded mean oscillation*, or BMO, if

$$\sup_I \frac{1}{|I|} \int_I |g - g_I|\, dm < \infty,$$

where the sup is taken over all arcs $I \subset \mathbb{T}$ and

$$g_I := \frac{1}{|I|} \int_I g\, dm$$

is the "average" of g on I.

With the norm

$$\|g\|_* := \left| \int g\, dm \right| + \sup_I \frac{1}{|I|} \int_I |g - g_I|\, dm,$$

one can check that BMO is a Banach space of functions on $\mathbb{T}$.

DEFINITION 3.5.9. The functions in $BMOA := BMO \cap H^1(\mathbb{T})$ are the analytic functions of bounded mean oscillation.

REMARK 3.5.10. A few facts about the space BMO are in order here.
1. The first summand in the definition of the norm $\|\cdot\|_*$ is necessary in order to distinguish the constants.
2. In the definition of BMO, it is not important that we subtract off g_I, the mean value of g on I. If there is a constant $M \geq 0$ such that for each interval I, there is a constant a_I so that
$$\frac{1}{|I|}\int_I |g - a_I|\, dm \leq M,$$
then $|g_I - a_I| \leq M$ and so
$$\sup_I \frac{1}{|I|}\int_I |g - g_I|\, dm \leq 2M.$$
From this, it follows from the inequality $||g|-|g_I||\leq |g-g_I|$ that $|g| \in BMO$ whenever $g \in BMO$.
3. It is clear that $L^\infty \subset BMO$. However this containment is strict since $f(e^{i\theta}) = \log|\theta|$ belongs to BMO but is unbounded.
4. If $g \in BMO$ then it follows from the John-Nirenberg theorem [39], p. 233, that for any $p \in (1, \infty)$
$$\sup_I \left\{\frac{1}{|I|}\int_I |g - g_I|^p\, dm\right\}^{1/p} \leq C_p \|g\|_*$$
and so, in particular, $BMO \subset \bigcap_{p>1} L^p$.
5. By looking at characteristic functions of arcs of the circle, it is not too difficult to see that BMO is not separable.

Recall from our previous discussion that $PL^\infty \not\subset H^\infty$. However, the following is true.

THEOREM 3.5.11 (Spanne [90], Stein [93]). *The Riesz projection operator P is continuous from L^∞ onto $BMOA$.*

Thus the continuous linear functional on H^1 defined by
$$\ell(f) = \int f\bar{g}\, dm, \ g \in L^\infty$$
can be written as (since $f_r(\zeta) := f(r\zeta)$ goes to f in H^1 as $r \to 1^-$, Theorem 3.2.3)
$$\ell(f) = \lim_{r \to 1^-} \int f_r \bar{g}\, dm = \lim_{r \to 1^-} \int (Pf_r)\bar{g}\, dm = \lim_{r \to 1^-} \int f_r \overline{(Pg)}\, dm$$
and so every $\ell \in (H^1)^*$ can be written in the form
$$\ell(f) = \lim_{r \to 1^-} \int f(r\zeta)\bar{g}(\zeta)\, dm(\zeta) \tag{3.5.12}$$
for some $g \in BMOA$.

The converse statement (that the linear functional as defined by eq.(3.5.12) is continuous on H^1 for a given $BMOA$ function g) is the content of a very deep theorem of Fefferman [34] and Fefferman and Stein [35].

THEOREM 3.5.13.   1. $(H^1)^*$ *is isometrically isomorphic to* $L^\infty/\overline{H^\infty_0}$.

2. *(Fefferman-Stein [34] [35])* $\ell \in (H^1)^*$ *if and only if there is a* $g \in BMOA$ *so that*
$$\ell(f) = \ell_g(f) := \lim_{r \to 1^-} \int_{\mathbb{T}} f(r\zeta)\overline{g}(\zeta)\, dm(\zeta).$$

3. *(Fefferman-Stein [34] [35]) The norm of this linear functional is comparable to the BMOA-norm of* g.

REMARK 3.5.14. Again, as with H^p, the mapping $g \to \ell_g$ is sesquilinear. If the pairing were changed to
$$\ell_g(f) = \lim_{r \to 1^-} \int_{\mathbb{T}} f(r\zeta)g(\overline{\zeta})\, dm(\zeta),$$
the mapping $g \to \ell_g$ would serve as an isomorphism between $(H^1)^*$ and $BMOA$. In either case, we say that $(H^1)^*$ "is" $BMOA$.

Actually this result follows from a more general result of Fefferman which shows that $(\Re H^1)^*$ can be identified with BMO. Here
$$\Re H^1 = \{\Re F : F \in H^1\}$$
is the $\mathbb{R}$-linear space with norm $\|\Re F\| := \|F\|_{H^1}$, where the imaginary part of F vanishes at the origin. The main result of Fefferman [34] [35] is the following.

THEOREM 3.5.15 (Fefferman duality theorem). *The dual of* $\Re H^1$ *can be identified with* BMO *in the following sense:*

1. *If* $g \in BMO$ *and real valued, the* $\mathbb{R}$-*linear functional*

(3.5.16) $$f \to \int f\overline{g}\, dm$$

   (initially defined on $\Re H^1 \cap L^2$ *which is dense in* $\Re H^1$*) can be extended to be a continuous linear functional on* $\Re H^1$.
2. *Every continuous* $\mathbb{R}$-*linear functional on* $\Re H^1$ *arises as in eq.(3.5.16) for some unique* $g \in BMO$.
3. *The norm of this linear functional is equivalent to the BMO norm of* g.

It is often useful, as we will see in Chapter 5, to represent the linear functional
$$\ell(f) = \lim_{r \to 1^-} \int f_r \overline{g}\, dm, \ f \in H^1,\ g \in BMOA,$$
in a slightly different form using the "truncation operator". The reader can consult a paper of C. Sundberg [97] for further results about this operator as well as other applications.

DEFINITION 3.5.17. For $g \in BMO$ and $\rho > 0$ define the *truncation* $h_\rho(g)$ of g to be
$$h_\rho(g) := \begin{cases} g & \text{if } |g| \leq \rho, \\ \rho g/|g| & \text{otherwise.} \end{cases}$$

Note that $h_\rho(g) \in BMO$ (since it is bounded) and since
$$|h_\rho(g) - h_\rho(g_I)| \leq |g - g_I|,$$

we have
$$\frac{1}{|I|} \int_I |h_\rho(g) - h_\rho(g)_I| \, dm$$
$$\leq \frac{1}{|I|} \int_I |h_\rho(g) - h_\rho(g_I)| \, dm + \frac{1}{|I|} \int_I |h_\rho(g)_I - h_\rho(g_I)| \, dm$$
$$\leq \frac{1}{|I|} \int_I |g - g_I| \, dm + |h_\rho(g)_I - h_\rho(g_I)|$$
$$\leq \|g\|_* + \left| \frac{1}{|I|} \int_I (h_\rho(g) - h_\rho(g_I)) \, dm \right|$$
$$\leq \|g\|_* + \frac{1}{|I|} \int_I |g - g_I| \, dm$$
$$\leq 2\|g\|_*$$

Thus, by the second remark following the definition of BMO,

(3.5.18) $$\|h_\rho(g)\|_* \leq C\|g\|_*$$

For $g \in BMOA$, the linear functional $\ell_g(f)$ can be written using the truncation operator in the following way.

PROPOSITION 3.5.19.   1. *For each $g \in BMO$, the linear functional*
$$f \to \lim_{\rho \to \infty} \int f \overline{h_\rho(g)} \, dm$$
is continuous on $\Re H^1$.
2. *For $g \in BMOA$ and $f \in H^1$*
$$\lim_{r \to 1^-} \int f_r \overline{g} \, dm = \lim_{\rho \to \infty} \int f \overline{h_\rho(g)} \, dm.$$

PROOF. Since $h_\rho(g)$ is a bounded BMO function, the linear functional
$$f \to \int f \overline{h_\rho(g)} \, dm$$
is continuous on $\Re H^1$. Next note that for $f \in \Re H^2$ we have
$$\int f \overline{h_\rho(g)} \, dm = \int_{|g| \leq \rho} f \overline{g} \, dm + \int_{|g| > \rho} f \frac{\rho \overline{g}}{|g|} \, dm$$
and so by the dominated convergence theorem (note that $f\overline{g} \in L^1$ since $BMO \subset L^p$ for all $p > 1$ and $f \in L^2$) we obtain

(3.5.20) $$\lim_{\rho \to \infty} \int f \overline{h_\rho(g)} \, dm = \int f \overline{g} \, dm.$$

To show that the following limit
$$\lim_{\rho \to \infty} \int f \overline{h_\rho(g)} \, dm$$
exists for every $f \in \Re H^1$, we let $\rho_n \to \infty$ and $\varepsilon > 0$ be given. Then using eq.(3.5.18) and the fact that $f_s \to f$ in H^1 we can fix $s \in (0,1)$ so that
$$\left\{ \sup_{\rho > 0} \|h_\rho(g)\|_* \right\} \|f_s - f\|_1 \leq \varepsilon/2.$$

From eq.(3.5.20)

(3.5.21) $$\left| \int f_s \overline{h_{\rho_n}(g)} \, dm - \int f_s \overline{h_{\rho_m}(g)} \, dm \right| \to 0, \quad n, m \to \infty.$$

Then
$$\left| \int f \overline{h_{\rho_n}(g)} \, dm - \int f \overline{h_{\rho_m}(g)} \, dm \right|$$
$$\leq \left| \int (f - f_s) \overline{h_{\rho_n}(g)} \, dm \right| + \left| \int (f - f_s) \overline{h_{\rho_m}(g)} \, dm \right|$$
$$+ \left| \int f_s \overline{h_{\rho_n}(g)} \, dm - \int f_s \overline{h_{\rho_m}(g)} \, dm \right|$$
$$\leq 2 \left\{ \sup_{\rho > 0} \|h_\rho(g)\|_* \right\} \|f_s - f\|_1 + \left| \int f_s \overline{h_{\rho_n}(g)} \, dm - \int f_s \overline{h_{\rho_m}(g)} \, dm \right|$$
$$\leq \varepsilon + \left| \int f_s \overline{h_{\rho_n}(g)} \, dm - \int f_s \overline{h_{\rho_m}(g)} \, dm \right|$$

Thus, from eq.(3.5.21),
$$\int f \overline{h_{\rho_n}(g)} \, dm$$
is a Cauchy sequence and so

(3.5.22) $$\lim_{\rho \to \infty} \int f \overline{h_\rho(g)} \, dm \quad \text{exists}.$$

From our previous discussion, the linear functional
$$f \to \int f \overline{h_\rho(g)} \, dm$$
is continuous on $\Re H^1$ for each fixed $\rho > 0$ and by eq.(3.5.22) and the Banach-Steinhaus theorem, the linear functional
$$L_g(f) = \lim_{\rho \to \infty} \int f \overline{h_\rho(g)} \, dm$$
is continuous on $\Re H^1$. This proves statement (1).

For $g \in BMOA$, the linear functional
$$\ell_g(f) := \lim_{r \to 1^-} \int f_r \overline{g} \, dm$$
is continuous on H^1. From the above and eq.(3.5.20), $L_g = \ell_g$ on H^2 (which is dense in H^1). Thus $L_g = \ell_g$ on H^1 which proves statement (2). □

3.5.4. Vanishing mean oscillation and H^1. A natural subspace of BMO is the space VMO, the functions of *vanishing mean oscillation*. These are functions $f \in BMO$ such that
$$\lim_{a \to 0} \left\{ \sup_{|I| \leq a} \frac{1}{|I|} \int_I |f - f_I| \, dm \right\} = 0.$$

A routine argument shows that VMO is a closed subspace of BMO. Furthermore, there are the following important equivalent characterizations of VMO.

THEOREM 3.5.23. *For $f \in BMO$, the following are equivalent.*
1. $f \in VMO$.
2. $\|T_t f - f\|_* \to 0$ as $t \to 0$, where $(T_t f)(\zeta) = f(\zeta e^{-it})$.

3. $\|f_r - f\|_* \to 0$ as $r \to 1^-$, where
$$f_r(\zeta) = \int_{\mathbb{T}} f(w) P_{r\zeta}(w) \, dm(w)$$
is the Poisson extension of f at the point $r\zeta$.

REMARK 3.5.24. Condition (3) in the previous theorem says that VMO is a separable space. In fact, the trigonometric polynomials are dense in VMO. One sees this by noting that for fixed $r \in (0,1)$, the function f_r is infinitely differentiable and so can be approximated in VMO-norm by a sequence of trigonometric polynomials. Now use condition (3).

DEFINITION 3.5.25. As we did for BMO, we consider the analytic functions of vanishing mean oscillation $VMOA := VMO \cap H^1(\mathbb{T})$.

The following result of Sarason is the analog of Stein's (Spanne's) result for $BMOA$.

THEOREM 3.5.26 (Sarason [80]). *The Riesz projection operator P is continuous from $C(\mathbb{T})$ onto $VMOA$.*

Sarason's theorem allows us to relate $VMOA$ with the pre-dual of H^1 as follows. Let A denote the *disk algebra* of functions in $C(\mathbb{T}) \cap H^1(\mathbb{T})$. By the Riesz representation theorem, $(C(\mathbb{T}))^*$ is isometrically isomorphic to $M(\mathbb{T})$ via the pairing
$$\int f d\mu.$$
Thus, from eq.(3.5.4), the dual of $C(\mathbb{T})/\overline{zA}$ is isometrically isomorphic to the space
$$(\overline{zA})^\perp = \left\{ \mu \in M(\mathbb{T}) : \int \overline{\zeta}^n d\mu = 0 \ \forall n \in \mathbb{N} \cup \{0\} \right\}$$
which, by the F. and M. Riesz theorem is equal to $\overline{H^1}$. Thus $(C(\mathbb{T})/\overline{zA})^* \simeq \overline{H^1}$.

In order to represent $C(\mathbb{T})/\overline{zA}$ as a space of analytic functions, we use the Riesz projection operator and Sarason's theorem. The Riesz projection $P : C(\mathbb{T}) \to VMOA$ is continuous and onto. Moreover the kernel of P is $\overline{zA}$ and so the map
$$f + \overline{zA} \to Pf$$
is an isomorphism from $C(\mathbb{T})/\overline{zA}$ onto $VMOA$. Putting this all together, we see that
$$(VMOA)^* \simeq \overline{H^1}.$$
To represent this duality as was done with the duality between H^1 and $BMOA$ we note that by Fefferman's duality theorem, the linear functional
$$f \to \lim_{r \to 1^-} \int_{\mathbb{T}} f_r \overline{g} \, dm$$
is continuous on $VMOA$ for each H^1. Conversely if $\ell \in (VMOA)^*$, the linear functional
$$h \to \ell(Ph)$$
is continuous on $C(\mathbb{T})$ and so there is a measure $\mu \in M(\mathbb{T})$ so that
$$\ell(Ph) = \int h \, d\mu$$

for all $h \in C(\mathbb{T})$. Notice that
$$\int h \, d\mu = 0$$
for all $h \in \overline{zA}$ and so by the F. and M. Riesz theorem, $d\mu = \bar{g} \, dm$ for some $g \in H^1$. Thus for $f \in VMOA$ we have $f = Ph$ for some $h \in C(\mathbb{T})$ and so

$$\begin{aligned}
\ell(f) &= \ell(Ph) \\
&= \int h\bar{g} \, dm \\
&= \lim_{r \to 1^-} \int h \overline{Pg_r} \, dm \\
&= \lim_{r \to 1^-} \int f \overline{g_r} \, dm \\
&= \lim_{r \to 1^-} \int f_r \bar{g} \, dm.
\end{aligned}$$

Combining all these results, we have the following theorem.

THEOREM 3.5.27.   1. *The dual of $C(\mathbb{T})/\overline{zA}$ is isometrically isomorphic to $\overline{H^1}$.*
2. $C(\mathbb{T})/\overline{zA} \simeq VMOA$.
3. $\ell \in (VMOA)^*$ *if and only if there is a $g \in H^1$ such that*
$$\ell(f) = \ell_g(f) := \lim_{r \to 1^-} \int_{\mathbb{T}} f_r(\zeta) \bar{g}(\zeta) \, dm(\zeta)$$
for all $f \in VMOA$.

We mention that an analog to Fefferman's duality theorem, Theorem 3.5.15 is true for VMO; that is, the dual of VMO can be identified with $\Re H^1$.

3.5.5. $L^1/\overline{H_0^1}$. Recall from the previous section on Cauchy transforms that the space $K = PM(\mathbb{T})$ of Cauchy transforms is endowed with the quotient space topology of $M(\mathbb{T})/H_0^1$; specifically,
$$\|P\mu\|_K = \inf\left\{\|\mu - \nu\| : \nu \in \overline{H_0^1}\right\}.$$
Also recall that K can be decomposed as
$$K = K_a \oplus K_s,$$
where $K_a = PM_a$ (the Cauchy transforms of measures $\mu \in M(\mathbb{T})$ with $\mu \ll m$) and $K_s = PM_s$ (the Cauchy transforms of measures $\mu \in M(\mathbb{T})$ with $\mu \perp m$). Moreover, K_a is isometrically isomorphic to $M_a/\overline{H_0^1}$ and K_s is isometrically isomorphic to M_s. From the F. and M. Riesz theorem, we have
$$\overline{H_0^1}^\perp = \left\{g \in L^\infty : \int \bar{f}g \, dm = 0 \; \forall f \in H_0^1\right\} = H^\infty$$
and so from eq.(3.5.4), $(M_a/\overline{H_0^1})^*$ is isometrically isomorphic to $\overline{H_0^1}^\perp = H^\infty$, where each $\ell \in (M_a/\overline{H_0^1})^*$ is of the form
$$\ell([f]) = \int f\bar{g} \, dm$$

for some $g \in H^\infty$. Thus the space $(M_a/\overline{H_0^1}) \simeq K_a$ is important since it is the pre-dual of H^∞. In the previous sentence, notice that we said that $K_a \simeq M_a/\overline{H_0^1}$ is "the" pre-dual of H^∞. This is not a careless use of language but a theorem of Ando [10].

THEOREM 3.5.28. *Let X be a complex Banach space such that X^* is isometrically isomorphic to H^∞. Then X is isometrically isomorphic to $M_a/\overline{H_0^1}$.*

The fact that H^∞ has a unique (isometric) pre-dual is not to be taken lightly since this is not the case for a general Banach space (for example l^1 [13]).

The dual of K_a can be identified with H^∞ directly. Indeed if $f = Ph$ for some $h \in L^1$, and $g \in H^\infty$, we use the fact that $h_r(\zeta) = \sum_{n=-\infty}^{\infty} r^{|n|} \zeta^n \hat{h}(n)$ (the Poisson extension of h) goes to h in L^1 to get

$$\begin{aligned}\int h\bar{g}\,dm &= \lim_{r \to 1^-} \int h_r \bar{g}\,dm \\ &= \lim_{r \to 1^-} \int h_r \overline{Pg}\,dm \\ &= \lim_{r \to 1^-} \int f_r \bar{g}\,dm\end{aligned}$$

Thus $\ell \in (K_a)^*$ if and only if

$$\ell(f) = \ell_g(f) := \lim_{r \to 1^-} \int f_r \bar{g}\,dm$$

for some $g \in H^\infty$.

3.5.6. H^p $(0 < p < 1)$. Even though Day [25] showed that L^p $(0 < p < 1)$ has no non-zero bounded linear functionals, the Hardy space H^p $(0 < p < 1)$ has many. For example, by Proposition 3.2.6, the linear functional $f \to f(z)$ is continuous for each fixed $z \in \mathbb{D}$. There are many others and their complete description is the content of the next result of Duren, Romberg and Shields [32] which we will describe after a few definitions.

DEFINITION 3.5.29. 1. Let $n \in \mathbb{N} \cup \{0\}$ and $\alpha \in (0,1)$. Define the *Lipschitz class* $\Lambda_\alpha^n(\mathbb{T})$ to be the space of functions $g(\theta) = g(e^{i\theta}) \in C^{(n)}(\mathbb{T})$ such that

$$\sup_{\theta \in \mathbb{R}, t \neq 0} \frac{|g^{(n)}(\theta + t) - g^{(n)}(\theta)|}{|t|^\alpha} < \infty.$$

Λ_α^n becomes a (non-separable) Banach space when given the norm

$$\|g\|_{\Lambda_\alpha^n} := \|g\|_\infty + \max_{0 \leq j \leq n} \sup_{\theta \in \mathbb{R}, t \neq 0} \frac{|g^{(j)}(\theta + t) - g^{(j)}(\theta)|}{|t|^\alpha}.$$

2. Define the Zygmund class $\Lambda_*^n(\mathbb{T})$ to be the space of functions $g \in C^{(n)}(\mathbb{T})$ such that

$$\sup_{\theta \in \mathbb{R}, t \neq 0} \frac{|g^{(n)}(\theta + t) - 2g^{(n)}(\theta) + g^{(n)}(\theta - t)|}{|t|} < \infty.$$

A norm on Λ_*^n is

$$\|g\|_{\Lambda_*^n} = \|g\|_\infty + \max_{0 \leq j \leq n} \sup_{\theta \in \mathbb{R}, t \neq 0} \frac{|g^{(j)}(\theta + t) - 2g^{(j)}(\theta) + g^{(j)}(\theta - t)|}{|t|}.$$

3. We remark that the Riesz projection operator acts well on these spaces[7] and in fact
$$P\Lambda_\alpha^n \subset \Lambda_\alpha^n, \quad P\Lambda_*^n \subset \Lambda_*^n.$$
We often refer to the spaces $P\Lambda_\alpha^n$ and $P\Lambda_*^n$ as the *analytic Lipschitz* and *analytic Zygmund* classes (respectively).

We also remark that an analytic function f on $\mathbb{D}$ belongs to $P\Lambda_\alpha^n$ if and only if
$$|f^{(n+1)}(z)| = O\left(\frac{1}{(1-|z|)^{1-\alpha}}\right)$$
and f belongs to $P\Lambda_*^n$ if and only if
$$|f^{(n+2)}(z)| = O\left(\frac{1}{1-|z|}\right).$$

DEFINITION 3.5.30. We set
$$\Lambda_{1/p-1} = \begin{cases} \Lambda_{1/p-n}^{n-1} & \text{if } p \in (1/(n+1), 1/n), \\ \Lambda_*^{n-1} & \text{if } p = 1/(n+1). \end{cases}$$
and $P\Lambda_{1/p-1}$ to be the corresponding space of analytic functions.

We are now ready to discuss the dual space of H^p ($0 < p < 1$).

THEOREM 3.5.31 (Duren, Romberg, Shields [32]). 1. *For $p \in (0,1)$, and $\ell \in (H^p)^*$ there exists a unique function g which is holomorphic in $\mathbb{D}$ and continuous on $\mathbb{D}^-$ such that*
$$(3.5.32) \qquad \ell(f) = \lim_{r \to 1^-} \int_\mathbb{T} f(r\zeta)\overline{g}(\zeta)\, dm(\zeta), \quad \forall f \in H^p.$$

2. *Moreover $g \in P\Lambda_{1/p-1}$ and the norm of the linear functional in eq.(3.5.32) is equivalent to the $\Lambda_{1/p-1}$ norm of g.*

Before proceeding to the next section, we make some remarks about the complexity of the topology of the H^p ($0 < p < 1$) spaces. As mentioned previously, H^p, with the metric
$$d(f,g) = \|f - g\|^p$$
is an *F-space* (a complete topological vector space whose topology is given by a translation invariant metric) and many of the results from Banach space theory such as the uniform boundedness principle, the open mapping theorem, and the closed graph theorem are still valid [30], Chapter II. Unfortunately, the Hahn-Banach separation theorem is not. At the center of problem is the fact that H^p ($0 < p < 1$) is neither normable nor locally convex. Recall that a topological vector space is *locally convex* if it has a base for its topology consisting of convex sets.

In L^p ($0 < p < 1$) it is well known that there are no non-trivial open convex subsets (so certainly L^p is neither normable [8] nor locally convex) and it follows [78], p. 37, that $(L^p)^* = (0)$. For H^p ($0 < p < 1$), there are non-trivial convex open subsets, e.g.,
$$\{\, f \in H^p : \Re f(0) > 0\, \},$$

---

[7] For the Lipschitz classes this was discovered by Privalov, [67]. See [51], p. 110 for a treatment of this. For the Zygmund classes, this was discovered by Zygmund [104].

[8] Recall a result of Kolmogorov [50] which says that a TVS has an equivalent norm topology if and only if the space contains a non-empty bounded open convex set. Here a set A is *bounded* if for every open neighborhood U of the origin, there is a number $\varepsilon > 0$ so that $\varepsilon A \subset U$.

but unfortunately, by a result of Livingston [56], none of them are bounded and so H^p is not locally convex.

In a locally convex F-space X, the Hahn-Banach separation theorem holds: Given a closed subspace E and a vector $x \notin E$, there is an $\ell \in X^*$ such that $\ell(E) = 0$ but $\ell(x) \neq 0$. Since H^p ($0 < p < 1$) is non-locally convex, one might suspect there might be a problem with the Hahn-Banach separation property. Our fears are realized since indeed there are examples of non-trivial subspaces E of H^p such that $E^\perp = \{\ell \in (H^p)^* : \ell(E) = 0\} = (0)$, see eq.(6.1.10). As it turns out though, this pathology is not particular to the Hardy spaces but is a reflection of the non-local convexity of the topology. A sharp result of Kalton [46] describes completely the relationship between the Hahn-Banach separation property and local convexity: An F-space is locally convex if and only if it has the Hahn-Banach separation property.

3.5.7. Summary table. The following table describes the dual pairings via the "Cauchy duality"

$$\int f\bar{g}\,dm, \quad f \in X, g \in X^* \tag{3.5.33}$$

(at least when this makes sense).

X	X^*	alternate pairing to eq.(3.5.33)
$H^p, p > 1$	H^q	—
H^1	$BMOA$	$\int f\overline{g_1}dm, g_1 \in L^\infty, Pg_1 = g$
$\Re H^1$	BMO	—
$VMOA \simeq C(\mathbb{T})/\overline{zA}$	H^1	$\int f_1\bar{g}dm, f_1 \in C(\mathbb{T}), Pf_1 = f$
VMO	$\Re H^1$	—
$K_a \simeq M_a/\overline{H_0^1}$	H^∞	$\int f_1\bar{g}dm, f_1 \in L^1, Pf_1 = f$
$H^p, 0 < p < 1$	$P\Lambda_{1/p-1}$	—

For future reference, we also list the behavior of the Riesz projection operator P.

X	PX
$L^p, 1 < p < \infty$	H^p
L^1	K_a
L^∞	$BMOA$
$C(\mathbb{T})$	$VMOA$
$M(\mathbb{T})$	K

3.6. The Nevanlinna class

3.6.1. The Nevanlinna class. From the factorization theorem for H^p functions, Theorem 3.2.4, any H^p function can be written in the form $bs_\mu F_f$, where b is the Blaschke product, s_μ is a singular inner function with positive singular measure μ, and F_f is an outer function formed from the L^p function f on $\mathbb{T}$ with $\log|f| \in L^1$. In a sense, the broadest class of functions one can form from these factors, for which the factors still make sense, are where μ is just a real measure (not necessarily positive) which is singular with respect to Lebesgue measure on $\mathbb{T}$, and $\log|f| \in L^1$ (f not necessarily L^p). Writing the real singular μ measure as

$\mu_1 - \mu_2$ as the difference of two positive measures, this class of functions will be of the form

(3.6.1) $$b\frac{S_{\mu_1}}{S_{\mu_2}}F_f$$

and form an important class of functions called the Nevanlinna class. A formal definition will be given below. We refer the reader to the papers [87] [103] and Chapter 2 of Duren's book [31] for more detailed information.

For $f \in \mathfrak{H}(\mathbb{D})$ the function $\log(1 + |f|)$ is a subharmonic function on $\mathbb{D}$ and so the integrals

$$T(r,f) = \int_{\mathbb{T}} \log(1 + |f(r\zeta)|) \, dm(\zeta)$$

define increasing functions of r on $[0,1)$.

DEFINITION 3.6.2. Define the *Nevanlinna class*, denoted by N, to be the $f \in \mathfrak{H}(\mathbb{D})$ for which

$$\|f\| = \lim_{r \to 1} T(r,f) < \infty.$$

Some authors define N to be the space of $f \in \mathfrak{H}(\mathbb{D})$ for which

$$\sup_{0<r<1} \int_{\mathbb{T}} \log^+ |f(r\zeta)| \, dm(\zeta) = \lim_{r \to 1^-} \int_{\mathbb{T}} \log^+ |f(r\zeta)| \, dm(\zeta) < \infty.$$

Using the inequality

$$\log^+ x \le \log(1+x) \le \log 2 + \log^+ x,$$

one sees that these two definitions are equivalent. Using the inequalities

$$\log(1 + |x+y|) \le \log(1 + |x|) + \log(1 + |y|)$$

$$\log(1 + |x||y|) \le \log(1 + |x|) + \log(1 + |y|),$$

it is routine to show that

(3.6.3) $$\rho(f,g) := \|f - g\|$$

defines a translation invariant metric on N and that N is a topological group under addition ($f_n \to f$, $g_n \to g$ in $N \Rightarrow f_n + g_n \to f + g$ in N) and N is an algebra under multiplication ($f,g \in N \Rightarrow f \cdot g \in N$). Unfortunately N is not a topological vector space since scalar multiplication is discontinuous [24] [9]. In fact, N is totally disconnected. Nevertheless, using the inequality

$$\log(1 + |f(z)|) \le 2\|f\|/(1-|z|), \quad |z| < 1$$

(which follows from the (area) sub-mean value property of the sub-harmonic function $\log(1 + |f|)$ as well as the fact that the integrals $\int \log(1 + |f(s\zeta)|) \, dm(\zeta)$ are increasing for $0 < s < 1$) along with Montel's theorem, one can show that N is a complete metric space. Furthermore, see below, every $f \in N$ can be written as the quotient of two bounded functions and thus, by the theorems of Fatou and Lusin-Privalov, f has non-tangential boundary values $f(\zeta)$ almost everywhere and $\log |f(\zeta)|$ is integrable. Finally, every $f \in N$ can be factored as in eq.(3.6.1) and every function of the form as in eq.(3.6.1) belongs to N.

---

[9]Note that for any g as in eq.(3.6.1), $\lim_{a \to 0} \rho(ag, 0) = \mu_2(\mathbb{T})$ which shows, by taking $\mu_2 \ne 0$, that N is not a topological vector space [87].

DEFINITION 3.6.4. Let $\mathfrak{N}(\mathbb{D})$ denote the meromorphic functions of *bounded type*, that is, the functions which can be written as the quotient of two bounded functions.

REMARK 3.6.5. An equivalent definition can be found in [**23**], p. 39.

Note that if $f \in \mathfrak{N}(\mathbb{D})$, then
$$f = \frac{I_1}{I_2} O,$$
where I_1, I_2 are inner and O is outer (i.e., $O = F_f$ for some log-integrable f). Another important result due to F. and R. Nevanlinna is that $N \subset \mathfrak{N}(\mathbb{D})$ [**31**], Theorem 2.1.

3.6.2. The Smirnov class. Although (N, ρ) is not a topological vector space, there is a natural closed subalgebra of N which is. Recall that
$$f \in N \Leftrightarrow \lim_{r \to 1^-} \int \log(1 + |f(r\zeta)|) \, dm(\zeta) < \infty.$$
Furthermore f has non-tangential boundary values which satisfy
$$\int \log(1 + |f(\zeta)|) \, dm(\zeta) < \infty.$$
This leads to the following definition which links these two properties.

DEFINITION 3.6.6. We will say f belongs to the *Smirnov class* N^+ if
$$\lim_{r \to 1^-} \int_{\mathbb{T}} \log(1 + |f(r\zeta)|) \, dm(\zeta) = \int_{\mathbb{T}} \log(1 + |f(\zeta)|) \, dm(\zeta).$$

Using the Nevanlinna factorization, one can prove the following useful characterization of the Smirnov class.

PROPOSITION 3.6.7. *Let $f \in N$. The following are equivalent:*
1. $f \in N^+$.
2.
$$\lim_{c \to \infty} \sup_{0 < r < 1} \int_{\{\log^+ |f_r| > c\}} \log^+ |f_r| dm = 0.$$

3. *In the canonical factorization eq.(3.6.1), $\mu_2 \equiv 0$.*

REMARK 3.6.8. Condition (2) above says that $\{\log^+ |f_r| : r \in (0, 1)\}$ is "uniformly integrable". An equivalent condition for this is
$$\lim_{|E| \to 0} \int_E \log^+ |f_r| dm = 0$$
uniformly in r (see [**37**], p. 120).

It turns out that (N^+, ρ) is a closed subspace of N and that, in contrast to N, N^+ is a topological vector space and hence an F-space. In fact (see below), N^+ is a topological algebra. As with H^p $(0 < p < 1)$, N^+ does not have the Hahn-Banach separation property and so, by Kalton's theorem, N^+ is not locally convex.

By the Nevanlinna and H^p factorization theorems, Theorem 3.2.4 and eq.(3.6.1), we have the following inclusions.
$$\bigcup_{p > 0} H^p \subset N^+ \subset N \subset \mathfrak{N}.$$

Also notice from the Nevanlinna factorization theorem, eq.(3.6.1), and the fact that $|F_f| = |f|$ almost everywhere, that if $f \in N^+$ and the boundary function $f(\zeta) \in L^p$, then $f \in H^p$, i.e.,

$$N^+ \cap L^p = H^p.$$

This is not necessarily true if $f \in N$. For example $f = 1/s_\mu$, where s_μ is a singular inner function, belongs to N, has L^p boundary values, but does not belong to H^p for any p.

One can also define, in a very similar way, the Nevanlinna, Smirnov, and functions of bounded type, for the exterior disk $\mathbb{D}_e$ (denoted by $N(\mathbb{D}_e)$, $N^+(\mathbb{D}_e)$, and $\mathfrak{N}(\mathbb{D}_e)$) with analogous results.

3.6.3. log L. On N^+, the metric is defined by

$$\rho(f, g) := \int_{\mathbb{T}} \log(1 + |f(\zeta) - g(\zeta)|)\, dm(\zeta)$$

which naturally leads us to consider the following class of functions.

DEFINITION 3.6.9. Let $\log L$ denote the class of complex measurable functions f on $\mathbb{T}$ for which $\rho(f, 0) < \infty$.

Using standard measure theory arguments, one can show that for a sequence $\{f_n : n \in \mathbb{N}\} \subset \log L$

$$\rho(f_n, f) \to 0 \Leftrightarrow f_n \to f \text{ in measure and } \int \log^+ |f_n|\, dm \to \int \log^+ |f|\, dm.$$

Now using this equivalent characterization of convergence in $\log L$, one can use an argument laid out in Gamelin's book [**37**], p. 122, to show that $\log L$ is a topological algebra ($f_n \to f$ and $g_n \to g$ in $\log L \Rightarrow f_n + g_n \to f + g$ and $f_n g_n \to fg$ in $\log L$).

For $f \in H^p$, $f_r \to f$ in H^p. A result of Yanagihara [**103**] says the same is true for N^+.

PROPOSITION 3.6.10. If $f \in N^+$, then $\rho(f_r, f) \to 0$ as $r \to 1^-$.

PROOF. Since $f_r \to f$ almost everywhere, we can apply Egorov's theorem to say that for a given $\varepsilon > 0$, there is a closed set $E \subset \mathbb{T}$ such that $|E| > 1 - \varepsilon$ (dm is normalized) and $f_r \to f$ uniformly on E. Write $\rho(f_r, f)$ as

$$\rho(f_r, f) = \int_{\mathbb{T}} \log(1 + |f_r - f|) dm = \int_E + \int_{\mathbb{T} \setminus E}$$

and notice that the first integral goes to zero as $r \to 1^-$. For the second integral,

$$\int_{\mathbb{T} \setminus E} \leq \int_{\mathbb{T} \setminus E} \log 2\, dm + \int_{\mathbb{T} \setminus E} \log^+ |f_r - f|\, dm$$

$$\leq \varepsilon \log 2 + \int_{\mathbb{T} \setminus E} \log^+ |f_r - f|\, dm$$

$$\leq \varepsilon \log 2 + c_\varepsilon,$$

where $c_\varepsilon \to 0$ as $\varepsilon \to 0$. Here we are using the uniform integrability of the family $\{\log^+ |f_r| : r \in (0, 1)\}$ (see Remark 3.6.8). □

3.6. THE NEVANLINNA CLASS

This result says that the Smirnov class N^+ can be thought of as the closure of the analytic polynomials in $\log L$ [10]. There is also an interesting result due to Aleksandrov which tells where N^+ lies in the space $\log L$.

DEFINITION 3.6.11. Define the space
$$N^- := \{\, f \in \log L : \overline{\zeta} f(\overline{\zeta}) \in N^+ \,\}.$$

In a way, N^- can be thought of as the space $\{\overline{f} : f \in N^+, f(0) = 0\}$. One can also think of N^- as the boundary functions of functions in $N^+(\mathbb{D}_e)$ which vanish at infinity. Aleksandrov's result [4] is
$$\log L = N^+ + N^-$$
and
$$\bigvee \left\{ \frac{1}{1 - \overline{\zeta}z} : \zeta \in \mathbb{T} \right\} = N^+ \cap N^-,$$
where the closed linear span is in the metric of $\log L$.

---

[10] For fixed $r \in (0, 1)$, the function f_r can be approximated by polynomials via a Taylor expansion in the metric of $\log L$. Now apply $\rho(f_r, f) \to 0$.

CHAPTER 4

The Hardy spaces of the upper-half plane

4.1. Motivation

Recall from the introduction that the object of Chapter 6 will be to show that a typical backward shift invariant subspace of H^p ($0 < p < 1$) takes the form $\mathcal{E}^p(I, F, k)$ [1] consisting of H^p functions f which satisfy the three conditions

1. $f \in H^p \cap I\overline{H_0^p}$.
2. f has an analytic continuation to a neighborhood of $\mathbb{T}\setminus F$.
3. At every $\zeta \in F_0$ (the isolated points of F), f has a pole of order at most $k(\zeta)$.

At some point in Chapter 6, we will need to solve the approximation problem

$$\mathcal{E}^p(I, F, k) = e^p(I, F, k),$$

where

$$e^p(I, F, k) := \left(H^2 \cap I\overline{H_0^2} \right) \bigvee \left\{ \frac{1}{(1-\overline{\zeta}z)^j} : j = 1, \cdots, k(\zeta); \zeta \in F \right\}.$$

As it turns out, we can reduce the problem to solving

$$E^p(1, F, k) = e^p(1, F, k) = \bigvee \left\{ \frac{1}{(1-\overline{\zeta}z)^j} : j = 1, \cdots, k(\zeta); \zeta \in F \right\}.$$

The problem is then to show that functions from $E^p(1, F, k)$; specifically, those functions $f \in H^p$ for which

1. $f \in H^p \cap \overline{H_0^p}$, i.e., f has a pseudocontinuation to a function in the Hardy space of the exterior disk which vanishes at infinity.
2. f has an analytic continuation across $\mathbb{T}\setminus F$.
3. At every point $\zeta \in F_0$, f has a pole of order no more than $k(\zeta)$,

can be approximated in H^p with *rational* functions satisfying these three properties.

As we shall see in Chapter 6, it will become necessary to replace the Hardy space $H^p(\mathbb{C}_\infty \setminus \mathbb{T})$ [2] (which is what we are looking at above) with the Hardy space

$$\mathcal{H}^p(\mathbb{C}\setminus\mathbb{R}) := \left\{ f \in \mathfrak{H}(\mathbb{C}\setminus\mathbb{R}) : \sup_{y \neq 0} \int_{-\infty}^{\infty} |f(x+iy)|^p dx < \infty \right\}$$

and then relate it to our original approximation problem by means of conformal mapping. The analogous problem in this new setting will be to show that every $f \in E^p(F, k)$[3] of functions in $\mathcal{H}^p(\mathbb{C}\setminus\mathbb{R})$ satisfying the conditions

1. $\lim_{y \to 0^+} f(x+iy) = \lim_{y \to 0^-} f(x+iy)$ almost everywhere.

---

[1] I is an inner function with spectrum $\sigma(I)$, F is a closed subset of the circle which contains $\sigma(I) \cap \mathbb{T}$, and $k : F \to \{1, \cdots, n_p\}$ ($n_p = [1/p]$) for which $k = n_p$ on $(F\setminus F_0) \cup (\sigma(I) \cap \mathbb{T})$.
[2] $f \in H^p(\mathbb{C}_\infty \setminus \mathbb{T})$ if $f \in \mathfrak{H}(\mathbb{C}_\infty \setminus \mathbb{T})$ and $\sup_{r \neq 1} \int |f(r\zeta)|^p dm(\zeta) < \infty$.
[3] Here F is s closed subset of $\mathbb{R}$ and $k : F \to \mathbb{N} \cap [1, n_p]$.

2. f has an analytic continuation across $\mathbb{R}\setminus F$.
3. At each point $x \in F_0$, f has a pole of order no more than $k(x)$.

can be approximated by rational functions from $E^p(F,k)$. Aleksandrov's ingenious trick is to view this rational approximation problem in terms of distributions and then to employ the tools of "real variable H^p theory" such as the atomic decomposition. Before launching into the technical details of this subject, we would like to spend a few moments to give the reader the "lay of the land" and mention where all this fits into our situation.

To each function $f \in \mathcal{H}^p(\mathbb{C}\setminus\mathbb{R})$, there is unique associated tempered distribution ℓ_f defined by

$$\ell_f(\psi) = \lim_{y \to 0^+} \int \{ f(x+iy) - f(x-iy) \} \psi(x)dx,$$

where ψ belongs to the space of infinitely differentiable rapidly decreasing functions. We will denote the set of distributions arising in this way by $H^p(\mathbb{R})$ (an alternate but equivalent definition will be given later). Conversely, given a distribution $\ell \in H^p(\mathbb{R})$, we can recover the unique function $f = f_\ell \in \mathcal{H}^p(\mathbb{C}\setminus\mathbb{R})$ such that $\ell_f = \ell$ by means of the "Cauchy integral formula"

$$(4.1.1) \qquad f_\ell(z) = \frac{1}{2\pi i} \ell\left(\frac{1}{\cdot - z}\right).$$

A reader familiar with the theory of tempered distributions might feel somewhat uncomfortable with the formula in eq.(4.1.1) since, for fixed $z \in \mathbb{C}\setminus\mathbb{R}$, the function $t \to (t-z)^{-1}$ is not a rapidly decreasing function and so the expression does not seem to be meaningful. However, with distributions belonging to $H^p(\mathbb{R})$, the function defined in eq.(4.1.1) is meaningful and even has several desirable properties.

By this identification of $\mathcal{H}^p(\mathbb{C}\setminus\mathbb{R})$ with $H^p(\mathbb{R})$, we can view $E^p(F,k)$ as the distributions $\ell \in H^p(\mathbb{R})$ such that

1.
$$\lim_{y \to 0^+} \ell\left(\frac{1}{\cdot - (x+iy)}\right) = \lim_{y \to 0^-} \ell\left(\frac{1}{\cdot - (x+iy)}\right)$$

   almost everywhere.
2. $\operatorname{supp} \ell \subset F$.
3. At each point $x \in F_0$,

$$\ell|U_x = \sum_{j=0}^{k(x)} c_j \delta_x^{(j)},$$

where U_x is any interval about x which does not contain any other point of F except x.

Our distributional "rational approximation" problem will be to show that the distributions satisfying the above three conditions can be approximated (in the metric of $H^p(\mathbb{R})$) with "rational" distributions from $E^p(F,k)$, namely those of the form

$$\ell = \sum_{n=1}^{N} \sum_{j=0}^{k(x_n)} c_{j,n} \delta_{x_n}^{(j)},$$

where $x_1, \cdots, x_N \in F$ and $c_{j,n} \in \mathbb{C}$. Indeed, a computation shows that the $\mathcal{H}^p(\mathbb{C}\setminus\mathbb{R})$ function corresponding to this distribution ℓ, via eq.(4.1.1), is

$$f_\ell(z) = \frac{1}{2\pi i}\,\ell\,\Big(\frac{1}{\cdot - z}\Big) = \frac{1}{2\pi i}\sum_{n=1}^{N}\sum_{j=0}^{k(x_n)}\frac{j!\,c_{j,n}}{(x_n - z)^{j+1}}$$

which is indeed a rational function.

Why this desire to view everything in terms of distributions? A powerful tool from the theory of $H^p(\mathbb{R})$ spaces will be the fact that every $\ell \in H^p(\mathbb{R})$ can be written in the form

(4.1.2) $$\ell = \sum_{k=1}^{\infty} \lambda_k\, b_k,$$

where λ_k are complex constants, b_k are certain compactly supported bounded functions called "atoms", and the sum converges in the sense of distributions, meaning

$$\ell(\psi) = \sum_{k=1}^{\infty} \lambda_k \int b_k(x)\,\psi(x)\,dx.$$

Again, ψ is an infinitely differentiable rapidly decreasing function. The above decomposition is called the "atomic decomposition" of ℓ and was discovered by Coifman [21]. In order to approximate functions in $E^p(F, k)$ with rational functions, we will need to control the singularities of functions from $E^p(F, k)$. This will be done by viewing this space in terms of distributions and then by defining certain operators first on finite linear combinations of atoms (which is a dense set of $H^p(\mathbb{R})$). The operators involved will be Hilbert transforms which act well on atoms since atoms are compactly supported, bounded, and satisfy certain vanishing moment conditions.

Why is this chapter called the Hardy spaces of the upper half-plane? Some of the key tools that will be used in Chapter 6, associating functions in $\mathcal{H}^p(\mathbb{C}\setminus\mathbb{R})$ with tempered distributions, the atomic decomposition, etc., were developed as a result of generalizing the Hardy spaces to several variables where the tools of complex analysis were often replaced by more real variable and harmonic analysis tools. These problems were originally posed for the Hardy spaces of analytic and harmonic functions for the half space in $\mathbb{R}^n$. In this chapter, we will discuss this material from its origins and clarify why the introduction of distributions is essential. We ask the reader's patience since we will often need to take many detours along the way into the various aspects of harmonic function theory. We feel these detours are necessary to both familiarize the reader with the necessary tools but also to put all the various (but equivalent) definitions of "H^p" in their proper context for later use.

4.2. Basic definitions

We begin with some definitions and well-known results from the theory of Hardy spaces of the upper half-plane where we follow [31] and [39].

DEFINITION 4.2.1. For $p \in (0, \infty)$, let $\mathcal{H}^p = \mathcal{H}^p(\mathbb{C}_+)$ denote the space of analytic functions f on the upper-half plane $\mathbb{C}_+ = \{z \in \mathbb{C} : \Im z > 0\}$ for which the quantity

$$\|f\| := \sup_{y>0}\Big\{\int_{\mathbb{R}} |f(x+iy)|^p\,dx\Big\}^{1/p}$$

is finite.

As was the case with $H^p = H^p(\mathbb{D})$, if $f \in \mathcal{H}^p$, the integral means
$$m_p(f,y) := \Big\{ \int_{\mathbb{R}} |f(x+iy)|^p\, dx \Big\}^{1/p}$$
are increasing as $y \downarrow 0$ [4] and so
$$\|f\| = \lim_{y \to 0^+} \Big\{ \int_{\mathbb{R}} |f(x+iy)|^p\, dx \Big\}^{1/p}.$$
One can also prove the following growth estimate for $f \in \mathcal{H}^p$
$$(4.2.2) \qquad |f(x+iy)| \le \Big(\frac{2}{\pi y}\Big)^{1/p} \|f\|,$$
from which, via Montel's theorem, one can prove that $\mathcal{H}^p$ is an F-space when $p \in (0,1)$ and a Banach space, when $p \in [1,\infty)$.

If ϕ is the usual conformal map $\phi : \mathbb{D} \to \mathbb{C}_+$,
$$\phi(z) := i\frac{1+z}{1-z}$$
and $f \in \mathfrak{H}(\mathbb{C}_+)$, then $f \circ \phi \in \mathfrak{H}(\mathbb{D})$. Furthermore, it is not difficult to show that
$$f \in \mathcal{H}^p \Leftrightarrow (f \circ \phi)(\phi')^{1/p} \in H^p(\mathbb{D}).$$

REMARK 4.2.3. The following elementary facts about $\mathcal{H}^p$ functions can be found in [31] and [39].

1. If $f \in \mathcal{H}^p$, then the boundary function
$$f(x) := \lim_{y \to 0^+} f(x+iy)$$
exists almost everywhere, f belongs to $L^p := L^p(\mathbb{R})$, and
$$\int_{\mathbb{R}} \frac{\log|f(x)|}{1+x^2}\, dx > -\infty.$$
Hence, just as with $H^p(\mathbb{D})$, the boundary values of $\mathcal{H}^p$ functions cannot vanish on a set of positive measure. We will denote the space of boundary functions of $\mathcal{H}^p$ functions by $\mathcal{H}^p(\mathbb{R})$.

2. For $f \in \mathcal{H}^p$,
$$\|f\|^p = \lim_{y \to 0^+} \int_{\mathbb{R}} |f(x+iy)|^p\, dx = \int_{\mathbb{R}} |f(x)|^p\, dx$$
and furthermore,
$$\lim_{y \to 0^+} \int_{\mathbb{R}} |f(x+iy) - f(x)|^p\, dx = 0.$$
Thus the map which takes $f \in \mathcal{H}^p$ to its L^p boundary function $f(x) \in \mathcal{H}^p(\mathbb{R})$ is an isometry.

---

[4]The hypothesis $f \in \mathcal{H}^p$ is important here since, unlike the analog for $H^p(\mathbb{D})$, this is not true if f is any analytic function on $\mathbb{C}_+$ [31], p. 192.

3. Similar to the $H^p(\mathbb{D})$ case, functions f belonging to $\mathcal{H}^p$ also have non-tangential limits almost everywhere. Here a typical "non-tangential approach region" is of the form

$$\{x + iy \in \mathbb{C}_+ : |x - t| < \alpha\, y\}$$

for some $\alpha > 0$, which is a cone with vertex $t \in \mathbb{R}$ and angle equal to $2\arctan \alpha$.

4. Though not needed in our further discussion, there is an analogous factorization theorem similar to the Nevanlinna factorization theorem for $H^p(\mathbb{D})$. That is to say, every $f \in \mathcal{H}^p$ can be factored as

$$f = bsF,$$

where b is a Blaschke product, s is a singular inner function, and F is an outer function, each for the upper-half plane. We refer the reader to [31], p. 193 for the details.

5. Suppose $f \in \mathcal{H}^{p_1}(\mathbb{R})$ and $f \in L^{p_2}$ for some $p_2 > p_1$. Then $f \in \mathcal{H}^{p_2}(\mathbb{R})$ [38], p. 234.

6. Functions in $\mathcal{H}^p(\mathbb{R}) \cap L^1$ for $p \in (0,1]$ satisfy the moment conditions

$$\int x^k f(x)\, dx = 0$$

whenever $k = 0, \cdots, [1/p] - 1$ and $x^k f \in L^1$ [95], p. 128, or Corollary 4.3.6.

7. Finding dense sets in $\mathcal{H}^p$ is often very useful. However, unlike $H^p(\mathbb{D})$ where the polynomials are dense, dense sets in $\mathcal{H}^p$ are often harder to come by. For example, the set $\mathcal{H}^p \cap \mathcal{H}^q$ is dense in $\mathcal{H}^p$ for any $q > p$, [38], p. 237. [5] For other examples of dense subsets of $\mathcal{H}^p$, see [95], p. 128.

4.3. Poisson and conjugate Poisson integrals

To further understand $\mathcal{H}^p$ as well as an important associated space of harmonic functions $\mathcal{H}_h^p$ (see definition below), we need to review some results from harmonic function theory. Recall the *Poisson kernel*

$$P_z(t) := \frac{1}{\pi}\Im\left(\frac{1}{t-z}\right) = \frac{1}{\pi}\frac{y}{(x-t)^2 + y^2}, \quad t \in \mathbb{R},\, z = x + iy \in \mathbb{C}_+$$

is harmonic on $\mathbb{C}_+$ for each fixed $t \in \mathbb{R}$. Differentiating under the integral sign, one can show that for $f \in L^p := L^p(\mathbb{R})$ ($1 \leq p \leq \infty$), the *Poisson integral* of f,

$$U_f(z) := \int_{\mathbb{R}} f(t) P_z(t)\, dt$$

belongs to $\mathrm{Har}(\mathbb{C}_+)$ (the harmonic functions on $\mathbb{C}_+$). Moreover, we have the following classical result of Fatou [33].

THEOREM 4.3.1. *If $f \in L^p$ ($1 \leq p \leq \infty$), then*

1. $U_f \in \mathrm{Har}(\mathbb{C}_+)$.

---

[5] Here is a brief outline. Let $f \in \mathcal{H}^p(\mathbb{R})$ and note that $f_\varepsilon := f(x + i\varepsilon)$ is a bounded C^∞ function belonging to $\mathcal{H}^p(\mathbb{R})$ with $f_\varepsilon \to f$ in L^p. Let $N > 1/q$ and set $G_n = (in/x + in)^N$. $G_n \in C^\infty$, $|G_n(x)| = O(|x|^{-N})$ (for large $|x|$), and so $G_n f_\varepsilon \in \mathcal{H}^q$. Furthermore, $G_n f_\varepsilon \to f_\varepsilon$ in L^p as $n \to \infty$.

2. (*Fatou* [33]) *If $x \in \mathbb{R}$ is a Lebesgue point of f* [6], *then*
$$\lim_{y \to 0^+} U_f(x+iy) = f(x).$$

3. *For all $y > 0$,*
$$\left\{ \int_{\mathbb{R}} |U_f(x+iy)|^p dx \right\}^{1/p} \leq \|f\|_{L^p}.$$

4. *For $p \in [1, \infty)$,*
$$\int_{\mathbb{R}} |U_f(x+iy) - f(x)|^p dx \to 0$$
as $y \to 0^+$.

One can also define the *Poisson integral of a measure* $\mu \in M(\mathbb{R})$ (the space of finite Borel measures on $\mathbb{R}$) by
$$U_\mu(z) := \int_{\mathbb{R}} P_z(t) d\mu(t).$$
With this definition, we have another classical result.

THEOREM 4.3.2. *If $\mu \in M(\mathbb{R})$, then*
1. $U_\mu \in \text{Har}(\mathbb{C}_+)$.
2. *For all $y > 0$,*
$$\int_{\mathbb{R}} |U_\mu(x+iy)| dx \leq \|\mu\|\ [7].$$

3. *For every $\phi \in C_0(\mathbb{R})$ (continuous functions on $\mathbb{R}$ which vanish at infinity)*
$$\lim_{y \to 0^+} \int_{\mathbb{R}} U_\mu(x+iy)\phi(x) dx = \int_{\mathbb{R}} \phi(x) d\mu(x),$$
that is to say $U_\mu(\cdot + iy)$ converges to μ wk-* as $y \to 0^+$.

The following is the converse of these two results.

THEOREM 4.3.3. *Let $p \in [1, \infty)$ and $u \in \text{Har}(\mathbb{C}_+)$ with*
$$\sup_{y>0} \int_{\mathbb{R}} |u(x+iy)|^p dx < \infty.$$

1. *If $p \in (1, \infty)$, then $u = U_f$ for some $f \in L^p$.*
2. *If $p = 1$, then $u = U_\mu$ for some $\mu \in M(\mathbb{R})$.*

Clearly $f \in \mathcal{H}^p$ ($1 \leq p < \infty$) satisfies the hypothesis of Theorem 4.3.3 and so $f = U_g$ (if $1 < p < \infty$) for some $g \in L^p$ or $f = U_\mu$ (if $p = 1$) for some $\mu \in M(\mathbb{R})$. Identifying the g (or μ) is possible with the following result.

THEOREM 4.3.4. *If $f \in \mathcal{H}^p$ ($1 \leq p < \infty$), then $f(z) = U_f(z)$* [8]. *Conversely, if $h \in L^p$ and $F(z) := U_h(z)$ is analytic on $\mathbb{C}_+$, then $F \in \mathcal{H}^p$ and its boundary function is h.*

---

[6]x is a Lebesgue point for f if $\lim_{h \to 0} h^{-1} \int_x^{x+h} |f(t) - f(x)|\, dt = 0$. Recall that almost every point is a Lebesgue point for f.

[7]$\|\mu\|$ is the total variation norm of μ.

[8]Here U_f means the Poisson integral of the boundary function for f.

From this result, one can derive the important "Cauchy integral formula" for $\mathcal{H}^p$.

COROLLARY 4.3.5 (Cauchy integral formula). *Let $f \in \mathcal{H}^p$ ($1 \leq p < \infty$). Then*

$$f(z) = \frac{1}{2\pi i} \int_{\mathbb{R}} \frac{f(t)}{t-z}\, dt, \quad z \in \mathbb{C}_+$$

and the integral vanishes for $\Im z < 0$. Conversely, if $h \in L^p$ ($1 \leq p < \infty$), and

$$\frac{1}{2\pi i} \int_{\mathbb{R}} \frac{h(t)}{t-z}\, dt = 0, \quad \Im z < 0,$$

then for $z \in \mathbb{C}_+$, the above integral is a function $f \in \mathcal{H}^p$ whose boundary values are equal to $h(x)$ almost everywhere.

This proof comes from [31], p. 195, but we will include here it since the ideas from this proof will be used later on to examine a distributional version of the above Cauchy integral formula for $\mathcal{H}^p$ ($0 < p < 1$) (see Theorem 4.10.1).

PROOF. If $f \in \mathcal{H}^p$, the Cauchy integral

$$F(z) := \frac{1}{2\pi i} \int_{\mathbb{R}} \frac{f(t)}{t-z}\, dt$$

is analytic on $\mathbb{C} \setminus \mathbb{R}$. Since

$$\frac{1}{t-z} - \frac{1}{t-\bar{z}} = 2\pi i P_z(t),$$

we see from Theorem 4.3.4 that for $\Im z > 0$, $F(\bar{z}) = F(z) - f(z)$ and so $F(\bar{z})$ is analytic on $\mathbb{C}_+$. But this means that F must be constant on the lower half plane. However, since $F(x+iy) \to 0$ as $y \to \pm \infty$ (for each fixed $x \in \mathbb{R}$), the constant must be zero and so

$$F(z) = \begin{cases} f(z) & \Im z > 0, \\ 0 & \Im z < 0. \end{cases}$$

Conversely, if

$$\frac{1}{2\pi i} \int \frac{h(t)}{t-z}\, dt = 0$$

for $\Im z < 0$, the function

$$f(z) := \frac{1}{2\pi i} \int \frac{h(t)}{t-z}\, dt$$

is analytic on $\mathbb{C}_+$ and

$$f(z) = \frac{1}{2\pi i} \int \frac{h(t)}{t-z}\, dt - \frac{1}{2\pi i} \int \frac{h(t)}{t-\bar{z}}\, dt = U_h(z), \quad \Im z > 0.$$

Applying Theorem 4.3.4 completes the proof. □

The following corollary will be useful in Chapter 6.

COROLLARY 4.3.6. *For $f \in \mathcal{H}^1$,*

$$\int f(t) dt = 0.$$

PROOF. By Corollary 4.3.5,
$$\int \frac{f(t)}{t+iy} dt = 0, \ y > 0$$
and so
$$\int \frac{yf(t)}{t+iy} dt = 0, \ y > 0.$$
The result follows by taking limits as $y \to \infty$. [9]

If $f \in L^p$ ($1 \leq p < \infty$), we define the *harmonic conjugate* of U_f, by
$$V_f(z) := \int_\mathbb{R} f(t) Q_z(t) \, dt,$$
where
$$Q_z(t) := \frac{1}{\pi} \Re \left(\frac{1}{z-t} \right) = \frac{1}{\pi} \frac{x-t}{(x-t)^2 + y^2}, \ t \in \mathbb{R}, z = x + iy \in \mathbb{C}_+,$$
is the *conjugate Poisson kernel*. It is routine to check, using the fact that $Q_z(t)$ is harmonic on $\mathbb{C}_+$ for fixed $t \in \mathbb{R}$, that V_f is harmonic on $\mathbb{C}_+$ and
$$U_f(z) + iV_f(z) = \frac{1}{\pi i} \int_\mathbb{R} \frac{f(t)}{t-z} dt$$
is analytic on $\mathbb{C}_+$. The following theorem of M. Riesz says even more.

THEOREM 4.3.7 (M. Riesz [73] [74]). *If $p \in (1, \infty)$ and $f \in L^p$, then*
$$\int |V_f(x+iy)|^p dx \leq C_p \|f\|_{L^p}, \ \forall \, y > 0$$
and so $U_f + iV_f \in \mathcal{H}^p$.

REMARK 4.3.8. Note that the Riesz theorem is false if $p = 1$.

We will say more about the conjugation operator in the following section.

4.4. Maximal functions

If f is a locally integrable function on $\mathbb{R}$, define the *Hardy-Littlewood maximal function* Mf by
$$(Mf)(x) := \sup_{r>0} \frac{1}{2r} \int_{x-r}^{x+r} |f(t)| \, dt$$
whenever this supremum exists. An important result of Hardy and Littlewood is the following.

THEOREM 4.4.1 (Hardy-Littlewood [42]). 1. *If $p \in [1, \infty)$, and $f \in L^p$, then Mf is finite almost everywhere.*
2. *If $f \in L^1$, then for every $\lambda > 0$,*

(4.4.2) $$\operatorname{meas} \{ x \in \mathbb{R} : (Mf)(x) > \lambda \} \leq \frac{C}{\lambda} \int |f(t)| \, dt,$$

where C does not depend on f or λ.
3. *If $p \in (1, \infty]$ and $f \in L^p$, then $Mf \in L^p$ with $\|Mf\|_{L^p} \leq C_p \|f\|_{L^p}$.*

---

[9]The proof presented here was found in Koosis' book [51], p. 122.

REMARK 4.4.3. 1. One sometimes sees the maximal function defined by
$$(\tilde{M}f)(x) := \sup_I \frac{1}{|I|} \int_I |f(t)|dt,$$
where the supremum is taken over *all* intervals which contain the point x. Clearly $(Mf)(x) \leq (\tilde{M}f)(x)$ and it turns out that the Hardy-Littlewood result above holds for this maximal operator $\tilde{M}$ as well [39], p. 22.

2. In the literature, a mapping $f \to M(f)$ satisfying eq.(4.4.2) is said to be of "weak-type $(1,1)$".

DEFINITION 4.4.4. For a harmonic function u on $\mathbb{C}_+$, define the *non-tangential maximal function* u^* by
$$u^*(t) := \sup_{|x-t|<y} |u(x+iy)|$$
whenever this supremum exists.

An important result is the following (see [39], p. 22, or [94], p. 197, for a proof).

PROPOSITION 4.4.5. *If $f \in L^1$, and U_f is the Poisson integral of f, then*
$$U_f^*(t) \leq C(Mf)(t), \quad t \in \mathbb{R},$$
where $C > 0$ is independent of f. Hence (from Theorem 4.4.1) there is a constant $C' > 0$ such that for each $\lambda > 0$,
$$\text{meas}\left\{ x \in \mathbb{R} : U_f^*(x) > \lambda \right\} \leq \frac{C'}{\lambda} \|f\|_{L^1}.$$

REMARK 4.4.6. In the above result, we looked at the non-tangential maximal function $u^*(t) = \sup\{|u(x+iy)| : |x-t| < y\}$. The region
$$\{x + iy \in \mathbb{C}_+ : |x-t| < y\}$$
could be replaced by a more general non-tangential region
$$\{x + iy \in \mathbb{C}_+ : |x-t| < \alpha\, y\}$$
for some $\alpha > 0$. In this situation, Proposition 4.4.5 is the same except that the constant C above is replaced by a constant C_α which depends on α.

The non-tangential maximal function also behaves well with respect to the conjugation operator
$$V_f(z) = \int f(t) Q_z(t)\, dt.$$

THEOREM 4.4.7. 1. (*Kolmogorov* [49]) *If $\in L^1$, then*
$$\text{meas}\left\{ x \in \mathbb{R} : V_f^*(x) > \lambda \right\} \leq \frac{C}{\lambda} \|f\|_{L^1},$$
where C is independent of f and λ.

2. (*M. Riesz* [73] [74]) *If $p \in (1, \infty)$, there is a constant c_p such that*
$$\|V_f^*\|_{L^p} \leq c_p \|f\|_{L^p}$$
for all $f \in L^p$.

The important connection between Hardy spaces and the non-tangential maximal function is the following result which serves as an equivalent characterization of $\mathcal{H}^p$. One direction is classical and due to Hardy and Littlewood [42]. The other direction is more recent and due to Burkholder, Gundy, and Silverstein [17].

THEOREM 4.4.8. *Let $p \in (0, \infty)$ and f be an analytic function on $\mathbb{C}_+$. Then f belongs to $\mathcal{H}^p$ if and only if $f^* \in L^p$. Moreover*
$$\|f\|_{\mathcal{H}^p} \asymp \|f^*\|_{L^p}.$$

REMARK 4.4.9. Again, in the above results, if the "standard" non-tangential approach region $\{x+iy \in \mathbb{C}_+ : |x-t| < y\}$ is replaced by $\{x+iy \in \mathbb{C}_+ : |x-t| < \alpha y\}$, then the constants C and c_p above would also depend on α.

4.5. The Hilbert transform

For $f \in L^p$ ($1 \leq p < \infty$), the radial boundary function of U_f is (almost everywhere) equal to f. What is the radial boundary function for V_f? The answer involves the Hilbert transform of f.

For $f \in L^p$ ($1 \leq p < \infty$), and $\varepsilon > 0$, define
$$(H_\varepsilon f)(x) := \frac{1}{\pi} \int_{|x-t| \geq \varepsilon} \frac{f(t)}{x-t}\, dt$$
and define the *Hilbert transform* to be the limit
$$(Hf)(x) := \lim_{\varepsilon \to 0^+} (H_\varepsilon f)(x) = \text{P.V.}\, \frac{1}{\pi} \int \frac{f(t)}{x-t}\, dt$$
whenever it exists. A deep and important result of Calderon and Zygmund [94], Chapter 2, says that the Hilbert transform (which is a "singular integral operator") exists almost everywhere for L^p functions and has some desirable properties.

THEOREM 4.5.1.   1. *For $f \in L^p$ ($1 \leq p < \infty$),*
$$\lim_{\varepsilon \to 0^+} (H_\varepsilon f)(x) = (Hf)(x)$$
exists almost everywhere.
2. *For $p \in (1, \infty)$, there is a constant $c_p > 0$ such that $\|H_\varepsilon f\|_{L^p} \leq c_p \|f\|_{L^p}$ for all $\varepsilon > 0$ and all $f \in L^p$.*
3. *For $p \in (1, \infty)$, $\lim_{\varepsilon \to 0^+} H_\varepsilon f = Hf$ in L^p.*
4. *The following weak-$(1,1)$ estimate holds: There is a $C > 0$ so that for all $f \in L^1$ and $\lambda > 0$,*
$$\text{meas}\{x : |(Hf)(x)| > \lambda\} \leq C \frac{\|f\|_{L^1}}{\lambda}.$$

REMARK 4.5.2.   1. The condition $p \in (1, \infty)$ is important here since $HL^1 \not\subset L^1$ and $HL^\infty \not\subset L^\infty$, see [94], Chapter 2. For example
$$(H\chi_{[a,b]})(x) = \frac{1}{\pi} \log \left| \frac{x-a}{x-b} \right|$$
is neither integrable nor bounded on $\mathbb{R}$.

Recall from Fatou's theorem, Theorem 4.3.1, that for $f \in L^p$ ($1 \leq p < \infty$), the Poisson integral
$$U_f(x + iy) = \int P_{x+iy}(t) f(t)\, dt$$

has a limit as $y \to 0^+$ at every Lebesgue point x of f equal to $f(x)$. This next result examines the same limit for the harmonic conjugate

$$V_f(x+iy) = \int Q_{x+iy}(t)f(t)dt.$$

THEOREM 4.5.3 ([**96**], p. 218). *Suppose $f \in L^p$ ($1 \leq p < \infty$), then at every Lebesgue point x of f*

$$\lim_{y \to 0^+} \{ V_f(x+iy) - (H_y f)(x) \} = 0.$$

REMARK 4.5.4. 1. Combining this with our previous result about Hilbert transforms, we can say that the boundary values of the conjugate function V_f are equal to the Hilbert transform Hf almost everywhere. Recall that the F. and M. Riesz theorem, Theorem 3.3.1, told us when a function $f \in L^p(\mathbb{T})$ belonged to $H^p(\mathbb{T})$. Using Corollary 4.3.5 and the above result, one can find a criterion for a function $f \in L^p$ to belong to $\mathcal{H}^p(\mathbb{R})$. We leave it to the reader to check that the result is: Let $f \in L^p$ ($1 \leq p < \infty$), then

$$f \in \mathcal{H}^p(\mathbb{R}) \Leftrightarrow H(\Re f) = \Im f.$$

In Chapter 6, we will develop an analog of this result (see Corollary 6.4.29), due to Aleksandrov, for $p \in (0,1)$ where we must finesse the fact that $\Re f$ may not be integrable and so $H(\Re f)$ does not make sense.

2. If $f \in L^p$ ($1 < p < \infty$), one can say even more. In this case $Hf \in L^p$ and

$$\int Q_z(t)f(t)\,dt = \int P_z(t)(Hf)(t)\,dt,$$

see [**94**], p. 65. From here it is easy to see, from Fatou's theorem (Theorem 4.3.1), that $V_f(x+iy) \to (Hf)(x)$ at every Lebesgue point x of f as $y \to 0^+$.

4.6. Some examples

For $p \in [1, \infty)$, the Hardy spaces are reasonably understood via the Poisson and conjugate Poisson integrals. For $p \in (0,1)$, these integrals are no longer defined and so we will construct some examples of Hardy space functions for $p \in (0,1)$. The construction of these examples will not only make our discussion more complete but will be vital for some of our constructions in our final chapter on rational approximation.

EXAMPLE 4.6.1. We begin with some examples of rational functions whose simple poles lie on the real axis; that is, functions f of the form

(4.6.2) $$f(z) = \sum_{j=1}^{N} \frac{c_j}{z - a_j}, \quad c_j \in \mathbb{C}, a_j \in \mathbb{R}.$$

Such functions never belong to $\mathcal{H}^p$ for $p \in [1, \infty)$ since they are not L^p integrable on the real axis. However, for $p \in (0,1)$, it is possible to judiciously choose the c_j and a_j so that such rational functions belong to $\mathcal{H}^p$. At the risk of being overly pedantic, we will demonstrate such a class of examples.

First, one can show that given an integer $K \geq 1$, an integer $N > K - 1$, and distinct real numbers $a_1, \cdots, a_N$, there are constants $c_1, \cdots, c_N$ (not all zero) so that

(4.6.3) $$|f(z)| = O\left(|z|^{-K}\right)$$

for large $|z|$. To see this, form the $(K-1) \times N$ matrix

$$A = \begin{pmatrix} 1 & 1 & \cdots & 1 \\ a_1 & a_2 & \cdots & a_N \\ a_1^2 & a_2^2 & \cdots & a_N^2 \\ \vdots & \vdots & \cdots & \vdots \\ a_1^{K-2} & a_2^{K-2} & \cdots & a_N^{K-2} \end{pmatrix}$$

and observe that since $N > K - 1$, there is a non-zero column vector $C = (c_j)_{j=1}^N$ such that $AC = 0$; equivalently,

$$\sum_{j=1}^N c_j a_j^k = 0 \ \forall \ k = 0, \cdots, K-2.$$

With this choice of c_j's, note that for large $|z|$,

$$f(z) = \sum_{j=1}^N c_j \sum_{n=0}^\infty \frac{a_j^n}{z^{n+1}} = \sum_{n=0}^\infty \frac{1}{z^{n+1}} \sum_{j=1}^N c_j a_j^n = \sum_{n=K-1}^\infty \frac{1}{z^{n+1}} \sum_{j=1}^N c_j a_j^n$$

and so $|f(z)| = O(|z|^{-K})$ as desired.

To show that a rational function f of the form eq.(4.6.2) can belong to $\mathcal{H}^p$, we fix $p \in (0,1)$ and an integer $k > 1/p$ and use the above argument to arrange the c_j and a_j so that

$$|f(z)| = O\left(\frac{1}{|z|^k}\right).$$

For this function,

$$\int |f(x+iy)|^p \, dx = \int_{|x|<R} + \int_{|x|>R},$$

where R is sufficiently large so that the interval $(-R, R)$ contains all the poles of f. Then

$$\int_{|x|<R} \left| \sum_{j=1}^N \frac{c_j}{x+iy-a_j} \right|^p dx \leq \sum_{j=1}^N |c_j|^p \int_{|x|<R} \frac{1}{((x-a_j)^2+y^2)^{p/2}} \, dx$$

$$\leq \sum_{j=1}^N |c_j|^p \int_{|x|<R} \frac{1}{|x-a_j|^p} \, dx$$

which is finite since $p \in (0,1)$. Moreover, the left-hand side of the above equation is uniformly bounded in $y > 0$. For the other integral, we use the fact that $|f(z)| =$

$O(|z|^{-k})$ to get

$$\int_{|x|>R} \Big| \sum_{j=1}^{N} \frac{c_j}{x+iy-a_j} \Big|^p \, dx \leq C \int_{|x|>R} \frac{1}{|x+iy|^{kp}} \, dx$$
$$\leq C \int_{|x|>R} \frac{1}{(x^2+y^2)^{kp/2}} \, dx$$
$$\leq C \int_{|x|>R} \frac{1}{|x|^{kp}} \, dx$$

which is finite since $kp > 1$. Furthermore, the left-hand side of the above equation is uniformly bounded in y. Combining these two results we get

$$\sup_{y>0} \int_{-\infty}^{\infty} \Big| \sum_{j=1}^{N} \frac{c_j}{x+iy-a_j} \Big|^p \, dx < \infty,$$

which says $f \in \mathcal{H}^p$.

EXAMPLE 4.6.4. By Theorem 4.3.7,

$$(Kf)(z) := \frac{1}{2\pi i} \int \frac{f(t)}{t-z} \, dt$$

belongs to $\mathcal{H}^p$ for $f \in L^p$ ($1 < p < \infty$). In general though, Kf does not belong to $\mathcal{H}^p$ for $p \in (0,1)$ even if f is bounded and has compact support. However, if f satisfies certain moment conditions, then Kf will indeed belong to $\mathcal{H}^p$. An example of such a result is the following: Let $p \in (0,1)$ be fixed and $f \in L^1(I)$ (I is a bounded interval) with

$$\int_I t^k f(t) \, dt = 0 \quad \forall \, k = 0, 1, \cdots, [1/p] - 1.$$

Then $Kf \in \mathcal{H}^p$.

To see this, choose $R > 0$ large enough so that $I \subset (-R, R)$. Then for each $y > 0$,

(4.6.5) $$\int |(Kf)(x+iy)|^p \, dx = \int_{|x|<R} + \int_{|x|>R}.$$

To estimate the first integral on the right-hand side of eq.(4.6.5), observe that

$$\int_{|x|<R} |(Kf)(x+iy)|^p \, dx$$
$$= \frac{1}{2^p} \int_{|x|<R} |U_f(x+iy) + iV_f(x+iy)|^p \, dx$$
$$\leq \frac{1}{2^p} \Big\{ \int_{|x|<R} |U_f(x+iy)|^p \, dx + \int_{|x|<R} |V_f(x+iy)|^p \, dx \Big\}$$
$$\leq \frac{1}{2^p} \Big\{ \int_{|x|<R} |U_f^*(x)|^p \, dx + \int_{|x|<R} |V_f^*(x)|^p \, dx \Big\}$$

Recall that if g is measurable, and $p > 0$, then

$$\int |g(t)|^p \, dt = \int_0^{\infty} p\lambda^{p-1} \text{meas}\{x \in \mathbb{R} : |g(x)| > \lambda\} \, d\lambda.$$

Applying this to the above, we get

$$\int_{|x|<R} |U_f^*(x)|^p \, dx = \int_0^\infty p\lambda^{p-1} \text{meas}\{|x| < R : U_f^*(x) > \lambda\} \, d\lambda$$

$$= \int_0^1 p\lambda^{p-1} \text{meas}\{|x| < R : U_f^*(x) > \lambda\} \, d\lambda$$

$$+ \int_1^\infty p\lambda^{p-1} \text{meas}\{|x| < R : U_f^*(x) > \lambda\} \, d\lambda$$

$$\leq \int_0^1 2Rp\lambda^{p-1} \, d\lambda + \int_1^\infty p\lambda^{p-1} \frac{C}{\lambda} \|f\|_{L^1} \, d\lambda.$$

Notice that in estimating the second integral above, we used Proposition 4.4.5. Both the integrals above converge since $p \in (0,1)$. In a very similar way (except use Theorem 4.4.7 in place of Proposition 4.4.5) one shows that

$$\int_{|x|<R} |V_f^*(x)|^p \, dx < \infty.$$

Thus the supremum in $y > 0$ of the first integral on the right-hand side of eq.(4.6.5) is finite.

To show the second integral on the right-hand side of eq.(4.6.5) has a uniform bound in $y > 0$, we use the moment conditions

$$\int_I t^k f(t) \, dt = 0 \quad k = 0, 1, \cdots, [1/p] - 1.$$

Indeed for $|z| > R$,

$$(Kf)(z) = -\frac{1}{2\pi i} \frac{1}{z} \sum_{n=0}^\infty \frac{1}{z^n} \int t^n f(t) \, dt = -\frac{1}{2\pi i} \sum_{n=[1/p]}^\infty \frac{1}{z^{n+1}} \int t^n f(t) \, dt.$$

Thus for $|z| > R$,

$$|(Kf)(z)| = O\left(\frac{1}{|z|^{[1/p]+1}}\right)$$

from which

$$|(Kf)(x+iy)| \leq C \frac{1}{(x^2+y^2)^{([1/p]+1)/2}} \leq C \frac{1}{|x|^{[1/p]+1}}, \quad \forall \, y > 0.$$

From this we have

$$\int_{|x|>R} |(Kf)(x+iy)|^p \, dx \leq C \int_{|x|>R} \frac{1}{|x|^{p[1/p]+p}} \, dx < \infty$$

since $1/p - 1 < [1/p]$ and so $p[1/p] + p > 1$. The bound in the above equation is uniform in y and so we can combine this with our estimate of the first integral in eq.(4.6.5) to conclude $Kf \in \mathcal{H}^p$.

EXAMPLE 4.6.6. Let $p \in (0,1)$ with $1/p \notin \mathbb{N}$. For an $f \in L^1(I)$ (where I is a compact interval in $\mathbb{R}$) which does not have the vanishing moment conditions in the previous example, we can still make Kf die off sufficiently fast at infinity by subtracting off a certain rational function

$$R(f,a)(z) := \sum_{j=1}^{[1/p]} \frac{c_j(f,a)}{(z-a)^j},$$

where a is any fixed real number and

$$c_j(f,a) := -\frac{1}{2\pi i}\int (t-a)^{j-1}f(t)\,dt.$$

Observe that for sufficiently large $|z|$,

$$\begin{aligned}
(Kf)(z) - R(f,a)(z) &= \frac{1}{2\pi i}\int \frac{f(t)}{t-z}\,dt - \sum_{j=1}^{[1/p]}\frac{c_j(f,a)}{(z-a)^j}\\
&= \frac{1}{2\pi i}\int \frac{f(t)}{(t-a)-(z-a)}\,dt - \sum_{j=1}^{[1/p]}\frac{c_j(f,a)}{(z-a)^j}\\
&= -\frac{1}{2\pi i}\sum_{n=0}^{\infty}\frac{1}{(z-a)^{n+1}}\int (t-a)^n f(t)\,dt - \sum_{j=1}^{[1/p]}\frac{c_j(f,a)}{(z-a)^j}\\
&= -\frac{1}{2\pi i}\sum_{n=[1/p]}^{\infty}\frac{1}{(z-a)^{n+1}}\int (t-a)^n f(t)\,dt.
\end{aligned}$$

Thus

$$|(Kf)(z)| = O\left(\frac{1}{|z|^{[1/p]+1}}\right)$$

which is what is required to make $|(Kf)(x+iy)|^p$ integrable for large x. We leave it to the reader to the make the appropriate adaptations to the previous example (keeping in mind that

$$\int_{|x|<R}\frac{1}{|x-a|^{jp}}\,dx < \infty \quad \forall\, j = 1,\cdots,[1/p]$$

since $1/p \notin \mathbb{N}$!) to show that $Kf - R(f,a)$ belongs to $\mathcal{H}^p$.

EXAMPLE 4.6.7. When $p \in (0,1)$ and $1/p \in \mathbb{N}$, one sees that for given $f \in L^1(I)$ we cannot subtract off a rational function r with a pole at $x = a$ (as in the previous example) to get

$$\frac{1}{2\pi i}\int \frac{f(t)}{t-z}\,dt - r(z) \in \mathcal{H}^p.$$

This is because the principal part of the rational function r must contain a term of the form $(z-a)^{-1/p}$ in order to make it die off fast enough at infinity so the difference above belongs to L^p. However the term $(x-a)^{-1/p}$ is not in L^p for any interval which contains the point $x = a$.

It is possible however to subtract off a rational function with poles at the two points $x = a_1$ and $x = a_2$ so that the function

$$\frac{1}{2\pi i}\int \frac{f(t)}{t-z}\,dt - \sum_{j=1}^{1/p-1}\frac{c_j}{(z-a_1)^j} - \frac{c_{1/p}}{z-a_2}$$

belongs to $\mathcal{H}^p$. This proof is slightly more complicated, but nevertheless important, and will be taken up in Chapter 6 (see Lemma 6.4.65).

4.7. The harmonic Hardy space

An important related space of harmonic functions is the harmonic Hardy space $\mathcal{H}_h^p = \mathcal{H}_h^p(\mathbb{C}_+)$ (which we will define in a moment). For our presentation here, we begin by following a landmark paper of Fefferman and Stein [35] and mention that the various definitions of the $\mathcal{H}_h^p$ spaces (Sometimes they are also called H^p spaces) are all equivalent. We begin with a definition.

DEFINITION 4.7.1. A (complex) function $u \in \text{Har}(\mathbb{C}_+)$ belongs to $\mathcal{H}_h^p$ ($0 < p < \infty$), if there exists a $v \in \text{Har}(\mathbb{C}_+)$ which satisfies the Cauchy-Riemann equations

$$(4.7.2) \qquad \frac{\partial u}{\partial x} = \frac{\partial v}{\partial y}, \quad \frac{\partial v}{\partial x} = -\frac{\partial u}{\partial y}.$$

and such that

$$(4.7.3) \qquad \sup_{y>0} \left\{ \int \left(|u(x+iy)|^2 + |v(x+iy)|^2 \right)^{p/2} dx \right\}^{1/p} < \infty.$$

REMARK 4.7.4. 1. The condition in eq.(4.7.2) is equivalent to the existence of a $F \in \text{Har}(\mathbb{C}_+)$ on $\mathbb{C}_+$ so that $F_y = u$ and $F_x = v$.
2. If $u \in \mathcal{H}_h^p$ and v is a harmonic function satisfying eq.(4.7.2) and eq.(4.7.3), then $u + iv$ is analytic. Moreover,

$$|u + iv|^p \leq c(|u|^2 + |v|^2)^{p/2},$$

and so, from the definition of $\mathcal{H}^p$, $u + iv \in \mathcal{H}^p$.
3. Suppose that for fixed $u \in \text{Har}(\mathbb{C}_+)$, the harmonic functions v_1 and v_2 both satisfy eq.(4.7.2) *and* eq.(4.7.3). From above, we conclude that $F_1 = u + iv_1$ and $F_2 = u + iv_2$ belong to $\mathcal{H}^p$. By the basic fact that $v_1 = v_2 + c$, we get $F_1 = F_2 + ic$. But since the boundary functions of F_1 and F_2 belong to L^p, c would belong to L^p which is only true when $c = 0$. Thus, for a given $u \in \text{Har}(\mathbb{C}_+)$, such a $v \in \text{Har}(\mathbb{C}_+)$ satisfying both eq.(4.7.2) *and* eq.(4.7.3) is unique and so we can "norm" $\mathcal{H}_h^p$ by

$$(4.7.5) \qquad \|u\| := \sup_{y>0} \left\{ \int \left(|u(x+iy)|^2 + |v(x+iy)|^2 \right)^{p/2} dx \right\}^{1/p} < \infty.$$

4. If $u \in \mathcal{H}_h^p$ ($0 < p < \infty$), then from eq.(4.7.3),

$$\sup_{y>0} \int |u(x+iy)|^p \, dx < \infty.$$

For $p \in (1, \infty)$, this will turn out to be an equivalent definition of $\mathcal{H}_h^p$ (see below) while for $p \in (0, 1]$ it is not. For example, if $u(x, y) = y/(x^2 + y^2)$, then $u \in \text{Har}(\mathbb{C}_+)$ and

$$\int_{\mathbb{R}} u(x,y) dy = \pi$$

for all $y > 0$. However the conjugate function to u is $v(x, y) = x/(x^2+y^2)+c$ which can never satisfy eq.(4.7.3) (with $p = 1$) for any c. Thus u does not belong to $\mathcal{H}_h^1$.
5. Functions u in $\mathcal{H}_h^p$ satisfy the growth condition

$$|u(x+iy)| \leq C y^{-1/p}, \quad x \in \mathbb{R}, \, y > 0$$

and so by Montel's theorem (for harmonic functions), $\mathcal{H}_h^p$, with the "norm" in eq.(4.7.5), is an F-space for $p \in (0,1)$ and a Banach space for $p \in [1, \infty)$.
6. From (4), if $u \in \mathcal{H}_h^p \cap \mathfrak{H}(\mathbb{C}_+)$, then $u \in \mathcal{H}^p$.
7. If $u \in \mathcal{H}^p$, the function $v = -iu$ satisfies the conditions eq.(4.7.2) and eq.(4.7.3). Thus $\mathcal{H}^p \subset \mathcal{H}_h^p$.

The $\mathcal{H}_h^p$ spaces can be alternatively understood by means of the following result which has its beginnings in Hardy and Littlewood [**42**] and brought to fruition by Burkholder, Gundy, and Silverstein [**17**]. The complete form we present here can be found (even for several variables) in Fefferman and Stein [**35**] (Theorem 8, Theorem 9, and Corollary 2). Recall that for $u \in \text{Har}(\mathbb{C}_+)$, the non-tangential maximal function u^* is defined by

$$u^*(t) = \sup_{|x-t|<y} |u(x+iy)|.$$

THEOREM 4.7.6. *Let $p \in (0, \infty)$ and $u \in \text{Har}(\mathbb{C}_+)$. Then the following are equivalent.*
1. $u \in \mathcal{H}_h^p$.
2. $u^* \in L^p$.
3. $u^+(x) := \sup_{y>0} |u(x+iy)| \in L^p$.
4. *The "Lusin area function"*

$$(Su)(t) := \left\{ \int_{|x-t|<y} |\nabla u(x+iy)|^2 \, dx \, dy \right\}^{1/2}$$

belongs to L^p and for each $x \in \mathbb{R}$, $u(x+iy) \to 0$ as $y \to +\infty$. Moreover, when any of the above conditions hold,

$$\|u^*\|_{L^p} \asymp \|u^+\|_{L^p} \asymp \|Su\|_{L^p} \asymp \|u\|.$$

REMARK 4.7.7. For $u \in \mathcal{H}_h^p$, the "Lusin area function" Su is finite almost everywhere and by (a harmonic version of) Theorem 2.1.5 (see [**93**], p. 206), u has non-tangential limits almost everywhere.

By the results from the previous section, Theorem 4.3.3, the space $\mathcal{H}_h^p$ ($1 < p < \infty$) is less mysterious.

THEOREM 4.7.8. *For $p \in (1, \infty)$ and $u \in \text{Har}(\mathbb{C}_+)$, the following are equivalent.*
1. $u \in \mathcal{H}_h^p$.
2. $u = U_f$ for some $f \in L^p$.
3. $\sup_{y>0} \int_{\mathbb{R}} |u(x+iy)|^p dx < \infty.$
4. $u^* \in L^p$.

4.8. Distributions

To motivate the need for distribution theory, we recall from Theorem 4.7.8 and Corollary 4.3.5 the formulas

$$f(z) = U_f(z) = \int P_z(t) f(t) \, dt, \ z \in \mathbb{C}_+, \ f \in \mathcal{H}_h^p(1 < p < \infty),$$

$$f(z) = \frac{1}{2\pi i} \int_{\mathbb{R}} \frac{f(t)}{t-z}\, dt, \quad z \in \mathbb{C}_+, \quad f \in \mathcal{H}^p (1 \le p < \infty).$$

When $p \in (0,1)$, these theorems no longer makes any sense since $f(t)$ may not integrable. However, if we are willing to think in terms of distributions, the Cauchy and Poisson integral formulas still holds. As we shall see below, for $f \in \mathcal{H}^p$ or $\mathcal{H}_h^p$, the limit

$$\lim_{y \to 0} f(\cdot + iy)$$

exists as a tempered distribution ℓ and moreover, we can recover f from its corresponding boundary distribution by the "Cauchy formula" (for $f \in \mathcal{H}^p$)

$$f(z) = \frac{1}{2\pi i} \ell \left(\frac{1}{\cdot - z} \right)$$

or the "Poisson formula"

$$f(z) = \ell(P_z(\cdot))$$

for $f \in \mathcal{H}_h^p$. This ability to think of $\mathcal{H}^p/\mathcal{H}_h^p$ functions as distributions for $p \in (0,1)$ will play a crucial role in Aleksandrov's characterization of the B-invariant subspaces of H^p.

4.8.1. General properties of distributions. This section contains a brief treatment of the theory of distributions. We refer the reader to [**78**], Chapters 6 and 7, as well as [**96**], Chapter 1, for a more in-depth discussion.

DEFINITION 4.8.1. Let $\mathcal{S}$ denote the class of *rapidly decreasing functions* (often called the "test functions") $\psi \in C^\infty(\mathbb{R})$ for which the semi-norms

$$\|\psi\|_{j,k} := \sup \left\{ |x^j \psi^{(k)}(x)| : x \in \mathbb{R} \right\}, \quad j,k \in \mathbb{N} \cup \{0\}$$

are finite.

With the above semi-norms, one can endow $\mathcal{S}$ with the metric

$$d(\phi, \psi) := \sum_{j,k \ge 0} 2^{-j-k} \frac{\|\phi - \psi\|_{j,k}}{1 + \|\phi - \psi\|_{j,k}}.$$

Some well known facts about $\mathcal{S}$ are recorded below.

PROPOSITION 4.8.2.
1. *With respect to the metric d, $\mathcal{S}$ is a complete metrizable locally convex topological vector space.*
2. *$\mathcal{S}$ is separable.*
3. *$C_0^\infty(\mathbb{R})$ (the infinitely differentiable functions with compact support) is a dense subset of $\mathcal{S}$.*

DEFINITION 4.8.3. The collection of all continuous linear functionals on $\mathcal{S}$ is called the space of *tempered distributions* and is denoted by $\mathcal{S}'$.

The following is a simple characterization of $\mathcal{S}'$.

PROPOSITION 4.8.4. *A linear functional ℓ on $\mathcal{S}$ is a tempered distribution if and only if there is a positive constant C and $m, n \in \mathbb{N} \cup \{0\}$ such that*

$$|\ell(\psi)| \le C \sum_{0 \le j \le m, 0 \le k \le n} \|\psi\|_{j,k}$$

for all $\psi \in \mathcal{S}$.

From this theorem, one can see that the following are tempered distributions.

EXAMPLE 4.8.5.   1. If $f \in L^p$ $(1 \leq p < \infty)$, the linear functional ℓ_f defined by
$$\ell_f(\psi) := \int_{\mathbb{R}} f(x)\psi(x)dx$$
is a tempered distribution. More generally, if f is a measurable function of "polynomial growth", that is $|f(x)| \leq C(1+|x|^2)^k$ for some $k \in \mathbb{N} \cup \{0\}$, then ℓ_f (defined as above) is also a tempered distribution.

2. If $\mu \in M(\mathbb{R})$ (the finite Borel measures on $\mathbb{R}$), the linear functional
$$\ell_\mu(\psi) := \int_{\mathbb{R}} \psi(x)d\mu(x)$$
is a tempered distribution. More generally, if μ is a Borel measure (not necessarily finite) with
$$\int \frac{d\mu(x)}{(1+|x|^2)^k} < \infty$$
for some $k \in \mathbb{N} \cup \{0\}$, then ℓ_μ (as above) is a tempered distribution.

3. If $a \in \mathbb{R}$, the linear functional
$$\delta_a(\psi) := \psi(a)$$
is a tempered distribution called the *Dirac distribution*.

REMARK 4.8.6. In the literature, one often drops the ℓ_f (ℓ_μ) in the first two examples above and equates the function f (measure μ) with the distribution ℓ_f (ℓ_μ).

There is a more general concept of a "distribution" which is obtained by replacing $\mathcal{S}$ (with its metric topology) with the space $\mathcal{D} = \mathcal{D}(\mathbb{R})$ of infinitely differentiable functions with compact support in $\mathbb{R}$ (with an inductive limit topology). The dual of $\mathcal{D}$ (with respect to this inductive limit topology) is denoted by $\mathcal{D}'$ and is called the space of distributions.

The topology on $\mathcal{D}$, although standard, is somewhat complicated and is discussed thoroughly in Rudin's book [**78**], Chapter 6. We take a moment to mention how sequences converge: A sequence $\{\psi_j : j \in \mathbb{N}\}$ converges to ψ in the topology of $\mathcal{D}$ if and only if (i) there is a compact set $K \subset \mathbb{R}$ which contains the supports of ψ_j for all j, (ii) for each $k \in \mathbb{N} \cup \{0\}$,
$$\sup\{|\psi_j^{(k)}(x) - \psi^{(k)}(x)| : x \in \mathbb{R}\} \to 0, \quad j \to \infty.$$
There are also the following useful characterizations of $\mathcal{D}'$.

PROPOSITION 4.8.7. *Let ℓ be a linear functional on $\mathcal{D}$. Then the following are equivalent.*

1. $\ell \in \mathcal{D}'$
2. ℓ *is "sequentially continuous", i.e., if $\{\psi_j : j \in \mathbb{N}\}$ is any sequence in $\mathcal{D}$ which converges to zero in $\mathcal{D}$, then $\ell(\psi_j) \to 0$.*
3. *To every compact set $K \subset \mathbb{R}$, there is an integer $N_K \in \mathbb{N} \cup \{0\}$ and a constant C_K such that*
$$|\ell(\psi)| \leq C_K \max_{0 \leq j \leq N_K}\left\{\sup_{x \in \mathbb{R}}|\psi^{(j)}(x)|\right\}$$
for every $\psi \in \mathcal{D}$ whose support lies in K.

DEFINITION 4.8.8. If there is an N so that the above holds for all compact sets K, we call the smallest such N the *order* of the distribution ℓ. If no such N exists, the distribution is said to have *infinite order*.

For example, the distributions mentioned in Example 4.8.5 have order zero; while the distribution $\psi \to \psi'(0)$ has order one. All tempered distributions have finite order. We will mention more about the order of a distribution below.

How do the spaces $\mathcal{D}'$ and $\mathcal{S}'$ compare? It turns out that the inclusion mapping $i: \mathcal{D} \to \mathcal{S}$ is continuous (note that the topology on $\mathcal{D}$ is the inductive limit topology while the topology on $\mathcal{S}$ is a metric topology). Thus if $\ell \in \mathcal{S}'$, then $\ell \circ i \in \mathcal{D}'$. This allows us to think of $\mathcal{S}'$ as a subspace of $\mathcal{D}'$.

DEFINITION 4.8.9.
1. We say a distribution ℓ *vanishes* on an open set $\Omega \subset \mathbb{R}$ if $\ell(\psi) = 0$ for every $\psi \in C_0^\infty(\Omega)$ (the infinitely differentiable functions with compact support in Ω).
2. If we let W be the union of all the open sets on which ℓ vanishes and let $S_\ell := \mathbb{R} \setminus W$, then S_ℓ is a closed set called the *support* of the distribution ℓ.

EXAMPLE 4.8.10.
1. If $\ell = \ell_f$ for some $f \in L^p$, then S_ℓ is equal to the support of f
2. If $\ell = \ell_\mu$ for some $\mu \in M(\mathbb{R})$, then S_ℓ is the support of μ
3. If $\ell = \delta_a$, then $S_\ell = \{a\}$.

An interesting result is that any $\ell \in \mathcal{D}'$ with compact support belongs to $\mathcal{S}'$.

With distributions, we have the following important operations.

DEFINITION 4.8.11.
1. *Multiplication*: If $\phi \in \mathcal{S}$ and $\ell \in \mathcal{S}'$, then the product $\phi\ell$ is a tempered distribution defined by
$$(\phi\ell)(\psi) := \ell(\phi\psi).$$

2. *Differentiation*: If $\ell \in \mathcal{S}'$ and $n \in \mathbb{N}$, then the *n-th derivative* of ℓ, $\ell^{(n)}$, is the tempered distribution defined by
$$\ell^{(n)}(\psi) := (-1)^n \ell(\psi^{(n)}).$$
For example, $\delta_a^{(n)}(\psi) = (-1)^n \psi^{(n)}(a)$.
3. *Convolutions*: For $\psi, \phi \in \mathcal{S}$, define the *convolution*
$$(\phi * \psi)(x) := \int \phi(y)\psi(x-y)dy$$
and observe that $\phi * \psi \in \mathcal{S}$ and that for $\psi_1, \psi_2, \psi_3 \in \mathcal{S}$, we have
$$(\psi_1 * \psi_2) * \psi_3 = \psi_1 * (\psi_2 * \psi_3)$$
$$\psi_1 * \psi_2 = \psi_2 * \psi_1.$$
If $\psi \in \mathcal{S}$, define
$$(\tau_x \psi)(y) := \psi(y-x)$$
$$\check{\psi}(y) := \psi(-y).$$
For $\ell \in \mathcal{S}'$, define the *convolution* of ℓ and ψ by
$$(\ell * \psi)(x) := \ell(\tau_x \check{\psi}).$$

It is known that $\ell * \psi \in C^\infty(\mathbb{R})$. In fact, $\ell * \psi$ is "slowly increasing" (has "polynomial growth"); more specifically, $(\ell * \psi)(x)/(1+|x|^2)^k$ is a bounded function for some $k \in \mathbb{N} \cup \{0\}$. We also have the formulas

$$(\ell * \psi)^{(n)}(x) = (\ell^{(n)} * \psi)(x) = (\ell * \psi^{(n)})(x).$$

Since $\ell * \phi$ has polynomial growth, it also can be regarded as a tempered distribution (via Example 4.8.5) and so one can also define $(\ell * \psi) * \phi$ and notice that

$$(\ell * \psi) * \phi = \ell * (\psi * \phi).$$

It turns out that if a distribution $\ell \in \mathcal{D}'$ has compact support, then it is tempered ($\ell \in \mathcal{S}'$) and thus has finite order. In particular, if a distribution ℓ has support in a finite set $\{a_1, \cdots, a_K\}$ and has order N, then ℓ is of the form

$$(4.8.12) \qquad \ell = \sum_{k=1}^{K} \sum_{j=0}^{N} c_{j,k} \delta_{a_k}^{(j)}$$

for some $c_{j,k} \in \mathbb{C}$.

4.8.2. Distributions and $\mathcal{H}_h^p$. As a first step towards viewing $\mathcal{H}_h^p$ or $\mathcal{H}^p(\mathbb{C}\backslash\mathbb{R})$ as a space of distributions, we have the following result of Fefferman and Stein [35].

THEOREM 4.8.13 (Fefferman-Stein). *Let $p \in (0, \infty)$ and $u \in \mathcal{H}_h^p$. Then the following are true.*
1.
$$\lim_{y \to 0^+} u(\cdot + iy) = \ell$$

exists in the sense of (tempered) distributions. That is to say, for all $\psi \in \mathcal{S}$, the limit

$$\ell(\psi) := \lim_{y \to 0^+} \int u(x+iy)\psi(x) dx$$

exists and defines an element of $\mathcal{S}'$.
2. *The function $u \in \mathcal{H}_h^p$ is uniquely determined by ℓ; namely, if $\ell(\psi) = 0$ for every $\psi \in \mathcal{S}$, then $u = 0$.*
3. *If $\ell = \lim_{y \to 0^+} u(\cdot + iy)$ in the sense of distributions, then*
   (a) *if $p \in (1, \infty)$,*

   $$|\ell(\psi)| \le A\|u\|\|\psi\|_{L^q} \;\; \forall \psi \in \mathcal{S};$$

   (b) *if $p = 1$,*

   $$|\ell(\psi)| \le A\|u\|\|\psi\|_{BMO} \;\; \forall \psi \in \mathcal{S};$$

   (c) *if $p \in (0, 1)$,*

   $$(4.8.14) \qquad |\ell(\psi)| \le A\|u\|\|\psi\|_{\Lambda_\alpha} \;\; \forall \psi \in \mathcal{S},$$

   where $\alpha = 1/p - 1$ and Λ_α is the Lipschitz or Zygmund class (see definition below).

REMARK 4.8.15. 1. Here BMO (bounded mean oscillation) is the space of locally integrable functions g on $\mathbb{R}$ for which

$$\|g\|_{BMO} := \sup_{I} \frac{1}{|I|} \int_{I} |g - g_I| dx$$

is finite. Here I is a bounded interval and $g_I := |I|^{-1} \int_I g dx$ is the average of g over I.

2. Let $n \in \mathbb{N} \cup \{0\}$ and $\alpha \in (0,1)$. Define the Lipschitz class $\Lambda_\alpha^n(\mathbb{R})$ to be the space of functions $g \in C^{(n)}(\mathbb{R}^-)$ [10] for which

$$\sup_{x \in \mathbb{R}, y \neq 0} \frac{|g^{(n)}(x+y) - g^{(n)}(x)|}{|y|^\alpha} < \infty.$$

$\Lambda_\alpha^n(\mathbb{R})$ becomes a Banach space when given the norm

$$\|g\|_{\Lambda_\alpha^n} := \|g\|_\infty + \max_{0 \leq j \leq n} \sup_{x \in \mathbb{R}, y \neq 0} \frac{|g^{(j)}(x+y) - g^{(j)}(x)|}{|y|^\alpha}.$$

Define the Zygmund class $\Lambda_*^n(\mathbb{R})$ to be the space of functions $g \in C^{(n)}(\mathbb{R}^-)$ for which

$$\sup_{x \in \mathbb{R}, y \neq 0} \frac{|g^{(n)}(x+y) - 2g^{(n)}(x) + g^{(n)}(x-y)|}{|y|} < \infty.$$

A norm on $\Lambda_*^n(\mathbb{R})$ is then

$$\|g\|_{\Lambda_*^n} = \|g\|_\infty + \max_{0 \leq j \leq n} \sup_{x \in \mathbb{R}, y \neq 0} \frac{|g^{(j)}(x+y) - 2g^{(j)}(x) + g^{(j)}(x-y)|}{|y|}.$$

In eq.(4.8.14) we set

$$\Lambda_{1/p-1}(\mathbb{R}) = \begin{cases} \Lambda_{1/p-n}^{n-1}(\mathbb{R}) & \text{if } p \in (1/(n+1), 1/n), \\ \Lambda_*^{n-1}(\mathbb{R}) & \text{if } p = 1/(n+1). \end{cases}$$

3. There is also the following equivalent definition of the spaces Λ_α due (independently) to Campanato [18] and Meyers [62]. A nice treatment of this can be found in [38], p. 292. For $\alpha > 0$, define L_α^∞ to be the set of locally bounded functions g for which there exists a constant $C > 0$ so that for every bounded interval $I \subset \mathbb{R}$, there is a polynomial $p_I(g)$ of degree at most $[\alpha]$ such that

$$|g - p_I(g)| \leq C|I|^\alpha \text{ on } I.$$

It turns out that $L_\alpha^\infty(\mathbb{R})$ can be normed in an appropriate way such that $L_\alpha^\infty(\mathbb{R}) = \Lambda_\alpha$ (equating functions which are equal almost everywhere) with equivalent norms.

4. From eq.(4.8.14) every distribution ℓ arising this way has order no more than $[1/p] - 1$.

5. As suggested by the inequality in eq.(4.8.14), we can extend the domain of ℓ to $\Lambda_{1/p-1}$ (see Remark 4.9.5 below).

Theorem 4.8.13 allows us to think of $\mathcal{H}_h^p$ as a "subset" of $\mathcal{S}'$. In order to distinguish between harmonic functions and distributions we introduce the following definition.

---

[10]$g \in C^{(n)}(\mathbb{R}^-)$ if $g \in C^{(n)}(\mathbb{R})$ and $\lim_{|x| \to \infty} g^{(n)}(x)$ exists and is finite.

DEFINITION 4.8.16. Let $H^p(\mathbb{R})$ $(0 < p < \infty)$ denote the space of distributions ℓ of the form

(4.8.17) $$\ell(\psi) = \lim_{y \to 0^+} \int u(x+iy)\psi(x)dx, \quad \psi \in \mathcal{S},$$

for some $u \in \mathcal{H}_h^p$.

The question now is, given $\ell \in \mathcal{S}'$, when does it arise as the distributional boundary values of an $\mathcal{H}_h^p$ function; equivalently, when is $\ell \in H^p(\mathbb{R})$?

EXAMPLE 4.8.18.   1. When $f \in L^p$ $(1 < p < \infty)$, Theorem 4.3.1 and Theorem 4.7.8 say that the function

$$U_f(x+iy) = \int f(t) P_z(t) dt$$

belongs to $\mathcal{H}_h^p$. Fatou's theorem, Theorem 4.3.1, says that

$$\lim_{y \to 0^+} U_f(\cdot + iy) = f \text{ in } L^p,$$

so certainly

$$U_f(\cdot + iy) \to \ell_f \text{ in the sense of distributions. }^{11}$$

Thus for $p \in (1, \infty)$, every L^p function arises as the distributional boundary values of an $\mathcal{H}_h^p$ function. Equating $f \in L^p$ with the distribution $\ell_f \in H^p(\mathbb{R})$, we can say $L^p \subset H^p(\mathbb{R})$.

2. Conversely, suppose that $u \in \mathcal{H}_h^p$ $(1 < p < \infty)$, then by Theorem 4.3.3 and Theorem 4.7.8, $u = U_f$ for some $f \in L^p$ and, by our above argument, the distributional boundary values of u are equal to ℓ_f. Thus for $p \in (1, \infty)$, $H^p(\mathbb{R}) = L^p$. We will see that when $p \in (0, 1]$, this is no longer the case. In fact for $p \in (0, 1)$, there are sums of point masses and their derivatives which belong to $H^p(\mathbb{R})$.

Following the model for $p \in (1, \infty)$, one is tempted to say that given an $\ell \in H^p(\mathbb{R})$, the $U_\ell \in \mathcal{H}_h^p$ which corresponds to this distribution, in the sense of eq.(4.8.17), should be

$$U_\ell(z) = \ell(P_z(\cdot)).$$

However, for fixed $z \in \mathbb{C}_+$, $P_z(t) \notin \mathcal{S}$ (it is not rapidly decreasing at infinity), and so the expression $\ell(P_z(\cdot))$ does not seems to be well defined. As it turns out, we can still define $\ell(P_z(\cdot))$ since distributions in $H^p(\mathbb{R})$ have several desirable properties. Our presentation will be one of Stein [95], p. 89, and will use properties of convolutions.

To set this up, we first write our Poisson kernel $P_z(t)$ in the form

$$P_{x+iy}(t) = P_y(x-t) = \frac{1}{\pi} \frac{y}{y^2 + (x-t)^2}$$

and so for $f \in L^1$,

$$U_f(x+iy) = (f * P_y)(x).$$

For a distribution $\ell \in H^p(\mathbb{R})$, we will now try to make some sense out of the expression $(\ell * P_y)(x)$.

---

[11] Recall that ℓ_f is the distribution defined by $\ell_f(\psi) = \int f(x)\psi(x)dx$.

We know that for $\ell \in \mathcal{S}'$ and $\psi \in \mathcal{S}$, the convolution $(\ell * \psi)(x)$ is an infinitely differentiable function on $\mathbb{R}$ which is "slowly increasing", that is

$$(\ell * \psi)(x)/(1 + |x|^2)^k \in L^\infty \text{ form some } k \in \mathbb{N} \cup \{0\}.$$

DEFINITION 4.8.19. We say that ℓ is *bounded* if $\ell * \psi \in L^\infty$ for all $\psi \in \mathcal{S}$.

Before attempting to define $\ell * P_y$, we review some basic facts about convolutions. If $1 \leq q_1, q_2 \leq \infty$ with $1/q_1 + 1/q_2 \geq 1$ and $f \in L^{q_1}$, $g \in L^{q_2}$, the Hausdorff-Young inequality says that the *convolution*

$$(f * g)(x) := \int f(x-t)g(t)\, dt$$

defines a function which belongs to L^r, where $1/r = 1/q_1 + 1/q_2 - 1$. If $f * g \in L^1$ (say $q_1 = q_2 = 1$), then $f * g$ also defines a distribution in the usual way

$$(f * g)(\psi) = \int (f * g)(x)\psi(x)dx$$

which is also equal to

$$((f * g) * \check{\psi})(0),$$

where $\check{\psi}(x) = \psi(-x)$.

DEFINITION 4.8.20. If ℓ is a bounded distribution and $h \in L^1$, we can define a tempered distribution $\ell \# h$ by

$$(\ell \# h)(\psi) := \int (\ell * \check{\psi})(x)\check{h}(x)dx, \quad \psi \in \mathcal{S}.$$

REMARK 4.8.21.   1. The above integral makes sense since ℓ is a bounded distribution.
2. The reader can use Proposition 4.8.4 to check that $\ell \# h$ is a tempered distribution and that $\ell \# h$ is in fact a bounded distribution.
3. Using the definition of $\ell \# h$, one can even show that

$$\ell \# (h_1 \# h_2) = (\ell \# h_1) \# h_2$$

   whenever $h_1, h_2 \in L^1$ [12].
4. This $\#$ operation might seem a bit mysterious but for the fact that if $\ell = \ell_f$ for some $f \in L^1$; namely, $\ell_f(\psi) = \int f\psi dx$, then

$$\begin{aligned}(\ell_f \# h)(\psi) &= \int (f * \check{\psi})(x)\check{h}(x)dx \\ &= ((f * \check{\psi}) * h)(0) \\ &= (f * (\check{\psi} * h))(0) \\ &= (f * (h * \check{\psi}))(0) \\ &= ((f * h) * \check{\psi})(0) \\ &= (f * h)(\psi).\end{aligned}$$

Thus in this case, $\ell_f \# h$ is the usual convolution $f * h$.

---

[12]From the Hausdorff-Young inequality, observe that any L^1 function defines a bounded distribution.

For fixed $y > 0$, the Poisson kernel $P_y(x) \in L^1$ and so $\ell \# P_y$ is a well defined bounded distribution. Moreover, for each $\psi \in \mathcal{S}$,

$$(\ell \# P_y)(\psi) = \int (\ell * \check{\psi})(x) P_y(x) dx$$

(note that $\check{P}_y = P_y$) which, by Theorem 4.3.1, goes to $(\ell * \check{\psi})(0) = \ell(\psi)$ as $y \to 0$ (since $\ell * \check{\psi}$ is a bounded C^∞ function). Thus

$$\ell \# P_y \to \ell$$

in the sense of distributions as $y \to 0^+$. What we will do next is to define, for fixed $y > 0$, $\ell \# P_y$ as a *function* $(\ell \# P_y)(x)$ and then show that $U_\ell(x + iy) := (\ell \# P_y)(x)$ is a harmonic function on $\mathbb{C}_+$ such that

$$\int U_\ell(x + iy) \psi(x) dx = (\ell \# P_y)(\psi), \quad \psi \in \mathcal{S},$$

and so $U_\ell(\cdot + iy) \to \ell$ in the sense of distributions.

For a function g and $y > 0$, define $g_y(x) = y^{-1} g(x/y)$. With this definition, $P_y = (P_1)_y$. We first claim that

$$P_y = \psi_y + \phi_y * P_y$$

for some $\phi, \psi \in \mathcal{S}$. This fact can be found in [95], p. 20, but we outline the proof here. Note that $(\mathcal{F} P_1)(x) = e^{-2\pi |x|}$ ($\mathcal{F}$ denotes the usual Fourier transform) is rapidly decreasing and is smooth at all points except $x = 0$. Let $\phi \in \mathcal{S}$ with $(\mathcal{F}\phi)(x) = 1$ for x near zero and set ψ to be a function in $\mathcal{S}$ with

$$(\mathcal{F}\psi)(x) = (1 - (\mathcal{F}\phi)(x))(\mathcal{F}P_1)(x).$$

Taking inverse Fourier transforms, we obtain the identity

$$P_1 = \psi + \phi * P_1$$

and so $P_y = \psi_y + \phi_y * P_y$.

The convolutions $\ell * \psi_y$ and $\ell * \phi_y$ are bounded C^∞ functions (since ℓ is a bounded distribution) and so $(\ell * \phi_y) * P_y$ is also a bounded C^∞ function. This allows us to make the following definition.

DEFINITION 4.8.22. For each $y > 0$, define $U_\ell(x + iy) = (\ell \# P_y)(x)$ by the following formula

$$U_\ell(x + iy) = (\ell \# P_y)(x) := (\ell * \psi_y)(x) + ((\ell * \phi_y) * P_y)(x).$$

From this definition, $\ell \# P_y$ is a bounded C^∞ function for each fixed $y > 0$. We remark that if $\ell = \ell_f$ for some $f \in L^1$, then $\ell_f \# P_y$ is equal to $f * P_y$ both as distributions *and* as functions. We also remark that by direct computation with the distributions and using the fact that the Poisson kernel is harmonic, one can show that

$$U_\ell(x + iy) := (\ell \# P_y)(x)$$

is harmonic on $\mathbb{C}_+$. When we get to the atomic decomposition in the next section, we will see another reason why this function is harmonic.

Finally, observe that if $\eta \in \mathcal{S}$, then

$$
\begin{aligned}
\int U_\ell(x+iy)\eta(x)dx &= \int (\ell \# P_y)(x)\eta(x)dx \\
&= \int (\ell * \psi_y)(x)\eta(x)dx + \int ((\ell * \phi_y) * P_y)(x)\eta(x)dx \\
&= ((\ell * \psi_y) * \check{\eta})(0) + (((\ell * \phi_y) * P_y) * \check{\eta})(0) \\
&= ((\ell * \check{\eta}) * \psi_y)(0) + (((\ell * \phi_y) * \check{\eta}) * P_y)(0) \\
&= ((\ell * \check{\eta}) * \psi_y)(0) + (((\ell * \check{\eta}) * \phi_y) * P_y)(0) \\
&= ((\ell * \check{\eta}) * \psi_y)(0) + ((\ell * \check{\eta}) * (\phi_y * P_y))(0) \\
&= \int (\ell * \check{\eta})(x) \check{\psi}_y(x) + (\ell * \check{\eta})(x)(\phi_y \check{*} P_y)(x)dx \\
&= \int (\ell * \check{\eta})(x) \check{P}_y(x)dx \\
&= \int (\ell * \check{\eta})(x) P_y(x)dx \\
&= (\ell \# P_y)(\eta).
\end{aligned}
$$

We leave it to the reader to justify that the above convolution calculations are valid by using the Hausdorff-Young inequality and the basic facts about convolutions from Definition 4.8.11. Thus, from the above calculation,

$$\int U_\ell(x+iy)\eta(x)dx = (\ell \# P_y)(\eta) \to \ell(\eta) \quad \text{as } y \to 0^+.$$

That is to say, $U_\ell(\cdot + iy) \to \ell$ in the sense of tempered distributions. We summarize all of this with the following result.

LEMMA 4.8.23 ([**95**], p. 89). *If ℓ is a bounded distribution, then the function*

$$U_\ell(x+iy) := (\ell \# P_y)(x)$$

is well defined and harmonic on $\mathbb{C}_+$. Moreover

$$\ell = \lim_{y \to 0^+} U_\ell(\cdot + iy)$$

in the sense of distributions.

REMARK 4.8.24.
1. The notation $\ell \# P_y$ was only used for pedagogical convenience and is not standard. Most of the time (see [**95**] for example) authors write $\ell * P_y$. In a sense, this notation is justified since if $\ell = \ell_f$ for some $f \in L^1$, then, as we saw earlier, $\ell \# P_y$ is really the ordinary convolution $f * P_y$.
2. Since, for $\phi \in \mathcal{S}$, $(\ell * \phi)(x) = \ell(\phi(\cdot - x))$, we will also write $(\ell \# P_y)(x) = \ell(P_{x+iy}(\cdot))$. In the next section on the atomic decomposition, it will turn out that $\ell(P_{x+iy}(\cdot))$ can be defined in a different way and, fortunately, these two definitions coincide.

Recall, a harmonic function belongs to $\mathcal{H}_h^p$ if and only if $u^* \in L^p$. This next result can be found in [**95**], p. 119, and is the extension of the Poisson integral theory for $\mathcal{H}_h^p$ ($1 \leq p < \infty$) developed in Theorem 4.7.8 to all $p \in (0, \infty)$.

PROPOSITION 4.8.25. *For $p \in (0, \infty)$ and $u \in Har(\mathbb{C}_+)$ the following are equivalent.*

1. $u \in \mathcal{H}_h^p$
2. $u(z) = U_\ell(z) = \ell(P_z(\cdot))$ for some $\ell \in H^p(\mathbb{R})$.

Moreover $\ell = \lim_{y \to 0^+} U_\ell(\cdot + iy)$ in the sense of distributions.

If $u \in \mathcal{H}_h^p$ ($1 < p < \infty$), then $u = U_f$ for some $f \in L^p$ and $\lim_{y \to 0} U_f(\cdot + iy) = f$ almost everywhere and in L^p. If $p = 1$ and $u \in \mathcal{H}_h^1$, then $u = U_\mu$ for some measure μ and $\lim_{y \to 0} U_\mu(\cdot + iy) = \mu$ weak-*. The above result is an extension of these results to all $p \in (0, \infty)$.

This next result of Fefferman and Stein [35] tells us precisely which distributions arise as the boundary distributions of $\mathcal{H}_h^p$ functions.

THEOREM 4.8.26 (Fefferman-Stein). *For a tempered distribution ℓ and $p \in (0, \infty)$, the following are equivalent.*

1. $\ell \in H^p(\mathbb{R})$, i.e., $\ell = \lim_{y \to 0^+} u(\cdot + iy)$ for some $u \in \mathcal{H}_h^p$.
2. ℓ is bounded and $U_\ell^* \in L^p$.
3. $\sup_{y > 0} |(\ell * \psi_y)(x)| \in L^p$ for some $\psi \in \mathcal{S}$ with $\int \psi = 1$. [13]
4. $\sup_{|x-t| < y} |(\ell * \psi_y)(x)| \in L^p$ for some $\psi \in \mathcal{S}$ with $\int \psi = 1$.
5. *The limit* $\lim_{\delta \to 0} \ell(e^{-\delta t^2} P_z(t)) := U(z)$ *exists for all* $z \in \mathbb{C}_+$ *and* $U^* \in L^p$.

REMARK 4.8.27. In the last condition in Theorem 4.8.26, note that $U_\delta(z) := \ell(e^{-\delta t^2} P_z(t))$ is harmonic and if the limit as $\delta \to 0$ exists for each $z \in \mathbb{C}_+$, it is not too difficult to show that the limit will be uniform on compact sets and so U will be harmonic. This allows us to define $\ell(P_z(\cdot))$ without making references to bounded distributions.

DEFINITION 4.8.28. From here, it is traditional to equate $\ell \in H^p(\mathbb{R})$ with $U_\ell \in \mathcal{H}_h^p$ and "norm" ℓ by

$$\|\ell\|_{H^p(\mathbb{R})} := \|U_\ell\|_{\mathcal{H}_h^p}.$$

The following are some examples of distributions which belong to $H^p(\mathbb{R})$ and can be found in [95], p. 129. Since for $p \in (1, \infty)$, $H^p(\mathbb{R})$ can be equated with L^p, our examples will be from the more interesting $p \in (0, 1)$ case.

EXAMPLE 4.8.29. Let $p \in (0, 1)$ be fixed.

1. A finite (complex) measure μ with compact support, regarded as a distribution [14], belongs to $H^p(\mathbb{R})$ if and only if

(4.8.30) $$\int x^k d\mu = 0 \quad \forall \, k = 0, 1, \cdots, [1/p] - 1.$$

2. If
$$\ell = \sum_{1 \leq k \leq N, 0 \leq j \leq M} c_{j,k} \delta_{a_k}^{(j)}, \quad a_k \in \mathbb{R}, c_{j,k} \in \mathbb{C},$$

then $\ell \in H^p(\mathbb{R})$ if and only if $M < [1/p] - 1$ and $\ell(x^s) = 0$ [15] for all $s = 0, 1, \cdots, [1/p] - 1$.

---

[13] Here $\psi_y(x) := y^{-1} \psi(x/y)$.
[14] i.e., $\mu(\psi) = \int \psi d\mu$, $\psi \in \mathcal{S}$.
[15] Here x^s is not in $\mathcal{S}$ and so $\ell(x^s)$ is not technically defined. However, ℓ has compact support and so if $\psi \in C_0^\infty$ which is one in a neighborhood of the support of ℓ, then $x^s \psi \in \mathcal{S}$ and $\ell(x^s)$ can be defined as $\ell(x^s \psi)$ and does not depend on the choice of $\psi \in C_0^\infty$ which is one near the support of ℓ.

4.9. The atomic decomposition

From the previous section, we saw that a compactly supported function which satisfies the moment conditions in eq.(4.8.30) belongs to $H^p(\mathbb{R})$. As is turns out, these functions are, in a sense, the building blocks for $H^p(\mathbb{R})$ ($0 < p \leq 1$). Hence they are given the appropriate name of "atoms".

DEFINITION 4.9.1. For fixed $p \in (0,1]$, an *atom* is a function b supported on a compact interval Δ_b such that

$$|b| \leq |\Delta_b|^{-1/p} \text{ almost everywhere}$$

$$\int x^j b(x) dx = 0 \ \forall \ j = 0, 1, \cdots, [1/p] - 1.$$

It is not too difficult to check that there is a universal constant $C_p > 0$, depending only on p, such that

(4.9.2) $$\|b\|_{H^p(\mathbb{R})} \leq C_p$$

for all atoms b [95], p. 129. This next result of Coifman is very useful in understanding $H^p(\mathbb{R})$ (or $\mathcal{H}_h^p$) and will be an important tool in Chapter 6. It essentially says that every $\ell \in H^p(\mathbb{R})$ can be written as an infinite linear combination of atoms.

THEOREM 4.9.3 (Coifman [21] [95], Chapter 3). *For any distribution ℓ belonging to $H^p(\mathbb{R})$ ($0 < p \leq 1$), there are atoms $\{b_k : k \in \mathbb{N}\}$ and complex numbers $\{\lambda_k : k \in \mathbb{N}\}$ so that*

$$\ell = \sum_{k=1}^{\infty} \lambda_k b_k,$$

where the above sum converges both in the sense of distributions and in $H^p(\mathbb{R})$. Moreover,

(4.9.4) $$C_1 \|\ell\|^p \leq \sum_{k=1}^{\infty} |\lambda_k|^p \leq C_2 \|\ell\|^p$$

for some positive constants $C_1, C_2 > 0$ which do not depend on ℓ.

Conversely, if $\ell \in \mathcal{S}'$ with $\ell = \sum_k \lambda_k b_k$, where $\{\lambda_k : k \in \mathbb{N}\}$ are complex constants with $\sum_k |\lambda_k|^p < \infty$ and $\{b_k : k \in \mathbb{N}\}$ are atoms, then $\ell \in H^p(\mathbb{R})$ with $\|\ell\|^p \asymp \sum_k |\lambda_k|^p$.

REMARK 4.9.5. 1. Note that the phrase "converges in the sense of distributions" means that for any $\psi \in \mathcal{S}$,

$$\ell(\psi) = \lim_{K \to \infty} \sum_{k=1}^{K} \lambda_k \int b_k(t) \psi(t) \, dt.$$

2. The decomposition of $\ell \in H^p(\mathbb{R})$ into atoms is by no way unique. Rather than being a technical annoyance, it is quite a blessing since we can often choose our atoms to satisfy certain useful properties. We mention a few that will be used later.

    (a) Given any positive integer $M \geq [1/p] - 1$, we can arrange the atoms $\{b_k : k \in \mathbb{N}\}$ in the above atomic decomposition to satisfy the further moment properties

    $$\int Q b_k \, dx = 0$$

for every polynomial Q whose degree is at most M [95], p. 105.

(b) If I is an open interval in $\mathbb{R}$ and $\ell \in H^p(\mathbb{R})$ for some $p \leq 1$, which vanishes on $\mathbb{R}\setminus I^-$, [16] then ℓ has an atomic decomposition whose atoms are supported in I [95], p. 137.

3. The atomic decomposition is also helpful in identifying the dual of $H^p(\mathbb{R})$ ($0 < p < 1$) with $\Lambda_{1/p-1}$ (the Lipschitz or Zygmund class)[17]. For example, for $g \in \Lambda_{1/p-1}$, the linear functional

$$L_g(u) = \int u(x)g(x)dx,$$

initially defined on finite linear combinations of atoms, has a bounded extension to $H^p(\mathbb{R})$ with $|L_g(u)| \leq A\|u\|\|g\|_{\Lambda_{1/p-1}}$. Furthermore, every continuous linear functional on $H^p(\mathbb{R})$ is of this form and the norm of this linear functional L_g is equivalent to the $\Lambda_{1/p-1}$ norm of g. This theorem was shown for H^p of the disk in [32]. For a formulation and proof of the result in the current situation, see [38], p. 293.

4. The above duality allows us to extend the class of "test functions" beyond $\mathcal{S}$. Indeed, if $\ell \in H^p(\mathbb{R})$ ($0 < p < 1$) and $\psi \in \mathcal{S}$, then

$$\ell(\psi) = \sum_{k=1}^{\infty} \lambda_k \int b_k(t)\psi(t)dt = L_\psi(\ell).$$

But since L_g is meaningful for $g \in \Lambda_{1/p-1}$, we can extend the domain of ℓ to $\Lambda_{1/p-1}$ by the relation

$$\ell(g) := L_g(\ell), \quad g \in \Lambda_{1/p-1}.$$

More specifically, we have

$$\ell(g) = \sum_{k=1}^{\infty} \lambda_k \int b_k(t)g(t)dt$$

and

$$|\ell(g)| \leq C_p \|g\|_{\Lambda_{1/p-1}} \|\ell\|.$$

Furthermore, the definition of $\ell(g)$ does not depend on the choice of atomic decomposition of ℓ.

5. Recall that if $\ell \in H^p(\mathbb{R})$ ($0 < p < 1$), the harmonic function

$$U_\ell(x+iy) = (\ell \# P_y)(x)$$

was the unique harmonic function in $\mathcal{H}_h^p$ for which

$$U_\ell(\cdot + iy) \to \ell$$

in the sense of distributions. Also recall that we abused notation slightly and often said $U_\ell(x+iy) = \ell(P_{x+iy}(\cdot))$ since when $\ell = \ell_f$ for some integrable function f,

$$\ell_f(P_{x+iy}(\cdot)) = \int P_{x+iy}(t)f(t)dt = (\ell \# P_y)(x).$$

---

[16] i.e., $\ell(\psi) = 0$ for every $\psi \in C_0^\infty(\mathbb{R}\setminus I^-)$.

[17] When $p = 1$, one can identify the dual of $H^1(\mathbb{R})$ with BMO.

It turns out that this is not really an abuse of notation after all since for fixed $x+iy \in \mathbb{C}_+$, $P_{x+iy}(t) \in \Lambda_{1/p-1}$ and so, by our work above, $\ell(P_{x+iy}(\cdot))$ is meaningfully defined via any atomic decomposition $\sum_k \lambda_k b_k$ of ℓ by

$$\ell(P_{x+iy}(\cdot)) = \sum_{k=1}^{\infty} \lambda_k \int b_k(t) P_{x+iy}(t) dt.$$

Notice that for any $K \in \mathbb{N}$, the function

$$\sum_{k=1}^{K} \lambda_k \int b_k(t) P_{x+iy}(t) dt$$

is harmonic on $\mathbb{C}_+$. Moreover, for large enough k, $|\lambda_k| \leq |\lambda_k|^p$ (since $p \in (0,1)$ and $\lambda_k \to 0$) and so, for $x+iy$ fixed in some compact subset of $\mathbb{C}_+$,

$$\begin{aligned}
\left| \sum_{k=1}^{K} \lambda_k \int b_k(t) P_{x+iy}(t) dt \right| &\leq C \sum_{k=1}^{K} |\lambda_k|^p \left| \int b_k(t) P_{x+iy}(t) dt \right| \\
&\leq C \sum_{k=1}^{K} |\lambda_k|^p \|b_k\| \|P_{x+iy}(\cdot)\|_{\Lambda_{1/p-1}} \\
&\leq C \sum_{k=1}^{K} |\lambda_k|^p \\
&\leq C \|\ell\|^p.
\end{aligned}$$

Thus the series

$$\sum_{k=1}^{\infty} \lambda_k \int b_k(t) P_{x+iy}(t) dt$$

converges uniformly on compact subsets of $\mathbb{C}_+$ to $\ell(P_{x+iy}(\cdot))$ which will necessarily be harmonic. We now claim that this harmonic function is indeed $U_\ell(x+iy) = (\ell \# P_y)(x)$. Indeed

$$\begin{aligned}
(\ell \# P_y)(x) &= (\ell * \psi_y)(x) + ((\ell * \phi_y) * P_y)(x) \\
&= \sum_k \lambda_k \int b_k(t) \psi_y(x-t)\, dt \\
&\quad + \int \left(\sum_k \lambda_k \int b_k(s) \phi_y(t-s) ds \right) P_y(x-t)\, dt \\
&= \sum_k \lambda_k (b_k * \psi_y)(x) \\
&\quad + \sum_k \lambda_k \int \left(\int b_k(s) \phi_y(t-s) ds \right) P_y(x-t)\, dt \\
&= \sum_k \lambda_k (b_k * \psi_y)(x) + \sum_k \lambda_k ((b_k * \phi_y) * P_y)(x) \\
&= \sum_k \lambda_k (b_k * \psi_y)(x) + \sum_k \lambda_k (b_k * (\phi_y * P_y))(x) \\
&= \sum_k \lambda_k (b_k * P_y)(x) \\
&= \ell(P_{x+iy}(\cdot))
\end{aligned}$$

Furthermore, from our estimates above, we have the pointwise estimate
$$|U_\ell(x+iy)| = |\ell(P_{x+iy}(\cdot))| \leq C_p \|P_{x+iy}(\cdot)\|_{\Lambda_{1/p-1}} \|\ell\|.$$

6. In a similar way, one can define for $\ell \in H^p(\mathbb{R})$ the functions
$$V_\ell(z) := \ell(Q_z(\cdot)), \quad z \in \mathbb{C}_+,$$

$$F_\ell(z) := \frac{1}{2\pi i} \ell\left(\frac{1}{\cdot - z}\right), \quad z \in \mathbb{C}_+,$$

(which are the distributional analogs of the conjugate Poisson integral and the Cauchy integral) and note that $V_\ell \in \operatorname{Har}(\mathbb{C}_+)$ and $F_\ell \in \mathfrak{H}(\mathbb{C}_+)$. Furthermore,

$$|V_\ell(z)| \leq C_p \|Q_z(\cdot)\|_{\Lambda_{1/p-1}} \|\ell\|, \quad |F_\ell(z)| \leq C_p \left\|\frac{1}{\cdot - z}\right\|_{\Lambda_{1/p-1}} \|\ell\|.$$

In particular, note that for fixed $x \in \mathbb{R}$,

(4.9.6) $$\lim_{y \to \pm\infty} F_\ell(x+iy) = 0.$$

We make the remark that one can use the techniques developed here (maximal functions, distributions, atomic decompositions, etc.) to look at the space "real $\mathcal{H}^p$", or even "real H^p" (for the disk). Two nice treatments of this are **[22]** and **[58]**.

4.10. Distributions and $\mathcal{H}^p$

In the previous two sections, we extended the Poisson integral theory for $\mathcal{H}^p_h$ to all $p \in (0, \infty)$ and saw that $u \in \mathcal{H}^p_h$ if and only if $u = U_\ell$ for some (unique) $\ell \in H^p(\mathbb{R})$. Furthermore, $\lim_{y \to 0^+} U_\ell(\cdot + iy) = \ell$ in the sense of distributions. In this section, we wish to extend the Cauchy integral formula, Corollary 4.3.5, for $\mathcal{H}^p$ to all $p \in (0, \infty)$.

For $g \in L^p$ ($1 \leq p < \infty$) define
$$F_g(z) := \frac{1}{2\pi i} \int \frac{g(t)}{t-z} dt$$

and recall that F_g is an analytic function on $\mathbb{C} \setminus \mathbb{R}$. Corollary 4.3.5 says that $f \in \mathcal{H}^p$ ($1 \leq p < \infty$) if and only if $f(z) = F_g(z)$ on $\mathbb{C}_+$ for some $g \in L^p$ with $F_g|\mathbb{C}_- = 0$ ($\mathbb{C}_-$ is the lower half plane $\{\Im z < 0\}$). Moreover, under these assumptions,

$$\lim_{y \to 0^+} F_g(x+iy) = g(x)$$

almost everywhere.

For $\ell \in H^p(\mathbb{R})$, we saw from the previous section, Remark 4.9.5 (6), that

$$F_\ell(z) := \frac{1}{2\pi i} \ell\left(\frac{1}{\cdot - z}\right)$$

is a well defined analytic function on $\mathbb{C} \setminus \mathbb{R}$. This next theorem is a Cauchy integral formula for $\mathcal{H}^p$ ($0 < p < 1$).

THEOREM 4.10.1. *Let $p \in (0,1)$. Then $f \in \mathcal{H}^p$ if and only if $f(z) = F_\ell(z)$ on $\mathbb{C}_+$ for some $\ell \in H^p(\mathbb{R})$ for which $F_\ell|\mathbb{C}_- = 0$. Moreover, under these assumptions,*

$$\lim_{y \to 0^+} f(\cdot + iy) = \ell$$

in the sense of distributions.

PROOF. The proof is pretty much the same as the proof of Corollary 4.3.5 except one thinks in terms of distributions. If $f \in \mathcal{H}^p$, then $f \in \mathcal{H}_h^p$ and by Proposition 4.8.25,

$$f(z) = U_\ell(z), \quad z \in \mathbb{C}_+,$$

for some $\ell \in H^p(\mathbb{R})$. By Remark 4.9.5, F_ℓ is analytic on $\mathbb{C}\backslash\mathbb{R}$ and by eq.(4.9.6), $\lim_{y \to \pm\infty} F_\ell(x + iy) = 0$ for fixed $x \in \mathbb{R}$. Using the identity

$$F_\ell(z) - F_\ell(\bar{z}) = U_\ell(z) = f(z),$$

argue as in the proof of Corollary 4.3.5 to show that

$$F(z) = \begin{cases} f(z) & \Im z > 0, \\ 0 & \Im z < 0. \end{cases}$$

Conversely, suppose that $\ell \in H^p(\mathbb{R})$ with $F_\ell|\mathbb{C}_- = 0$. For $z \in \mathbb{C}_+$, note that

$$F_\ell(z) = F_\ell(z) - F_\ell(\bar{z}) = U_\ell(z)$$

which belongs to $\mathcal{H}_h^p$. But F_ℓ is analytic on $\mathbb{C}_+$ and so $F_\ell \in \mathcal{H}^p$ (see Remark 4.7.4). Now apply Lemma 4.8.23 to complete the proof. □

4.11. The space $\mathcal{H}^p(\mathbb{C}\backslash\mathbb{R})$

Perhaps the most important space of functions that will be used in some approximation results later, is the space $\mathcal{H}^p(\mathbb{C}\backslash\mathbb{R})$ ($0 < p < \infty$).

DEFINITION 4.11.1. For fixed $p \in (0, \infty)$ let $\mathcal{H}^p(\mathbb{C}\backslash\mathbb{R})$ denote the analytic functions f on $\mathbb{C}\backslash\mathbb{R}$ for which

$$\|f\| := \sup_{y \neq 0} \left\{ \int_\mathbb{R} |f(x+iy)|^p dx \right\}^{1/p} < \infty.$$

One can think of an $f \in \mathcal{H}^p(\mathbb{C}\backslash\mathbb{R})$ as

$$f(z) = \begin{cases} F_1(z) & \text{if } \Im z > 0 \\ \overline{F_2(\bar{z})} & \text{if } \Im z < 0 \end{cases},$$

where $F_1, F_2 \in \mathcal{H}^p = \mathcal{H}^p(\mathbb{C}_+)$ and so the theory of $\mathcal{H}^p$ spaces applies to this space as well. With the above realization of $\mathcal{H}^p(\mathbb{C}\backslash\mathbb{R})$ functions one can see that

(4.11.2) $$\|f\| \asymp \max\{\|F_1\|, \|F_2\|\} \asymp \|F_1\| + \|F_2\|.$$

What will be very useful to us will be the fact that we can identify $\mathcal{H}^p(\mathbb{C}\backslash\mathbb{R})$ with $H^p(\mathbb{R})$ in the following way: Given $\ell \in H^p(\mathbb{R})$, the functions $F_\ell(z)$ and $F_\ell(\bar{z})$ are analytic (anti-analytic) functions on $\mathbb{C}\backslash\mathbb{R}$. Furthermore, $F_\ell(z) - F_\ell(\bar{z})$ is equal to $U_\ell(z)$ which, by Proposition 4.8.25, belongs to $\mathcal{H}_h^p$. From the definition of $\mathcal{H}_h^p$ and Remark 4.7.4 (2), there is a harmonic function V on $\mathbb{C}_+$ which satisfies the Cauchy-Riemann system

$$(U_\ell)_x = V_y, \quad V_x = -(U_\ell)_y$$

4.11. THE SPACE $\mathcal{H}^p(\mathbb{C}\setminus\mathbb{R})$

and such that
$$U_\ell + iV \in \mathcal{H}^p.$$
The first condition implies
$$V = -i(F_\ell(z) + F_\ell(\bar{z})) + c$$
for some constant c. The second condition ($U_\ell + iV \in \mathcal{H}^p$) says that for fixed $x \in \mathbb{R}$,
$$\lim_{y\to+\infty} \{ U_\ell(x+iy) + iV(x+iy) \} = 0.$$
But from Remark 4.9.5, $\lim_{y\to\pm\infty} F_\ell(x+iy) = 0$ and so c must be zero. Thus
$$U_\ell(z) + iV(z) = F_\ell(z) - F_\ell(\bar{z}) + i\{ -i(F_\ell(z) + F_\ell(\bar{z})) \} = 2F_\ell(z)$$
belongs to $\mathcal{H}^p = \mathcal{H}^p(\mathbb{C}_+)$. A similar argument shows that $F_\ell \in \mathcal{H}^p(\mathbb{C}_-)$ and so $F_\ell \in \mathcal{H}^p(\mathbb{C}\setminus\mathbb{R})$.

Moreover,
$$\lim_{y\to 0+} \{ F_\ell(\cdot + iy) - F_\ell(\cdot - iy) \} = \lim_{y\to 0+} U_\ell(\cdot + iy) = \ell$$
in the sense of distributions. Finally, observe that
$$U_\ell(z) := F_\ell(z) - F_\ell(\bar{z}), \quad V_\ell(z) = -i(F_\ell(z) + F_\ell(\bar{z}))$$
and so
$$\|\ell\|$$
$$= \|u_\ell\|$$
$$= \sup_{y>0} \left\{ \int_\mathbb{R} (|F_\ell(x+iy) - F_\ell(x-iy)|^2 + |F_\ell(x+iy) + F_\ell(x-iy)|^2)^{p/2} dx \right\}^{1/p}$$
$$= \sup_{y>0} \left\{ \int_\mathbb{R} (|F_\ell(x+iy)|^2 + |F_\ell(x-iy)|^2)^{p/2} dx \right\}^{1/p}$$
$$\asymp \|F_\ell\| \quad \text{from eq.(4.11.2)}.$$

Thus, for a given $\ell \in H^p(\mathbb{R})$, F_ℓ is the unique function in $\mathcal{H}^p(\mathbb{C}\setminus\mathbb{R})$ with
$$\lim_{y\to 0+} \{ F_\ell(\cdot + iy) - F_\ell(\cdot - iy) \} = \ell.$$
Furthermore, $\|\ell\|_{H^p(\mathbb{R})} \asymp \|F_\ell\|_{\mathcal{H}^p(\mathbb{C}\setminus\mathbb{R})}$.

Conversely, suppose $F \in \mathcal{H}^p(\mathbb{C}\setminus\mathbb{R})$. Then
$$F(z) = \begin{cases} F_1(z) & \text{if } \Im z > 0 \\ \overline{F_2(\bar{z})} & \text{if } \Im z < 0 \end{cases},$$
where $F_1, F_2 \in \mathcal{H}^p$. The distribution ℓ_F defined by
$$\ell_F := \lim_{y\to 0+} \{ F(\cdot + iy) - F(\cdot - iy) \} = \lim_{y\to 0+} \{ F_1(\cdot + iy) - \overline{F_2}(\cdot + iy) \}$$
belongs to $H^p(\mathbb{R})$ and, as before, $u := F_1 - \overline{F_2} \in \mathcal{H}_h^p$ and the pair u and $v := -i(F_1 + \overline{F_2})$ satisfy eq.(4.7.2) and eq.(4.7.3). Thus, as in the previous paragraph,
$$\|\ell_F\| = \|F_1 - \overline{F_2}\| \asymp \|F_1\| + \|F_2\| \asymp \|F\|.$$
Hence, for a given $F \in \mathcal{H}^p(\mathbb{C}\setminus\mathbb{R})$, there is a unique distribution ℓ_F such that
$$\ell_F := \lim_{y\to 0+} \{ F(\cdot + iy) - F(\cdot - iy) \}.$$
Moreover, this distribution also satisfies $\|\ell_F\|_{H^p(\mathbb{R})} \asymp \|F\|_{\mathcal{H}^p(\mathbb{C}\setminus\mathbb{R})}$.

EXAMPLE 4.11.3. Let $p \in (0,1)$ be fixed.

1. A finite (complex) measure μ with compact support, regarded as a distribution, belongs to $H^p(\mathbb{R})$ if and only if

$$\int x^k d\mu = 0 \ \forall \ k = 0, 1, \cdots, [1/p] - 1. \tag{4.11.4}$$

The corresponding $\mathcal{H}^p(\mathbb{C}\backslash\mathbb{R})$ function is

$$F_\mu(z) = \frac{1}{2\pi i} \int \frac{d\mu(t)}{t - z}.$$

Conversely, suppose that μ is a finite compactly supported measure on $\mathbb{R}$. Then $F_\mu \in \mathcal{H}^p(\mathbb{C}\backslash\mathbb{R})$ if and only if μ satisfies the moment conditions eq.(4.11.4).

2. If

$$\ell = \sum_{1 \le k \le N, 0 \le j \le M} c_{j,k} \delta_{a_k}^{(j)},$$

then $\ell \in H^p(\mathbb{R})$ if and only if $M < 1/p - 1$ and $\ell(x^s) = 0$ for all integers $0 \le s \le 1/p - 1$. The corresponding $\mathcal{H}^p(\mathbb{C}\backslash\mathbb{R})$ function is

$$F_\ell(z) = \frac{1}{2\pi i} \sum_{1 \le k \le N, 0 \le j \le M} \frac{j! c_{j,k}}{(a_k - z)^{j+1}}.$$

Conversely, the rational function

$$F(z) = \sum_{1 \le k \le N, 0 \le j \le M} \frac{C_{j,k}}{(a_k - z)^{j+1}}$$

belongs to $\mathcal{H}^p(\mathbb{C}\backslash\mathbb{R})$ if and only if $M < 1/p - 1$ and the associated distribution

$$\ell_F = \sum_{1 \le k \le N, 0 \le j \le M} 2\pi i \frac{C_{j,k}}{j!} \delta_{a_k}^{(j)}$$

satisfies the moment conditions $\ell(x^s) = 0$ for all $s = 0, 1, \cdots, [1/p] - 1$.

As one might have noticed from the previous examples, there seems to be a connection between the support of the distribution $\ell \in H^p(\mathbb{R})$ and the places where the corresponding function $F_\ell \in \mathcal{H}^p(\mathbb{C}\backslash\mathbb{R})$ has an analytic continuation. The precise connection is the following.

COROLLARY 4.11.5. *A distribution $\ell \in H^p(\mathbb{R})$ has support which omits an open interval $I \subset \mathbb{R}$ if and only if the corresponding analytic function*

$$F_\ell(z) = \frac{1}{2\pi i} \ell\left(\frac{1}{\cdot - z}\right)$$

in $\mathcal{H}^p(\mathbb{C}\backslash\mathbb{R})$ has an analytic continuation across I.

PROOF. Suppose that the support of ℓ omits I. Without loss of generality (i.e., taking a smaller interval), we can assume that suppt $\ell \cap I^- = \emptyset$. Let $\phi \in C_0^\infty(\mathbb{R})$ be zero on the support of ℓ and equal to one near I. For each fixed $z \in \mathbb{C}\backslash\mathbb{R}$, the function $\phi(t)(t - z)^{-1}$ is zero on the support of ℓ and so

$$\ell\left(\frac{1}{\cdot - z}\right) = \ell\left(\frac{1 - \phi(\cdot)}{\cdot - z}\right).$$

Notice that
$$\frac{1-\phi(t)}{t-z} \to \frac{1-\phi(t)}{t-z_0}$$
in Lipschitz (or Zygmund) norm as $z \to z_0$, where $z_0 \in I$ (note that $1 - \phi = 0$ near I). Thus F_ℓ has a continuous extension to I. An application of Morera's theorem shows that F_ℓ has an analytic continuation across I.

Conversely, suppose that F_ℓ has an analytic continuation across I. Let $\phi \in C_0^\infty(I)$. Then
$$\ell(\phi) = \lim_{y \to 0^+} \int \left\{ F_\ell(x+iy) - F_\ell(x-iy) \right\} \phi(x) dx = 0.$$
Thus the support of ℓ omits I. □

REMARK 4.11.6. One could have shown the first part of this theorem by using Remark 4.9.5 (2 b). We leave the details to the interested reader.

CHAPTER 5

The backward shift on H^p for $p \in [1, \infty)$

5.1. The case $p > 1$

Throughout this chapter, $p \in [1, \infty)$ and $q := p/(p-1)$ will be the conjugate index to p. For a function $f = \sum_n a_n z^n \in H^p$, Bf will represent the analytic function

$$(Bf)(z) = \frac{f(z) - f(0)}{z} = a_1 + a_2 z + a_3 z^2 \cdots.$$

Is is clear from the definition of H^p that B (the *backward shift operator*) is a linear transformation from H^p to H^p.

PROPOSITION 5.1.1. *For each* $s \in (0, \infty)$, B *is continuous on* H^s.

PROOF. Clearly $Bf \in H^s$ whenever $f \in H^s$. Moreover,

$$|Bf| \le |f| + |f(0)| \le |f| + c_s \|f\|$$

almost everywhere on $\mathbb{T}$. The last inequality follows from Proposition 3.2.6. From here, it follows that

$$\|Bf\| \le c_s \|f\|$$

and so B is continuous. □

From Theorem 3.5.6, $(H^p)^*$ ($1 < p < \infty$) can be identified with H^q via the pairing

$$<f, g> = \int_{\mathbb{T}} f(\zeta) \overline{g}(\zeta) \, dm(\zeta)$$

and a routine computation shows

$$<Bf, g> = <f, Sg>, \quad f \in H^p, g \in H^q,$$

where

$$(Sg)(z) := zg(z)$$

is the *forward shift operator* on H^q. Thus, for a subspace $\mathcal{M}$ of H^p,

$$B\mathcal{M} \subset \mathcal{M} \Leftrightarrow S\mathcal{M}^\perp \subset \mathcal{M}^\perp,$$

where

$$\mathcal{M}^\perp = \{ g \in H^q : <f, g> = 0 \; \forall f \in \mathcal{M} \}$$

is the annihilator of $\mathcal{M}$ (see Definition 3.5.1). Beurling gives a precise description of such S-invariant subspaces $\mathcal{M}^\perp$.

THEOREM 5.1.2 (Beurling's theorem [14]). *Let* $s \in (0, \infty)$ *and* $\mathcal{K}$ *be a non-zero* S-*invariant subspace of* H^s. *Then the following statements are true.*

1. $\mathcal{K} = IH^s$ *for some inner function* I.

2. For any $f \in H^s$

$$\bigvee \{ z^k f : k \in \mathbb{N} \cup \{0\} \} = I_f H^s \text{ }^1.$$

3. If I_1 and I_2 are inner functions, then $I_1 H^s \subset I_2 H^s$ if and only if $I_1/I_2 \in H^\infty$; that is, I_2 "divides" I_1.

REMARK 5.1.3.   1. Beurling's theorem was originally proved by Beurling for the Hilbert space H^2 and then generalized by others to H^s ($0 < s < \infty$), see [**37**], p. 132, and [**51**], p. 79.

2. From the Nevanlinna factorization theory, one can show that for any family $\mathcal{F}$ of inner functions, there is an inner function $I_\mathcal{F}$ with the property that (i) $I_\mathcal{F}$ "divides" every $I \in \mathcal{F}$, specifically, $I/I_\mathcal{F} \in H^\infty$ for all $I \in \mathcal{F}$; and (ii) if J is any inner function which divides every $I \in \mathcal{F}$, then J divides $I_\mathcal{F}$. The inner function $I_\mathcal{F}$ is called the *greatest common divisor* of $\mathcal{F}$ [**44**], p. 85 (or [**39**], p. 84). If $\mathcal{K}$ is an S-invariant subspace of H^s and I is the greatest common divisor of the inner factors of non-zero functions from $\mathcal{K}$, then $\mathcal{K} = I H^s$.

By the Hahn-Banach separation theorem (Theorem 3.5.2), note that for our B-invariant subspace $\mathcal{M}$ of H^p, we have $\mathcal{M} = (\mathcal{M}^\perp)^\perp = (IH^q)^\perp$ and so we can recover $\mathcal{M}$ by characterizing the annihilator

$$(IH^q)^\perp = \{ f \in H^p : <f, Ig> = 0 \ \forall g \in H^q \}.$$

This is done with the following theorem which is the main result of this chapter.

THEOREM 5.1.4. *Let $p \in (1, \infty)$ and I be an inner function.*

1. $(IH^q)^\perp = H^p \cap I\overline{H_0^p}$, where $H_0^p = \{f \in H^p : f(0) = 0\}$.
2. *Moreover, $f \in H^p \cap I\overline{H_0^p}$ if and only if f/I has a pseudocontinuation belonging to $H^p(\mathbb{D}_e)$ which vanishes at infinity.*

REMARK 5.1.5.   1. An important remark needs to be made here to avoid confusion. The space $H^p \cap I\overline{H_0^p}$ is understood on the *circle* as the space $H^p(\mathbb{T}) \cap I\overline{H_0^p(\mathbb{T})}$.

2. Recall from Chapter 2, Definition 2.2.1, that $T_f \in \mathfrak{M}(\mathbb{D}_e)$ is a pseudocontinuation of $f \in \mathfrak{M}(\mathbb{D})$ if the non-tangential limits of f and T_f exist and agree almost everywhere.

To emphasize the different aspects of the theory and to generalize this result to other spaces of holomorphic functions, we will give three different proofs of Theorem 5.1.4.

5.2. The first and most straightforward proof

This first proof of Theorem 5.1.4 is the easiest, but since it depends heavily on Beurling's theorem, it does not readily generalize to other function spaces.

FIRST PROOF OF THEOREM 5.1.4. First we will prove statement (1),

$$(IH^q)^\perp = H^p \cap I\overline{H_0^p}.$$

---

[1] I_f is the inner factor of f, see Theorem 3.2.4.

Indeed if $f \in H^p \cap \overline{IH_0^p}$, then $f = I\overline{h}$ almost everywhere on $\mathbb{T}$ for some $h \in H_0^p$. Thus if $g \in H^q$,

$$<f, Ig> = \int (I\overline{h})(\overline{Ig})\, dm = \int \overline{hg}\, dm = 0$$

since $gh \in H_0^1$ and I is an inner function (see the F. and M. Riesz theorem, Theorem 3.3.1). Hence $f \in (IH^q)^\perp$.

Conversely, if f belongs to

$$(IH^q)^\perp = \left(\bigvee \{z^k I : k \in \mathbb{N} \cup \{0\}\}\right)^\perp,$$

then

$$\int f\overline{I}\overline{z}^k\, dm = 0 \quad \forall k \in \mathbb{N} \cup \{0\}.$$

Again, by the F. and M. Riesz theorem, $f\overline{I} \in \overline{H_0^p}$ and so $f \in \overline{IH_0^p}$. From this follows $(IH^q)^\perp = H^p \cap \overline{IH_0^p}$, which proves statement (1).

To prove statement (2), we must show that $f \in H^p \cap \overline{IH_0^p}$ if and only if f/I has a pseudocontinuation belonging to $H^p(\mathbb{D}_e)$ which vanishes at infinity. If $f \in H^p \cap \overline{IH_0^p}$ then $f = I\overline{h}$ almost everywhere on $\mathbb{T}$ for some $h \in H_0^p$. From eq.(3.2.10), $\overline{h}(1/\overline{z}) \in H^p(\mathbb{D}_e)$ and vanishes at infinity. Thus if

$$T_{f/I}(z) := \overline{h}(1/\overline{z}), \quad z \in \mathbb{D}_e,$$

then $(f/I)(\zeta) = T_{f/I}(\zeta)$ almost everywhere on $\mathbb{T}$ (the non-tangential limit functions are equal almost everywhere) and $T_{f/I} \in H^p(\mathbb{D}_e)$ and vanishes at infinity. Conversely if f/I has a pseudocontinuation $h \in H^p(\mathbb{D}_e)$ with $h(\infty) = 0$, then from eq.(3.2.10), $\overline{h}(1/\overline{z}) \in H_0^p$. Thus, almost everywhere on $\mathbb{T}$,

$$f = Ih = I\overline{(\overline{h})} \in \overline{IH_0^p}.$$

This completes the proof of statement (2). $\square$

REMARK 5.2.1. 1. For any $s \in (0, \infty)$ and inner function I, the space $H^s \cap \overline{IH_0^s}$ is a closed B-invariant subspace of H^s. Furthermore, the same proof as above shows that $f \in H^s$ belongs to $H^s \cap \overline{IH_0^s}$ if and only if f/I has a pseudocontinuation to a function in $H^s(\mathbb{D}_e)$ which vanishes at infinity.

2. We leave it to the reader to use the Hahn-Banach and F. and M. Riesz theorems to show that for $p \in (1, \infty)$,

$$\bigvee \{B^n(BI) : n \in \mathbb{N} \cup \{0\}\} = H^p \cap \overline{IH_0^p}$$

and so $B|H^p \cap \overline{IH_0^p}$ is a cyclic operator with cyclic vector BI.

A function belonging to $H^p \cap \overline{IH_0^p}$ ($1 \leq p < \infty$) may have an analytic continuation across certain portions of $\mathbb{T}$. For an inner function $I = bs_\mu$ [2] recall, from Remark 3.2.5(1 d), the definition of the spectrum

$$\sigma(I) := \operatorname{clos} b^{-1}(\{0\}) \cup \operatorname{suppt}(\mu).$$

Also recall that

$$\sigma(I) = \{\lambda \in \mathbb{D}^- : \liminf_{z \to \lambda} I(z) = 0\}$$

---

[2] b is a Blaschke product with zeros $\{a_n : n \in \mathbb{N}\} \subset \mathbb{D}$ and s_μ is a singular inner function with positive singular measure μ on $\mathbb{T}$.

and that I has an analytic continuation to

$$\mathbb{C}_\infty \setminus \{ 1/\overline{z} : z \in \sigma(I) \}.$$

COROLLARY 5.2.2. *If $p \in [1, \infty)$, then every $f \in H^p \cap I\overline{H^p_0}$ has an analytic continuation to*

$$\mathbb{C}_\infty \setminus \{ 1/\overline{z} : z \in \sigma(I) \}.$$

REMARK 5.2.3.   1. The above result is, in a sense, optimal since by Remark 3.2.5(1 d), if $1/\overline{z} \in \sigma(I)$, then I does not have an analytic continuation to a neighborhood of $1/\overline{z}$. Since $BI \in H^p \cap I\overline{H^p_0}$ (in fact by Remark 5.2.1, it generates this space), we see that BI is a vector in $H^p \cap I\overline{H^p_0}$ which does not have an analytic continuation to $1/\overline{z}$.

2. It is important to point out that even though $H^1 \cap I\overline{H^1_0}$ has the above analytic continuation property, we still need to show that every B-invariant subspace of H^1 is of this form. This will be taken up in Section 8 of this chapter.

3. As mentioned in the introduction, for each $s \in (0,1)$, $H^s \cap I\overline{H^s_0}$ is a B-invariant subspace of H^s and $f \in H^s$ belongs to $H^s \cap I\overline{H^s_0}$ if and only if f/I has a pseudocontinuation to a function in $H^s(\mathbb{D}_e)$ which vanishes at infinity. However, Corollary 5.2.2 is false in this case. In fact, for *any* inner function I and $\zeta \in \mathbb{T}$, the function $(1 - \overline{\zeta}z)^{-1}$ belongs to H^s ($0 < s < 1$). Moreover, for almost every $z \in \mathbb{T}$,

$$\frac{1}{1 - \overline{\zeta}z} = I(z)\frac{\overline{z}\overline{I(z)}}{\overline{z} - \overline{\zeta}} \in I\overline{H^s_0}.$$

Thus, given any point ζ on the circle, there is a function from $H^s \cap I\overline{H^s_0}$ which is not analytic near ζ.

The proof of this corollary depends on a certain version of Morera's theorem (found in [**39**]) which we state here and leave the proof to the reader as an exercise.

LEMMA 5.2.4. *Suppose that $f \in H^1(\mathbb{D})$ and $F \in H^1(\mathbb{D}_e)$ such that F is a pseudocontinuation of f across an arc $J \subset \mathbb{T}$. If*

$$G(z) = \begin{cases} f(z) & \text{if } |z| < 1, \\ F(z) & \text{if } |z| > 1, \end{cases}$$

then for $\zeta_0 \in J$ and "small" $\delta > 0$ [3],

$$G(z) = \frac{1}{2\pi i}\int_{|\zeta - \zeta_0| = \delta} \frac{G(\zeta)}{\zeta - z}d\zeta, \quad \{z : |z - \zeta_0| < \delta\} \setminus \mathbb{T}$$

and so F is an analytic continuation of f across the arc J.

REMARK 5.2.5.   1. It is important that both f and F belong to H^1 since the result is false without it. For example, $(1-z)^{-1}$ satisfies all but the first hypothesis of Morera's theorem, but does not have an analytic continuation to a neighborhood of $\zeta = 1$. Note that $(1-z)^{-1} \in H^p$ ($0 < p < 1$) but is not in H^1.

---

[3]Here, we mean a δ such that the non-tangential limits of f and F exist on the set $\{z : |z - \zeta_0| = \delta\} \cap \mathbb{T}$. Such δ are of full measure.

2. Actually, the version we will use below is a "local" version of Morera's theorem: If $f \in \mathfrak{H}(\mathbb{D})$ and $F \in \mathfrak{M}(\mathbb{D}_e)$ with

$$\sup_{r_1 < r < 1} \int_J |f(r\zeta)|\, dm(\zeta) < \infty, \quad \sup_{1 < r < r_2} \int_J |F(r\zeta)|\, dm(\zeta) < \infty$$

(f and F are H^1 "near" J), and if F is a pseudocontinuation of f across J, then F is an analytic continuation of f across J.

PROOF OF COROLLARY 5.2.2. If $f \in H^p \cap I\overline{H_0^p}$, then $f = I\overline{h}$ ($h \in H_0^p$) almost everywhere on $\mathbb{T}$ and so, by the proof of Theorem 5.1.4, f has a (meromorphic) pseudocontinuation T_f given by the formula

$$T_f(z) = \frac{1}{\overline{I(1/\overline{z})}} \overline{h}(1/\overline{z}), \quad |z| > 1.$$

The inner function I is analytic on $\mathbb{C}_\infty \setminus \{1/\overline{z} : z \in \sigma(I)\}$ and $\overline{h}(1/\overline{z}) \in H^p(\mathbb{D}_e)$. Thus f and T_f are both H^1 "near" any arc J of $\mathbb{T}$ which does not intersect $\sigma(I)$. By the local version of Morera's theorem, T_f is an analytic continuation of f across J. □

5.3. The second proof - using Fatou's jump theorem

This second proof of Theorem 5.1.4 is a little longer but the payoff is that it can be slightly altered to examine the backward shift invariant subspaces of Banach spaces of analytic functions whose norm is given by

$$\|f\| = \left\{ \int |f(z)|^p d\mu(z) \right\}^{1/p},$$

where μ is a positive measure on the closed unit disk. Examples of such spaces include the Bergman spaces which will be discussed in the next section.

SECOND PROOF OF THEOREM 5.1.4. Let $f \in H^p$ have the property: f/I has a pseudocontinuation $T_{f/I} \in N^+(\mathbb{D}_e)$ [4] which vanishes at infinity. Then for almost every $\zeta \in \mathbb{T}$, we have

$$f(\zeta)\overline{I}(\zeta) = T_{f/I}(\zeta)$$

is a boundary function of an $H^p(\mathbb{D}_e)$ function which vanishes at infinity. Thus, by the F. and M. Riesz theorem,

$$\int_{\mathbb{T}} f(\zeta)\overline{I}(\zeta)\overline{\zeta}^k\, dm(\zeta) = \int_{\mathbb{T}} T_{f/I}(\zeta)\overline{\zeta}^k\, dm(\zeta) = 0 \quad \forall k \in \mathbb{N} \cup \{0\}.$$

From here it follows that f belongs to

$$\left(\bigvee \{z^k I : k \in \mathbb{N} \cup \{0\}\} \right)^\perp,$$

which, by Beurling's theorem, is equal to $(IH^q)^\perp$.

Conversely, suppose $f \in (IH^q)^\perp$. We must show that f/I has a pseudocontinuation $T_{f/I}$ which belongs to $H^p(\mathbb{D}_e)$ and vanishes at infinity. For fixed $\lambda \in \mathbb{D}$

$$\frac{zf - \lambda f(\lambda)}{z - \lambda} \in H^p, \quad \forall f \in H^p.$$

---

[4] Note that if f/I has a pseudocontinuation to a function in $N^+(\mathbb{D}_e)$ then, since f/I has L^p boundary values, $f/I \in H^p(\mathbb{D}_e)$. Thus the theorem could equivalently read $f \in (IH^q)^\perp$ if and only if f/I has a pseudocontinuation to a function in $N^+(\mathbb{D}_e)$ which vanishes at infinity.

Furthermore,
$$\left\|\frac{zf - \lambda f(\lambda)}{z - \lambda}\right\| \leq \frac{1}{1 - |\lambda|} \|zf - \lambda f(\lambda)\| \leq c_\lambda \|f\|$$
and so the operator
$$f \to \frac{zf - \lambda f(\lambda)}{z - \lambda}$$
is continuous on H^p. Furthermore, a routine computation yields
$$(I_d - \lambda B)^{-1} f = \frac{zf - \lambda f(\lambda)}{z - \lambda}.$$
Thus the spectrum $\sigma(B)$ is contained in $\mathbb{D}^-$. Another simple computation reveals
$$B \frac{1}{1 - \lambda z} = \lambda \frac{1}{1 - \lambda z} \quad \forall \lambda \in \mathbb{D}$$
and so $\sigma(B) = \mathbb{D}^-$.

Recall from elementary operator theory [**30**], p. 567, that for a bounded linear operator A on a Banach space X, the spectral radius is defined to be
$$r(A) := \sup \{ |z| : z \in \sigma(A) \}$$
and can be computed using the spectral radius formula
$$r(A) = \lim_{n \to \infty} \|A^n\|^{1/n}.$$
Applying this to the backward shift B, we see that $r(B) = 1$ and that for a B-invariant subspace $\mathcal{M}$, we have $\|B^n|\mathcal{M}\| \leq \|B^n\|$ for every $n \in \mathbb{N}$. By the spectral radius formula, $r(B|\mathcal{M}) \leq 1$ and so $\sigma(B|\mathcal{M}) \subset \mathbb{D}^-$. Another routine computation shows that
$$(I_d - \lambda B|\mathcal{M})^{-1} B|\mathcal{M} f = \frac{f - f(\lambda)}{z - \lambda} \quad \forall \lambda \in \mathbb{D}$$
and thus
$$\frac{f - f(\lambda)}{z - \lambda} \in \mathcal{M} \quad \forall f \in \mathcal{M}, \lambda \in \mathbb{D}.$$
This observation will be important for what follows.

By the previous paragraph,
$$\frac{f - f(\lambda)}{z - \lambda} \in \mathcal{M} \quad \forall f \in \mathcal{M}, \lambda \in \mathbb{D}$$
and so for a non-zero vector $g \in IH^q = \mathcal{M}^\perp$ we have
$$\int \frac{f(\zeta) - f(\lambda)}{\zeta - \lambda} \overline{g}(\zeta) \, dm(\zeta) = 0 \quad \forall \lambda \in \mathbb{D}. \tag{5.3.1}$$
Recall from Chapter 3, that if $h \in L^1$, the Cauchy transform
$$(Ch)(z) := \int_\mathbb{T} \frac{h(\zeta)}{\zeta - z} dm(\zeta)$$
is analytic on $\mathbb{C}_\infty \setminus \mathbb{T}$ and by Theorem 3.4.1, $Ch \in H^p(\mathbb{C}_\infty \setminus \mathbb{T})$ (i.e. belongs to $H^p(\mathbb{D})$ and $H^p(\mathbb{D}_e)$). Rewriting eq.(5.3.1) in terms of Cauchy transforms, we observe
$$(Cf\overline{g})(\lambda) = f(\lambda)(C\overline{g})(\lambda).$$
Notice, since $g \in H^q$, that $(C\overline{g})(\lambda) = P(\overline{\zeta g})(\lambda) = 0$ for all $\lambda \in \mathbb{D}$ and so
$$(Cf\overline{g})(\lambda) = 0 \quad \forall \lambda \in \mathbb{D}. \tag{5.3.2}$$

We now define the function which will serve as the pseudocontinuation of f. Define

(5.3.3) $$F(\lambda) := \frac{(Cf\overline{g})(\lambda)}{(C\overline{g})(\lambda)}, \quad \lambda \in \mathbb{D}_e.$$

By the Cauchy integral formula,

(5.3.4) $$(C\overline{g})(\lambda) = -\frac{1}{\lambda}\overline{g(1/\overline{\lambda})}, \quad \lambda \in \mathbb{D}_e,$$

which is not the zero function on $\mathbb{D}_e$, and so $F \in \mathfrak{M}(\mathbb{D}_e)$. Moreover F is the quotient of two Cauchy transforms, each of which belong to $H^s(\mathbb{D}_e)$ ($0 < s < 1$). Since $H^s(\mathbb{D}_e) \subset \mathfrak{N}(\mathbb{D}_e)$ (the functions of bounded type, or equivalently the quotient of two bounded analytic functions), F is also of bounded type on $\mathbb{D}_e$. Applying Fatou's jump theorem (Theorem 3.4.8) along with eq.(5.3.2) yields, for almost every $\zeta \in \mathbb{T}$,

$$\begin{aligned} F(\zeta) &= \lim_{r \to 1^+} F(r\zeta) \\ &= \lim_{r \to 1^+} \frac{(Cf\overline{g})(r\zeta)}{(C\overline{g})(r\zeta)} \\ &= \frac{0 + \overline{\zeta}f(\zeta)\overline{g}(\zeta)}{0 + \overline{\zeta}\overline{g}(\zeta)} \\ &= f(\zeta). \end{aligned}$$

Thus f has a pseudocontinuation $F \in \mathfrak{N}(\mathbb{D}_e)$.

In order to finish the proof, we need to show that f/I has a pseudocontinuation to a function in $N^+(\mathbb{D}_e)$ which vanishes at infinity. The function F, as defined in eq.(5.3.3), which serves as the pseudocontinuation of f, is independent of the choice of $g \in IH^q$ (pseudocontinuations are unique). Thus, in the definition of F, we will agree to take g to be the inner function I and so the pseudocontinuation of f is given by the formula

$$F(\lambda) = \frac{C(f\overline{I})(\lambda)}{C(\overline{I}(\lambda))} = -\lambda \frac{C(f\overline{I})(\lambda)}{\overline{I(1/\overline{\lambda})}}.$$

From here, one can see that f/I has a pseudocontinuation given by $-\lambda C(f\overline{I})(\lambda)$ which belongs to $H^p(\mathbb{D}_e) \subset N^+(\mathbb{D}_e)$. Furthermore, evaluating this pseudocontinuation at infinity yields

$$-\int f\overline{I}\,dm$$

which equals zero since $f \in (IH^q)^\perp$. $\square$

5.4. Application: Bergman spaces

The Cauchy transform/Fatou jump theorem proof of Theorem 5.1.4 can be used to examine the B-invariant subspaces of other spaces of analytic functions as was done in [9]. One particular example is the Bergman space. To get started, we list some of the basic properties of the Bergman spaces and refer the reader to [11] and [48] for further information. We also mention that since this section is not the main purpose of the book, but only an application, we will be a bit "skimpy" on the details.

DEFINITION 5.4.1. Let $p \in (0, \infty)$ and define the *Bergman space* L_a^p as the space of $f \in \mathfrak{H}(\mathbb{D})$ for which
$$\|f\|_{L_a^p} := \left\{ \int_{\mathbb{D}} |f(z)|^p \, dm_2(z) \right\}^{1/p} < \infty,$$
where $dm_2 = dxdy/\pi$ is normalized two-dimensional Lebesgue measure on $\mathbb{D}$.

A straightforward estimate using the mean value property for harmonic functions, yields
$$|f(z)| \le \frac{1}{(1-|z|)^{2/p}} \|f\|_{L_a^p}, \quad z \in \mathbb{D},$$
and an argument using Montel's theorem shows that L_a^p is a Banach space when $p \in [1, \infty)$ and an F-space when $p \in (0,1)$. Modifying the Hardy space arguments of the previous sections, one can show that the backward shift B is continuous on L_a^p with spectrum equal to $\mathbb{D}^-$.

For $p \in (1, \infty)$, the dual of L_a^p can be identified with the L^q-Dirichlet space D_q of $g \in \mathfrak{H}(\mathbb{D})$ with
$$\|g\|_{D_q} = \left\{ \int_{\mathbb{D}} |(zg)'|^q \, dm_2(z) \right\}^{1/q} < \infty$$
via the pairing
$$(5.4.2) \qquad <f, g> := \int_{\mathbb{D}} f(z) \overline{(zg)'(z)} \, dm_2(z).$$

REMARK 5.4.3.
1. If $g \in D_q$, one can show that $g \in H^q$ [**94**], Chapter 5, section 5. In fact the above Dirichlet norm is equivalent to
$$\|g\|_{H^q} + \left\{ \int_{|\zeta|=1} \int_{|w|=1} \left| \frac{g(\zeta)-g(w)}{\zeta-w} \right|^q dm(w) \, dm(\zeta) \right\}^{1/q}.$$

2. If $f = \sum_j a_j z^j \in L_a^2$, we can integrate in polar coordinates to obtain the identity
$$\int_{\mathbb{D}} |f|^2 \, dm_2 = \sum_j \frac{|a_j|^2}{j+1}.$$
Thus for $g = \sum_j b_j z^j \in D_2$, equivalently, $(zg)' \in L^2$, we observe
$$\int_{\mathbb{D}} |(zg)'|^2 \, dm_2 = \sum_j (j+1)|b_j|^2.$$
So in this case, one can show, again by computing in polar coordinates, that the pairing $<f,g>$ in eq.(5.4.2) satisfies the identity
$$(5.4.4) \qquad <f,g> = \lim_{r \to 1^-} \int_{\mathbb{T}} f(r\zeta) \overline{g}(\zeta) \, dm(\zeta).$$

3. It turns out the above formula is valid for $f \in L_a^p$ and $g \in D_q$. This ability to represent the area integral pairing as a Cauchy pairing will be very useful in what follows.

Using the Cauchy pairing in eq.(5.4.4), we see that if $\mathcal{M}$ is a B-invariant subspace of L_a^p, then $\mathcal{M}^\perp$ is an S-invariant (forward shift invariant) subspace of D_q. Our main result about Bergman space parallels Theorem 5.1.4.

5.4. APPLICATION: BERGMAN SPACES

THEOREM 5.4.5. *Let $p \in (1, \infty)$ and $\mathcal{M}$ be a non-trivial B-invariant subspace of L_a^p. Then*

1. *If $g \in \mathcal{M}^\perp$ and $f \in \mathcal{M}$, then $fg \in H^s$ for all $s \in (0,1)$. Thus $\mathcal{M}$ is contained in the Nevanlinna class N.*
2. *Every $f \in \mathcal{M}$ has a pseudocontinuation to a function belonging to $\mathfrak{N}(\mathbb{D}_e)$.* [5]
3. *If $\mathcal{M}^\perp = \bigvee\{z^n g : n \in \mathbb{N} \cup \{0\}\}$ where $g \in D_{q+\varepsilon}$ for some $\varepsilon > 0$, then $f \in \mathcal{M}$ if and only if the following conditions are satisfied.*
   (a) *$fg \in H^1$*
   (b) *f/I_g has a pseudocontinuation to a function belonging to $N^+(\mathbb{D}_e)$ which vanishes at infinity.*

REMARK 5.4.6. 1. The condition $\mathcal{M} \subset N(\mathbb{D})$ is significant here since, in general, Bergman functions are not in the Nevanlinna class. For example, Bergman functions need not have non-tangential limits on any set of positive measure. In fact, the radial limits of (non-zero) Bergman functions can be zero almost everywhere on the circle [45].

2. For the simplicity of our proof (below), we assume in (3) that $g \in D_{q+\varepsilon}$ which is equivalent to the condition that $g'' \cdot (1 - |z|) \in L^{q+\varepsilon}$. In fact, the same conclusion is true if g satisfies the weaker condition

$$g'' \cdot (1 - |z|) \log \frac{1}{1 - |z|} \in L^q,$$

see [8].

3. For $p \in (1,2)$, one can apply deep results of Shirokov [88] and Khrushchev [47], concerning the ideals of the Dirichlet space D_q, to show the condition that $\mathcal{M}^\perp$ is "singly generated" by a "slightly smoother" function g (in statement (3) of the theorem) always holds (one can even take g to be smooth on $\mathbb{D}^-$) and so for these p, one has a complete characterization of the B-invariant subspaces of L_a^p: Given $p \in (1,2)$, and a B-invariant subspace $\mathcal{M}$ of L_a^p, there is a $g \in D_q$ such that $f \in \mathcal{M}$ if and only $fg \in H^1$ and f/I_g has a pseudocontinuation to a function belonging to $N^+(\mathbb{D}_e)$ which vanishes at infinity. See [8] for details.

4. Parts (1) and (2) of this theorem were shown for $p = 2$ in a paper of Richter and Sundberg [70] by using a different technique (than we will use below) which involved a computation with the "local Dirichlet integral".

5. An analog of this theorem is true for more general Bergman spaces (for example weighted Bergman spaces), see [9].

The proof of this theorem will involve some of the same techniques used in the second proof of Theorem 5.1.4 as well as a delicate analysis of the Cauchy transform of a measure on $\mathbb{D}$. To get started, we review some topics from potential theory.

For $u \in C(\mathbb{T})$, we let $\hat{u}$ denote the solution to the Dirichlet problem on $\mathbb{D}$. That is to say $\hat{u} \in C(\mathbb{D}^-)$ and harmonic on $\mathbb{D}$ with $\hat{u}|\mathbb{T} = u$. For $\mu \in M(\mathbb{D}^-)$, one can use the maximum principle for harmonic functions to conclude that the mapping

$$u \to \int \hat{u} \, d\mu$$

---

[5] Recall the functions of bounded type from Section 6 of Chapter 3.

is a continuous linear functional on $C(\mathbb{T})$. By the Riesz representation theorem, there is a (unique) measure $\hat{\mu} \in M(\mathbb{T})$ such that

$$\int_{\mathbb{D}^-} \hat{u} d\mu = \int_{\mathbb{T}} u d\hat{\mu} \;\; \forall \, u \in C(\mathbb{T}).$$

DEFINITION 5.4.7. The measure $\hat{\mu}$ obtained in this way is called the *sweep* of μ and is often known in other potential theory settings by its French equivalent "balayage", see [**15**].

From the definition of sweep and the Riesz representation theorem, it is straightforward to show that the map

(5.4.8) $$\mu \to \hat{\mu}$$

is a contractive linear map from $M(\mathbb{D}^-)$ to $M(\mathbb{T})$ which is the identity if μ is supported on $\mathbb{T}$.

DEFINITION 5.4.9. For $f \in L^{\infty}_{loc}(\mathbb{D}) \cap L^1(\mathbb{D}, dm_2)$ [6], define the *Cauchy transform* of f to be the function

$$(Vf)(\lambda) := \int \frac{f(z)}{z - \lambda} \, dm_2(z) = \int \frac{f(z + \lambda)}{z} \, dm_2(z).^7$$

Using the fact that $1/z \in L^1_{loc}(dm_2)$, it is easy to show that Vf is a well defined function on $\mathbb{C} \setminus \mathbb{T}$ which is analytic on $\mathbb{D}_e$. By using the continuity of translation along with the dominated convergence theorem, it is routine to show that Vf is continuous on $\mathbb{D}$.

For $\lambda \in \mathbb{D}_e$, the function $z \to (z - \lambda)^{-1}$ is continuous on $\mathbb{D}^-$ and is harmonic on $\mathbb{D}$. If $\widehat{f \, dm_2}$ is the sweep of $f \, dm_2|\mathbb{D}$ to $\mathbb{T}$, then we have

$$(Vf)(\lambda) = (C\widehat{f \, dm_2})(\lambda) = \int_{\mathbb{T}} \frac{1}{\zeta - \lambda} d\widehat{f \, dm_2}(\zeta), \;\; \lambda \in \mathbb{D}_e.$$

By Kolmogorov's theorem (Theorem 3.4.1), along with eq.(5.4.8), we observe that $Vf \in H^p(\mathbb{D}_e)$ for all $p \in (0,1)$ and

(5.4.10) $$\|Vf\| = \|C(\widehat{f \, dm_2})\| \le c_p \|\widehat{f \, dm_2}\| \le c_p \|f\|_{L^1(\, dm_2)}.$$

Although Vf is continuous on $\mathbb{D}$, it is not necessarily analytic. However, we still have the following "H^p"-type estimates.

LEMMA 5.4.11. 1. *For each $p \in (0,1)$, there is a constant $c_p > 0$ such that*

$$\int_{\mathbb{T}} |(Vf)(r\zeta)|^p \, dm(\zeta) \le c_p \|f\|^p_{L^1(\, dm_2)} \;\; \forall \, r \in (0,1)$$

for every $f \in L^{\infty}_{loc}(\mathbb{D}) \cap L^1(\mathbb{D}, dm_2)$.

2. *For fixed $\varepsilon > 0$, then there is a constant $c_\varepsilon > 0$ so that*

$$\int_{\mathbb{T}} |(Vf)(r\zeta)| \, dm(\zeta) \le c_\varepsilon \|f\|_{L^{1+\varepsilon}(\, dm_2)} \;\; \forall \, r \in (0,1)$$

for every $f \in L^{\infty}_{loc} \cap L^{1+\varepsilon}(\mathbb{D}, dm_2)$.

---

[6] $f \in L^{\infty}_{loc}(\mathbb{D})$ if for each compact set $K \subset \mathbb{D}$, there is a constant C_K such that $|f||K \le C_K$.

[7] We are using the letter V here to distinguish this Cauchy transform from the Cauchy transform of measures on the circle described in Chapter 3. We also assume that $f(z + \lambda) = 0$ when $z + \lambda$ is outside the disk.

REMARK 5.4.12. Note that all the integrals above make sense since Vf is continuous on $\mathbb{D}$.

PROOF. For each $r \in (0,1)$
$$f\, dm_2 = f\, dm_2|\{|z| < r\} + f\, dm_2|\{|z| > r\} = d\mu_1 + d\mu_2.$$
By sweeping the measures μ_1 and μ_2 to $r\mathbb{T}$ as above, we notice that for each $p \in (0,1)$, $V\mu_1 = C\widehat{\mu_1} \in H^p(r\mathbb{D}_e)$ and $V\mu_2 = C\widehat{\mu_2} \in H^p(r\mathbb{D})$ with
$$\int_{\mathbb{T}} |(V\mu_1)(r\zeta)|^p\, dm(\zeta) \leq \|C\widehat{\mu_1}\|^p_{H^p(r\mathbb{D}_e)} \leq c_p \|\widehat{\mu_1}\|^p \leq c_p \|f\|^p_{L^1(\, dm_2)}\ ^8.$$
Similarly
$$\int_{\mathbb{T}} |(V\mu_2)(r\zeta)|^p\, dm(\zeta) \leq \|C\widehat{\mu_2}\|^p_{H^p(r\mathbb{D})} \leq c_p \|\widehat{\mu_2}\|^p \leq c_p \|f\|^p_{L^1(\, dm_2)}.$$
The estimate in statement (1) of the lemma now follows.

The proof of statement (2) is standard since Cauchy transforms Vf of $L^p(dm_2)$ ($p > 1$) functions are Sobolev functions [**101**], Chapter 2. However, we will give a self contained proof. For $r \in (0,1)$, notice that
$$\begin{aligned}\int_{\mathbb{T}} |(Vf)(r\zeta)|\, dm(\zeta) &\leq \int_{\mathbb{T}} \int_{\mathbb{D}} \frac{|f(z)|}{|z-r\zeta|}\, dm_2(z)\, dm(\zeta)\\ &= \int_{\mathbb{D}} |f(z)| \left\{ \int_{\mathbb{T}} \frac{dm(\zeta)}{|z-r\zeta|} \right\} dm_2(z)\\ &= \int_{|z|<r} + \int_{|z|>r}\end{aligned}$$
Using the well known estimate that
$$\int_{\mathbb{T}} \frac{dm(\zeta)}{|1-s\zeta|} = O\left(\log\frac{1}{1-s}\right),\ s \in (0,1),$$
yields
$$\int_{\mathbb{T}} \frac{dm(\zeta)}{|z-r\zeta|} = O\left(\log\frac{1}{||z|-r|}\right),\ |z| \neq r.$$
Thus, an application of Hölder's inequality as well as the fact that for any fixed $t \geq 1$, the integrals
$$\int_{|z|<r} \left(\log\frac{1}{||z|-r|}\right)^t dm_2(z) \text{ and } \int_{|z|>r} \left(\log\frac{1}{||z|-r|}\right)^t dm_2(z)$$
are uniformly bounded in r, yields statement (2) of the lemma. □

REMARK 5.4.13. The result (1) is true for more general measures and can also be shown by using the Littlewood subordination theorem, see [**9**].

This next result is a "L^p-Fatou's jump theorem" for Cauchy transforms whose measures are not supported in $\mathbb{T}$.

COROLLARY 5.4.14. Let $f \in L^\infty_{loc}(\mathbb{D}) \cap L^1(\mathbb{D}, dm_2)$. Then for each $p \in (0,1)$,
$$\lim_{r \to 1^-} \int_{\mathbb{T}} |(Vf)(r\zeta) - (Vf)(\zeta/r)|^p\, dm(\zeta) = 0.$$

---

[8]Here, we are using a slight variation of Theorem 3.4.1 for Cauchy transforms of measures on $r\mathbb{T}$. One can prove this by making the obvious modifications to the proof of Theorem 3.4.1 and the lemma that precedes it. The key here is that the constants involved are independent of r.

PROOF. Let $\varepsilon > 0$ be given and choose $\delta > 0$ with
$$\int_{\{1-\delta < |z| < 1\}} |f|\, dm_2 < \varepsilon.$$
Writing
$$f = f|\{1-\delta < |z| < 1\} + f|\{|z| < 1-\delta\} = f_1 + f_2,$$
we have, for fixed $p \in (0,1)$, the inequality
$$\int_{\mathbb{T}} |(Vf)(r\zeta) - (Vf)(\zeta/r)|^p \, dm(\zeta)$$
$$\leq \int_{\mathbb{T}} |(Vf_1)(r\zeta) - (Vf_1)(\zeta/r)|^p \, dm(\zeta) + \int_{\mathbb{T}} |(Vf_2)(r\zeta) - (Vf_2)(\zeta/r)|^p \, dm(\zeta).$$
Since Vf_2 is continuous near $\mathbb{T}$ (f_2 is not supported near $\mathbb{T}$), the second integral above converges to zero as $r \to 1^-$. For the first integral, observe that
$$\int_{\mathbb{T}} |(Vf_1)(r\zeta) - (Vf_1)(\zeta/r)|^p \, dm(\zeta)$$
$$\leq \int_{\mathbb{T}} |(Vf_1)(r\zeta)|^p \, dm(\zeta) + \int_{\mathbb{T}} |(Vf_1)(\zeta/r)|^p \, dm(\zeta).$$
By Lemma 5.4.11, the first integral is bounded by $c_p \|f_1\|_{L^1}^p \leq c_p \varepsilon^p$, while by eq.(5.4.10), the second integral is also bounded by $c_p \|f_1\|_{L^1}^p \leq c_p \varepsilon^p$. The proof now follows. $\square$

PROOF OF THEOREM 5.4.5. The spectrum of B is equal to $\mathbb{D}^-$ and a computation shows
$$(I_d - \lambda B)^{-1} B f = \frac{f - f(\lambda)}{z - \lambda} \quad \forall f \in L_a^p, \lambda \in \mathbb{D}.$$
Also recall from the second proof of Theorem 5.1.4 (using the spectral radius formula), that $\sigma(B|\mathcal{M}) \subset \mathbb{D}^-$ and so
$$\frac{f - f(\lambda)}{z - \lambda} \in \mathcal{M} \quad \forall \lambda \in \mathbb{D}.$$
Let g be a non-zero element of $\mathcal{M}^\perp$ and observe that
(5.4.15) $$< \frac{f - f(\lambda)}{z - \lambda}, g > = 0 \quad \forall \lambda \in \mathbb{D}.$$
Since the dual pairing $< f, g >$ can be written as an area integral
$$< f, g > = \int_{\mathbb{D}} f(z) \overline{(zg)'(z)} \, dm_2(z),$$
we can use eq.(5.4.15) to write
$$\int \frac{f\overline{(zg)'}}{z - \lambda} \, dm_2 = f(\lambda) \int \frac{\overline{(zg)'}}{z - \lambda} \, dm_2 \quad \lambda \in \mathbb{D}.$$
A routine power series computation shows that
$$\int \frac{\overline{(zg)'}}{z - \lambda} \, dm_2 = -\overline{\lambda g(\lambda)}$$
and so, using the language of Cauchy transforms,
(5.4.16) $$-\overline{\lambda g(\lambda)} f(\lambda) = V(\overline{f(zg)'})(\lambda), \quad \lambda \in \mathbb{D}.$$

We now apply Lemma 5.4.11 to conclude that $fg \in H^s$ for all $s \in (0,1)$. It follows from the fact that $g \in D_q \subset H^q$ (and is not the zero function), that f can be written as the quotient of two Hardy space functions. This means that $f \in N(\mathbb{D})$ which proves statement (1) of the theorem.

To prove statement (2), define the meromorphic function T_f on $\mathbb{D}_e$ by

$$(5.4.17) \qquad T_f(\lambda) := -\frac{\lambda V(\overline{f(zg)'})(\lambda)}{\overline{g}(1/\overline{\lambda})}.$$

By Lemma 5.4.11, T_f is the quotient of two Hardy space functions and so (since the Hardy spaces are contained in the functions of bounded type, see Chapter 3, Section 6), $T_f \in \mathfrak{N}(\mathbb{D}_e)$. Functions of bounded type have finite non-tangential limits almost everywhere. Thus, the (exterior) non-tangential boundary values of T_f exist and are given by

$$(5.4.18) \qquad -\frac{1}{\overline{\zeta g(\zeta)}} \lim_{r \to 1^-} V(\overline{f(zg)'})(\zeta/r) \quad \text{a.e.}$$

Since f is a function from the Nevanlinna class, we can apply the same reasoning as above along with eq.(5.4.16) to conclude that f has finite non-tangential boundary values given by

$$(5.4.19) \qquad -\frac{1}{\overline{\zeta g(\zeta)}} \lim_{r \to 1^-} V(\overline{f(zg)'})(r\zeta) \quad \text{a.e.}$$

By Corollary 5.4.14 and Chebychev's inequality,

$$V(\overline{f(zg)'})(r\zeta) - V(\overline{f(zg)'})(\zeta/r) \to 0 \quad \text{in measure as } r \to 1^-.$$

By standard measure theory arguments (the limits in eq.(5.4.18) and eq.(5.4.19) exist almost everywhere and hence in measure),

$$\lim_{r \to 1^-} V(\overline{f(zg)'})(r\zeta) = \lim_{r \to 1^-} V(\overline{f(zg)'})(\zeta/r) \quad \text{a.e.}$$

Combining this with eq.(5.4.18) and eq.(5.4.19) we see that T_f is a pseudocontinuation of f across $\mathbb{T}$. This proves statement (2) of the theorem.

We will now prove statement (3). Suppose that $f \in L_a^p$, $g \in D_q$ with $fg \in H^1$ and such that f/I_g has a pseudocontinuation $T_{f/I_g} \in N^+(\mathbb{D}_e)$ which vanishes at infinity. Then

$$T_{f/I_g}(z)\overline{O_g}(1/\overline{z}) \in H^p(\mathbb{D}_e)$$

(where O_g is the outer part of g) and vanishes at infinity. Furthermore the (exterior) boundary values of this function are equal to $f(\zeta)\overline{g}(\zeta)$ almost everywhere. Thus for all $k \in \mathbb{N} \cup \{0\}$ we can apply the F. and M. Riesz theorem to get,

$$<f, z^k g> = \lim_{r \to 1^-} \int_{\mathbb{T}} f(r\zeta)\overline{(z^k g)(r\zeta)}\, dm(\zeta) = \int_{\mathbb{T}} f(\zeta)\overline{g}(\zeta)\overline{\zeta}^k\, dm(\zeta) = 0.^9$$

This means that f annihilates

$$\bigvee \{z^k g : k \in \mathbb{N} \cup \{0\}\} = \mathcal{M}^\perp$$

and so, by the Hahn-Banach theorem, $f \in \mathcal{M}$.

---

[9]Since we are assuming $fg \in H^1$, then $f_r g_r \to fg$ in L^1. By the dominated convergence theorem, $f_r \overline{g_r} - f\overline{g} = (f_r g_r - fg)\overline{g_r}/g_r + fg(\overline{g_r}/g_r - \overline{g}/g)$ converges to zero in L^1 as $r \to 1^-$. Also note, by eq.(5.4.4), that the area integral pairing $<f,g>$ can also be represented as a Cauchy pairing on the circle.

Conversely, suppose
$$f \in \mathcal{M} = {}^\perp\left(\bigvee\{z^k g : k \in \mathbb{N} \cup \{0\}\}\right).$$
Then by eq.(5.4.16) and Lemma 5.4.11 [10], we obtain $fg \in H^1$. By the reasoning used to prove statement (2) of the theorem, we observe from eq.(5.4.17) that f/I_g has a pseudocontinuation T_{f/I_g} given by the formula
$$T_{f/I_g}(\lambda) = -\frac{\lambda V(f\overline{(zg)'})(\lambda)}{\overline{O_g(1/\overline{\lambda})}}, \quad \lambda \in \mathbb{D}_e.$$
Since the numerator in the above expression belongs to $H^s(\mathbb{D}_e)$ for $s \in (0,1)$ and the denominator is an outer function, we conclude (from Chapter 3, Section 6) that $T_{f/I_g} \in N^+(\mathbb{D}_e)$. To finish, we need to show that $T_{f/I_g}(\infty) = 0$. By the dominated convergence theorem,
$$T_{f/I_g}(\infty) = \frac{1}{\overline{O_g(0)}}\int_\mathbb{D} f(z)\overline{(zg)'}\,dm_2(z) = \frac{1}{\overline{O_g(0)}}<f,g>=0.$$
This proves statement (3) of the theorem. $\square$

5.5. Application: spectral properties

If $p \in [1,\infty)$, the subspace $\mathcal{M} = H^p \cap I\overline{H^p_0}$ is a B-invariant subspace of H^p and moreover, $f \in H^p$ belongs to $\mathcal{M}$ if and only if f/I has a pseudocontinuation belonging to $H^p(\mathbb{D}_e)$ which vanishes at infinity. When $p \in (1,\infty)$, these are all of the B-invariant subspaces of H^p and, as we shall see in Section 8 of this chapter, the same is true when $p = 1$. We also know from our previous work that $\sigma(B|\mathcal{M}) \subset \mathbb{D}^-$. In this section, we identify precisely the spectrum of $B|\mathcal{M}$. Before proceeding to the main theorem, which was originally shown by Moeller [63] in the Hilbert space setting H^2, we make a few general remarks.

Suppose that λ belongs to the resolvent of $B|\mathcal{M}$. To compute a formula for $(\lambda I_d - B|\mathcal{M})^{-1}f$, we observe that if
$$(\lambda I_d - B|\mathcal{M})^{-1}f = g,$$
then
$$f = (\lambda I_d - B|\mathcal{M})g.$$
A computation yields the formula
(5.5.1) $$g = \frac{zf - g(0)}{\lambda z - 1}.$$
Setting $c_\lambda(f) = g(0)$, we obtain the formula
$$(\lambda I_d - B|\mathcal{M})^{-1}f = \frac{zf - c_\lambda(f)}{\lambda z - 1}.$$
If $\lambda \in \mathbb{D}_e$, then $\lambda \notin \sigma(B|\mathcal{M})$ and so from eq.(5.5.1)
$$(\lambda z - 1)g = zf - g(0).$$
Evaluating this expression at the point $z = 1/\lambda$ says
$$g(0) = c_\lambda(f) = \frac{1}{\lambda}f(\frac{1}{\lambda}).$$

---
[10] $f\overline{(zg)'} \in L^{1+\varepsilon}(dm_2)$ for some $\varepsilon > 0$ since $g \in D_{q+\varepsilon}$ for some, possibly different, $\varepsilon > 0$.

As we shall see below, if $\lambda \in \mathbb{D}$ and belongs to the resolvent of $B|\mathcal{M}$, then

$$c_\lambda(f) = \frac{1}{\lambda} T_f\left(\frac{1}{\lambda}\right),$$

where T_f is the meromorphic pseudocontinuation of f. The main theorem, which exploits these ideas, is the following.

THEOREM 5.5.2. *For $p \in [1, \infty)$ and $\mathcal{M} = H^p \cap I\overline{H_0^p}$,*

$$\sigma(B|\mathcal{M}) = \overline{\sigma(I)}.$$

Moreover

1. *$\lambda \in \sigma(B|\mathcal{M}) \cap \mathbb{D}$ if and only if $(1 - \lambda z)^{-1} \in \mathcal{M}$.*
2. *$\overline{\zeta} \notin \sigma(B|\mathcal{M}) \cap \mathbb{T}$ if and only if there is an open neighborhood U of ζ such that every $f \in \mathcal{M}$ has an analytic continuation to U.*

PROOF OF THEOREM 5.5.2. To prove statement (1), note that if $\lambda \in \overline{\sigma(I)} \cap \mathbb{D}$, then $I(\overline{\lambda}) = 0$ and so the function

$$g_\lambda(z) := \frac{zI(z)}{z - \overline{\lambda}} \in H_0^p.$$

It follows that for almost every $\zeta \in \mathbb{T}$

$$\frac{1}{1 - \zeta\lambda} = I(\zeta)\overline{g_\lambda(\zeta)} \in H^p \cap I\overline{H_0^p} = \mathcal{M}.$$

It is easy to see that this argument can be reversed and so

$$\overline{\sigma(I)} \cap \mathbb{D} = \left\{ \lambda \in \mathbb{D} : \frac{1}{1 - \lambda z} \in \mathcal{M} \right\}.$$

We now want to show that $\sigma(B|\mathcal{M}) \cap \mathbb{D} = \overline{\sigma(I)} \cap \mathbb{D}$. From above and the observation that

$$B \frac{1}{1 - \lambda z} = \lambda \frac{1}{1 - \lambda z},$$

we have the inclusion one way, namely $\overline{\sigma(I)} \cap \mathbb{D} \subset \sigma(B|\mathcal{M}) \cap \mathbb{D}$. The other inclusion requires more work.

Assume that $\lambda \in \mathbb{D} \setminus \overline{\sigma(I)}$. Then the inner function I has a pseudocontinuation given by the formula

$$T_I(z) = \frac{1}{\overline{I(1/\overline{z})}}, \quad z \in \mathbb{D}_e,$$

which is analytic near the point $z = 1/\lambda$. If $f \in \mathcal{M} = H^p \cap I\overline{H_0^p}$, then $f = I\overline{h}$ almost everywhere on $\mathbb{T}$ and so f has a pseudocontinuation given by the formula

$$T_f(z) = T_I(z)\overline{h}(1/\overline{z}), \quad z \in \mathbb{D}_e.$$

Using the fact that T_I is analytic near $1/\lambda$ and that the mapping $h \to \overline{h}(\overline{\lambda})$ is continuous on H^p (see Proposition 3.2.6), we see that the linear functional

$$f \to T_f\left(\frac{1}{\lambda}\right)$$

is also continuous on H^p. Hence, the operator

$$R_\lambda f := \frac{zf - \lambda^{-1}T_f(\lambda^{-1})}{\lambda z - 1}$$

is a continuous linear operator from $\mathcal{M}$ to H^p.

Since $f = I\overline{h}$ almost everywhere on $\mathbb{T}$,

$$\begin{aligned}(R_\lambda f)(\zeta) &= \frac{\zeta f(\zeta) - \lambda^{-1}T_f(\lambda^{-1})}{\lambda\zeta - 1} \\ &= \frac{\zeta I(\zeta)\overline{h}(\zeta) - \lambda^{-1}T_f(\lambda^{-1})}{\lambda\zeta - 1} \\ &= I(\zeta)\left\{\frac{\overline{h}(\zeta) - \overline{\zeta I(\zeta)}\lambda^{-1}T_f(\lambda^{-1})}{\lambda - \overline{\zeta}}\right\} \\ &\in I\overline{H_0^p}\end{aligned}$$

and so $R_\lambda \mathcal{M} \subset \mathcal{M}$. Finally, a straightforward computation shows that R_λ is actually the inverse of $(\lambda I_d - B|\mathcal{M})^{-1}$. Hence $(\lambda I_d - B|\mathcal{M})$ is invertible and so $\sigma(B|\mathcal{M}) \cap \mathbb{D} \subset \overline{\sigma(I)} \cap \mathbb{D}$. From the above, we conclude that

$$\sigma(B|\mathcal{M}) \cap \mathbb{D} = \overline{\sigma(I)} \cap \mathbb{D} = \left\{\lambda \in \mathbb{D} : \frac{1}{1 - \lambda z} \in \mathcal{M}\right\},$$

which proves statement (1).

To prove statement (2), we will show that

(5.5.3) $$\sigma(B|\mathcal{M}) \cap \mathbb{T} = \mathbb{T}\setminus S,$$

where S is the set of $\overline{\zeta} \in \mathbb{T}$ with the property that there is a open neighborhood U of ζ such that every $f \in \mathcal{M}$ has an analytic continuation to U. To this end, suppose $\zeta \in \mathbb{T}$ and $\overline{\zeta} \notin \sigma(B|\mathcal{M})$. As mentioned at the beginning of this section, the function

$$w \to (wI - B|\mathcal{M})^{-1}$$

is an (operator-valued) analytic function on some neighborhood U of $\overline{\zeta}$ and so $w \to ((wI - B|\mathcal{M})^{-1}f)(0)$ is analytic on U for every $f \in \mathcal{M}$. Also, as we mentioned at the beginning of this section, for $w \in \mathbb{D}_e$

$$((wI - B|\mathcal{M})^{-1}f)(0) = wf(1/w).$$

Thus every $f \in \mathcal{M}$ has a analytic continuation to U and so $\overline{\zeta} \in S$.

Conversely, suppose that $\overline{\zeta} \in S$. Then there is a neighborhood U of $\overline{\zeta}$ such that every $f \in \mathcal{M}$ has an analytic continuation to U. Using the dominated convergence theorem, it is routine to show that the function

$$\frac{zf - \zeta^{-1}f(1/\zeta)}{\zeta z - 1}$$

is the norm limit of the functions

$$\frac{zf - w_n^{-1}f(1/w_n)}{w_n z - 1} = (w_n I - B|\mathcal{M})^{-1}f \in \mathcal{M}$$

for some sequence $\{w_n : n \in \mathbb{N}\} \subset \mathbb{D}_e$ with $w_n \to \zeta$ (note that $\sigma(B|\mathcal{M}) \subset \mathbb{D}^-$). Hence the mapping

(5.5.4) $$f \to \frac{zf - \zeta^{-1}f(1/\zeta)}{\zeta z - 1}$$

is a well defined linear transformation from $\mathcal{M}$ to itself. At least formally we have

$$(\zeta I - B|\mathcal{M})^{-1}f = \frac{zf - \zeta^{-1}f(1/\zeta)}{\zeta z - 1}.$$

To show that the linear transformation defined by eq.(5.5.4) is continuous, and hence $\zeta \notin \sigma(B|\mathcal{M})$, we use the closed graph theorem. Indeed, suppose that $\{f_n : n \in \mathbb{N}\}$ is a sequence in $\mathcal{M}$ with

$$f_n \to 0 \quad \text{and} \quad \frac{zf_n - \zeta^{-1}f_n(1/\zeta)}{\zeta z - 1} \to h \quad \text{in norm.}$$

Since the above limits also hold pointwise in $\mathbb{D}$, we conclude that $f_n(1/\zeta) \to c$ and so

$$h = \frac{-c\overline{\zeta}^2}{z - \overline{\zeta}}$$

which does not belong to H^p ($1 \leq p < \infty$), unless c is zero. So by the closed graph theorem, the operator defined by eq.(5.5.4) is continuous and hence $\zeta \notin \sigma(B|\mathcal{M})$.

The equality $\sigma(B|\mathcal{M}) \cap \mathbb{T} = \overline{\sigma(I)} \cap \mathbb{T}$ follows from eq.(5.5.3) and Remark 5.2.3. □

We conclude this section by mentioning the paper [**9**] which examines the spectral properties of $B|\mathcal{M}$, where $\mathcal{M}$ is an invariant subspace, for various classes of functions such as the weighted Bergman and Dirichlet spaces.

5.6. The third proof - using the Nevanlinna theory

This final proof of Theorem 5.1.4 is due to Aleksandrov [**5**] and can be used to describe the backward shift invariant subspaces of many other spaces of analytic functions.

Recall from Chapter 3, the Smirnov class N^+ is a closed subalgebra of the topological algebra $\log L$ and the polynomials are dense in N^+. Also recall the associated space

$$N^- := \{ f \in \log L : \overline{\zeta} f(\overline{\zeta}) \in N^+ \}$$

and note that one can think of functions in N^- as the boundary values of functions in the Smirnov class of the exterior disk which vanish at infinity.

Using the F. and M. Riesz theorem along with the fact that $N^+ \cap L^p = H^p$, one can prove the following (we assume that $p \in (1, \infty)$ and $q = p/(p-1)$ is the conjugate index to p).

LEMMA 5.6.1. *Let $f \in H^p$ and $g \in H^q$. Then*

$$f\overline{g} \in N^- \Leftrightarrow <B^n f, g> = 0 \quad \forall n \in \mathbb{N} \cup \{0\}.$$

REMARK 5.6.2. Since $N^+ \cap L^p = H^p$, the condition $f\overline{g} \in N^-$ could be written equivalently as $f\overline{g} \in \overline{H_0^p}$. This formalism of writing $f\overline{g} \in N^-$ might seems a bit much in our current situation. However, when we examine the B-invariant subspaces of $VMOA$ and $L^1/\overline{H_0^1}$, this notation will be more relevant.

Using the fact that if F is an outer function then $\overline{F(\overline{z})}$ is also an outer function, it is not difficult to prove the following.

LEMMA 5.6.3. *For $f \in H^p$ and $g \in H^q$,*

$$f\overline{g} \in N^- \Leftrightarrow f\overline{I_g} \in N^-,$$

where I_g is the inner part of g.

Also needed will be "Beurling's theorem" for the Smirnov class.

THEOREM 5.6.4. *Let $\mathcal{K}$ be a non-zero S-invariant subspace of N^+ and I the greatest common divisor of $\{I_g : g \in \mathcal{K}, g \not\equiv 0\}$. Then $\mathcal{K} = IN^+$.*

PROOF. We first claim that if g is a non-zero vector in $\mathcal{K}$, then I_g also belongs to $\mathcal{K}$. Indeed, let g be a non-zero vector in $\mathcal{K}$ and write $g = I_g O_g$ (the inner-outer factorization of g). Since $1/O_g \in N^+$ and polynomials are dense in N^+ (Proposition 3.6.10), there is a sequence $\{p_n : n \in \mathbb{N}\}$ of polynomials with $p_n \to 1/O_g$ in N^+. Since $\mathcal{K}$ is S-invariant, then $p_n g \in \mathcal{K}$ and using the fact that N^+ is a topological algebra, we obtain $p_n g \to I_g$ in N^+. Hence $I_g \in \mathcal{K}$.

Since the metric of H^2 is stronger than that of N^+, we can argue that $\mathcal{K} \cap H^2$ is a non-zero S-invariant subspace of H^2 and so, by Beurling's theorem for the Hardy space (Theorem 5.1.2), $\mathcal{K} \cap H^2 = IH^2$ for some inner function I. In fact, since $I_g \in \mathcal{K} \cap H^2$ for all $g \in \mathcal{K}$, I can be taken to be the greatest common divisor of $\{I_g : g \in \mathcal{K}\}$.

Finally, we claim that $\mathcal{K} = IN^+$. From the above,
$$I_g \in \mathcal{K} \cap H^2 = IH^2 \subset IN^+ \quad \forall g \in \mathcal{K}.$$

Now approximate O_g with polynomials in the metric of N^+ and use the fact that N^+ is a topological algebra (and the S-invariance of $\mathcal{K}$) again to conclude $I_g O_g = g \in IN^+$. Thus $\mathcal{K} \subset IN^+$. For the other direction, note that if $g = Ih \in IN^+$, then $h_r \in H^2$ ($h_r(\zeta) := h(r\zeta)$, $\zeta \in \mathbb{T}, r \in (0,1)$) and so $Ih_r \in IH^2 = \mathcal{K} \cap H^2$ for all $r \in (0,1)$. But since $h_r \to h$ in the metric of N^+ (Proposition 3.6.10) we conclude $Ih = g \in \mathcal{K}$. Thus $IN^+ \subset \mathcal{K}$ and equality follows. □

THIRD PROOF OF THEOREM 5.1.4. For an inner function I, define
$$(5.6.5) \qquad I^*(H^p) := \{ f \in H^p : f\bar{I} \in N^- \}.$$

Using Remark 5.6.2, one can show that $I^*(H^p) = H^p \cap \overline{IH_0^p}$. By the Hahn-Banach theorem, a function $f \in H^p$ belongs to $\mathcal{M}$ if and only if
$$<B^n f, g> = 0 \quad \forall g \in \mathcal{M}^\perp.$$

Combining this with Lemma 5.6.1 and Lemma 5.6.3, we observe
$$(5.6.6) \qquad f \in \mathcal{M} \Leftrightarrow \overline{I_g}f \in N^- \quad \forall g \in \mathcal{M}^\perp.$$

Letting I be the greatest common divisor of the family of inner functions
$$\{ I_g : g \in \mathcal{M}^\perp, g \not\equiv 0 \},$$
we now argue, using eq.(5.6.6), that $\mathcal{M} = I^*(H^p)$.

Indeed if $f \in I^*(H^p)$, that is, $f\bar{I} \in N^-$, then for a non-zero $g \in \mathcal{M}^\perp$,
$$f\overline{I_g} = f\bar{I}\frac{I}{I_g}.$$

By assumption, $f\bar{I} \in N^-$ while $\overline{I/I_g}$ belongs to $\overline{N^+}$ (since $I_g/I \in H^\infty$ and so $I/I_g = \overline{(I_g/I)} \in \overline{H^\infty}$) and so $f\overline{I_g} \in N^-$ for all non zero $g \in \mathcal{M}^\perp$. Thus, from eq.(5.6.6), $I^*(H^p) \subset \mathcal{M}$. To prove the other direction, we first notice from the choice of I as the greatest common inner divisor of $\mathcal{M}^\perp$, that $\mathcal{M}^\perp \subset IN^+$. By "Beurling's theorem" for N^+ (Theorem 5.6.4), the closure of $\mathcal{M}^\perp$ in N^+ is in fact equal to IN^+. Thus, there is a sequence of $\{g_n : n \in \mathbb{N}\} \subset \mathcal{M}^\perp$ with $g_n \to I$ in the metric of N^+ (which is also the metric of $\log L$). If $f \in \mathcal{M}$, then $f\overline{g_n} \in N^-$. But $\log L$ is a topological algebra, and so $f\overline{g_n} \to f\bar{I}$ in $\log L$ from which, $f\bar{I} \subset N^-$. Thus $\mathcal{M} \subset I^*(H^p)$ and equality follows. □

5.7. Application: $VMOA$, $BMOA$, and $L^1/\overline{H_0^1}$

As mentioned in the previous section, a paper of Aleksandrov [5] gives a unified approach to characterizing the B-invariant subspaces of certain Banach spaces of analytic functions which are contained in the Nevanlinna class. In this section we will give two examples of Aleksandrov's technique and characterize the B-invariant subspaces of $VMOA$ and $K_a \simeq L^1/\overline{H_0^1}$ (the space of Cauchy transforms of L^1 functions).

THEOREM 5.7.1 (Aleksandrov [5]). *Let $X = VMOA$ or K_a. If $\mathcal{M}$ is a B-invariant subspace of X, then there is an inner function I so that*

$$\mathcal{M} = I^*(X) = \{f \in X : f\overline{I} \in N^-\}.$$

REMARK 5.7.2. Note that $f\overline{I} \in N^-$ if and only if f/I has a pseudocontinuation belonging to $N^+(\mathbb{D}_e)$ which vanishes at infinity.

When $X = H^p$ ($1 < p < \infty$), the key to proving this result was Lemma 5.6.1. Unfortunately, the proof relied on the F. and M. Riesz theorem, a luxury not afforded us in the cases $X = VMOA$ or K_a.[11] The lemma is still indeed true but the proof needs to be changed.

LEMMA 5.7.3. *Let $X = VMOA$ or K_a and let X^* denote the corresponding dual space (H^1 or H^∞). Then for $f \in X$ and $g \in X^*$,*

$$f\overline{g} \in N^- \Leftrightarrow <B^n f, g> = 0 \ \forall n \in \mathbb{N} \cup \{0\}.$$

We will defer the proof of this lemma (which is at the heart of the matter) momentarily and note that the following lemma is the analog for Lemma 5.6.3 and the proof is the same.

LEMMA 5.7.4. *Let $X = VMOA$ or K_a and X^* be the corresponding dual space. Then, for $f \in X$ and $g \in X^*$,*

$$f\overline{g} \in N^- \Leftrightarrow f\overline{I_g} \in N^-.$$

Now combine Lemma 5.7.3 and Lemma 5.7.4 along with the proof of Theorem 5.1.4, making the obvious modifications, to prove Theorem 5.7.1.

PROOF OF LEMMA 5.7.3. We will prove Lemma 5.7.3 for the case $X = VMOA$ and point out (at the end of the proof) the changes that need to be made for K_a. For $f \in VMOA$, notice from Theorem 3.5.26 that $f = PF$ (P is the Riesz projection operator) for some (not necessarily unique) $F \in C(\mathbb{T})$. Despite the fact that F is not unique, one can check, using the definition of the Riesz projection operator, that the mapping $R : X \times X^* \to N^+$

(5.7.5) $$R(f, g) = P(F\overline{g})$$

is well defined.[12] We will now prove Lemma 5.7.3 by proving the following stronger statement: The following are equivalent.

---

[11]In each of these cases, the duality is given by $\lim_{r \to 1^-} \int f_r \overline{g} dm$, $f \in X, g \in X^*$, which can not always be written as $\int f\overline{g} dm$, since $f\overline{g}$ may not always be integrable. Our use of the F. and M. Riesz theorem requires that the duality be written as $\int f\overline{g} dm$.

[12]Suppose that $f = PF_1 = PF_2$. Then $\widehat{F_j \overline{g}}(n) = \sum_{k=-\infty}^{0} \hat{\overline{g}}(k) \hat{F}_j(n-k)$. Thus for $n \geq 0$ we have $n - k \geq 0$ and so (since $PF_1 = PF_2$) $\hat{F}_1(n-k) = \hat{F}_2(n-k)$. Hence, for $n \geq 0$, $\widehat{F_1 \overline{g}}(n) = \sum_{k=-\infty}^{0} \hat{\overline{g}}(k) \hat{F}_1(n-k) = \sum_{k=-\infty}^{0} \hat{\overline{g}}(k) \hat{F}_2(n-k) = \widehat{F_2 \overline{g}}(n)$. Hence $P(F_1 \overline{g}) = P(F_2 \overline{g})$.

1. $R(f,g) = 0$.
2. $f\bar{g} \in N^-$.
3. $< B^n f, g > = 0 \ \forall n \in \mathbb{N} \cup \{0\}$.

Suppose that $R(f,g) = 0$. Then, by the F. and M. Riesz theorem, $F\bar{g} \in \overline{H_0^1}$. Note that
$$F = PF + (I-P)F \in VMOA + \{\, \bar{g} : g \in VMOA, g(0) = 0 \,\}$$
and so
$$f\bar{g} = F\bar{g} - [(I-P)F]\bar{g} \in \overline{H_0^1} + N^- \subset N^-.$$
Thus (1) implies (2). To prove that (2) implies (1), simply reverse the argument.

We will now prove (2) $\Leftrightarrow$ (3). Since (1) $\Leftrightarrow$ (2), then
$$f\bar{g} \in N^- \quad \Leftrightarrow \quad P(F\bar{g}) = 0$$
$$\Leftrightarrow \quad \int_{\mathbb{T}} F\bar{g}\bar{\zeta}^n \, dm(\zeta) = 0 \ \forall n \in \mathbb{N} \cup \{0\}.$$

Recall that the dual pairing between $VMOA$ and H^1 can be written equivalently as
$$< f, g > = \lim_{r \to 1^-} \int f_r \bar{g} \, dm = \int F\bar{g} \, dm.$$
So, for $n \in \mathbb{N} \cup \{0\}$,
$$< B^n f, g > = < f, S^n g > = \int_{\mathbb{T}} F\bar{\zeta}^n \bar{g} dm,$$
which, from the above, we get (2) $\Leftrightarrow$ (3). The proof of Lemma 5.7.3 in the case $X = VMOA$ is thus complete.

For the proof of Lemma 5.7.3 when $X = K_a$, one changes the operator R in eq.(5.7.5) by $R(f,g) = P(F\bar{g})$, where $F \in L^1$ with $P(F) = f$. The proof follows as before, with the obvious modifications and the fact, from Kolmogorov's theorem (Theorem 3.4.1), that $(I-P)F \in \overline{H_0^p}$ $(0 < p < 1)$. $\square$

Using the methods developed in this section, one can characterize the wk-* closed B-invariant subspaces of $BMOA$. Since $BMOA$ with the norm topology is not separable, there is no hope of obtaining a reasonable characterization of the norm closed B-invariant subspaces. However, when one considers $BMOA$ with the wk-* topology, arising from the pairing
$$\lim_{r \to 1^-} \int f\bar{g_r} \, dm, \quad f \in BMOA, g \in H^1,$$
$BMOA$ is indeed separable. Moreover, since the (forward) shift operator S is continuous on H^1, its adjoint, B, is wk-* continuous on $BMOA$. Making a small change to the proof of Lemma 5.7.3; namely, define
$$R: X^* \times X = BMOA \times H^1 \to N^+, \quad R(f,g) := P(F\bar{g}),$$
where $F \in L^\infty$ with $P(F) = f$ (recall Theorem 3.5.11), one can prove the following:

THEOREM 5.7.6. *Let $\mathcal{M}$ be a wk-* closed B-invariant subspace of $BMOA$. Then, there is an inner function I such that*
$$\mathcal{M} = \{\, f \in BMOA : f\bar{I} \in N^- \,\}.$$

5.8. The case $p=1$

The invariant subspaces of H^1 are handled by a different technique of Aleksandrov which will also be used in part of the analysis of the H^p $(0 < p < 1)$ case. The result is the same as in H^p $(1 < p < \infty)$, but the proof is different.

THEOREM 5.8.1. *Let $\mathcal{M}$ be a B-invariant subspace of H^1. Then $\mathcal{M} = H^1 \cap \overline{IH_0^1}$ for some inner function I. Hence $f \in \mathcal{M}$ if and only if f/I has a pseudocontinuation belonging to $H^1(\mathbb{D}_e)$ which vanishes at infinity.*

The proof, although very elegant, is somewhat involved. For now, we give the reader a brief outline of the argument. The details will follow.

Outline: If f is a non-zero vector in $\mathcal{M}$, recall from Chapter 4 that the non-tangential maximal function $f^*(\zeta) = \sup_{z \in \Gamma_\alpha(\zeta)} |f(z)|$ belongs to L^1 (α is fixed). Since $\exp(-f^*) \in L^\infty$, we can use the Nevanlinna factorization theorem (Theorem 3.2.4) to construct an H^∞ outer function ϕ with $|\phi(\zeta)| = \exp(-f^*(\zeta))$ almost everywhere on $\mathbb{T}$. By our construction, $f\overline{\phi} \in L^\infty$ and so $P(\overline{\phi}f)$ is a non-zero vector in H^2, where P is the Riesz projection operator. Using a Fourier series argument, one proves $P(\overline{\phi}f) \in \mathcal{M} \cap H^2$. By Theorem 5.1.4, $\mathcal{M} \cap H^2 = H^2 \cap \overline{IH_0^2}$ for some inner function I. Next, we show $H^2 \cap \overline{IH_0^2}$ is dense in $H^1 \cap \overline{IH_0^1}$ using the duality between H^1 and $BMOA$ and the truncation operator in Proposition 3.5.19. Thus $H^1 \cap \overline{IH_0^1} \subset \mathcal{M}$. To show the other containment we let $f \in \mathcal{M}$ and set $\{\phi_n : n \in \mathbb{N}\}$ to be the H^∞ outer functions with $|\phi_n(\zeta)| = \exp(-f^*(\zeta)/n)$ almost everywhere on $\mathbb{T}$. Exactly as before, one can show $P(\overline{\phi_n}f) \in \mathcal{M} \cap H^2$. Since $\phi_n \to 1$ almost everywhere, a limit argument using Kolmogorov's theorem (Theorem 3.4.1) shows that $f \in H^1 \cap \overline{IH_0^1}$ and so $\mathcal{M} = H^1 \cap \overline{IH_0^1}$.

We begin our proof of Theorem 5.8.1 with the following lemma.

LEMMA 5.8.2. *For $f \in H^1$, $\|B^n f\|_1 \leq \pi(n+1)\|f\|_1 \quad \forall\, n \in \mathbb{N}$.*

PROOF. First notice that if $f = \sum a_n z^n$, then

$$(B^n f)(\zeta) = \frac{1}{\zeta^n}\left(f - \sum_{k=0}^{n-1} a_k \zeta^k\right) \quad \text{a.e}$$

Thus we have

$$\begin{aligned}
\|B^n f\|_1 &\leq \|f\|_1 + \sum_{k=0}^{n-1} |a_k| \\
&\leq \|f\|_1 + \sum_{k=0}^{n-1}(k+1)\frac{1}{k+1}|a_k| \\
&\leq \|f\|_1 + n\sum_{k=0}^{n-1}\frac{|a_k|}{k+1} \\
&\leq \|f\|_1 + n\pi\|f\|_1 \quad \text{from Hardy's inequality} \\
&\leq \pi(n+1)\|f\|_1.
\end{aligned}$$

In the second to last step above, we used Hardy's inequality which says that for $f = \sum_n a_n z^n \in H^1$,

(5.8.3) $$\sum_n |a_n|/(n+1) \leq \pi \|f\|_1.$$

□

LEMMA 5.8.4. *If $\{g^r : 0 < r < 1\}$ is a family of functions in H^2 with $g^r \to g$ in H^2 as $r \to 1^-$, then $(g^r)_r \to g$ in H^2 as $r \to 1^-$.*

PROOF. If $g^r(z) = \sum a_n(r) z^n$ and $g(z) = \sum a_n z^n$, then using the well known identity

$$\|\sum b_n z^n\|_2^2 = \sum_n |b_n|^2, \quad \sum b_n z^n \in H^2,$$

we have

$$\begin{aligned}
\|(g^r)_r - g\|_2^2 &= \sum |a_n(r) r^n - a_n|^2 \\
&\leq 2 \sum |a_n(r) r^n - a_n r^n|^2 + 2 \sum |a_n r^n - a_n|^2 \\
&\leq 2 \sum |a_n(r) - a_n|^2 + 2 \sum |a_n|^2 (1 - r^n)^2 \\
&= 2\|g^r - g\|_2^2 + 2 \sum |a_n|^2 (1 - r^n)^2.
\end{aligned}$$

The first summand goes to zero as $r \to 1^-$ (by hypothesis) while the second goes to zero by the dominated convergence theorem. □

PROPOSITION 5.8.5. *If $\mathcal{M}$ is a non-zero B-invariant subspace of H^1, then*

$$\mathcal{M} \cap H^2 \neq (0).$$

PROOF. Let f be a non-zero element of $\mathcal{M}$ (which we will assume is non-constant, else the result is trivial) and recall from Theorem 4.4.8 that the non-tangential maximal function

$$f^*(\zeta) = \sup_{z \in \Gamma_\alpha(\zeta)} |f(z)|$$

belongs to L^1 (α is fixed). By the Nevanlinna factorization theory (Theorem 3.2.4) there is a bounded outer function ϕ with

(5.8.6) $$|\phi| = e^{-f^*}$$

almost everywhere $\mathbb{T}$. Also notice that $\phi f^* \in L^\infty$ and so $\overline{\phi} f \in L^\infty$ as well.

From Fatou's theorem, $f_r \to f$ almost everywhere and moreover

$$|\overline{\phi} f_r - \overline{\phi} f| \leq 2|\phi| f^*.$$

By the dominated convergence theorem

(5.8.7) $$\overline{\phi} f_r \to \overline{\phi} f \text{ in } L^2.$$

and so, from Theorem 3.3.5,

(5.8.8) $$P(\overline{\phi} f_r) \to P(\overline{\phi} f) \text{ in } L^2 \text{ (and hence in } L^1\text{)}.$$

Here P is the Riesz projection operator.

Again by Fatou's theorem, $\phi_r \to \phi$ almost everywhere and moreover, from the inequality $|\overline{\phi_r} f - \overline{\phi} f| \le 2|f|$ (note that $|\phi_r|$, $0 < r < 1$, and $|\phi|$ are bounded by one, see eq.(5.8.6)), and the dominated convergence theorem,

$$\overline{\phi_r} f \to \overline{\phi} f \text{ in } L^1. \tag{5.8.9}$$

Using the identity

$$P(\overline{\phi} f) = \overline{\phi} f - (I_d - P)(\overline{\phi} f)$$

along with eq.(5.8.7), eq.(5.8.8), eq.(5.8.9), and Lemma 5.8.4, we observe that

$$P(\overline{\phi} f) = \lim_{r \to 1^-} \left\{ \overline{\phi_r} f - ((I_d - P)\overline{\phi} f_r)_r \right\}, \tag{5.8.10}$$

where the limit is in L^1.

If $f = \sum_n a_n z^n$ and $\phi = \sum_n b_n z^n$, we will now compute, for fixed $r \in (0, 1)$, the Fourier coefficients of the L^1 function $\overline{\phi_r} f - ((I_d - P)\overline{\phi} f_r)_r$. Indeed for $p \in \mathbb{Z}$,

$$\widehat{\overline{\phi_r} f}(p) = \int \overline{\phi_r} f \overline{\zeta}^p \, dm = \sum_{n=0}^\infty \overline{b_n} r^n \int f \overline{\zeta}^{p+n} \, dm = \sum_{n=0}^\infty \overline{b_n} r^n a_{n+p}. \tag{5.8.11}$$

In a similar way,

$$\widehat{\overline{\phi} f_r}(p) = \sum_{n=0}^\infty \overline{b_n} r^{n+p} a_{n+p}$$

and so the p-th Fourier coefficient of $(I_d - P)\overline{\phi} f_r$ is equal to

$$\sum_{n=0}^\infty \overline{b_n} r^{n+p} a_{n+p}, \ p < 0.$$

From here, it follows that the p-th Fourier coefficient of $((I_d - P)\overline{\phi} f_r)_r$ is equal to

$$\sum_{n=0}^\infty \overline{b_n} r^{|p|} r^{n+p} a_{n+p} = \sum_{n=0}^\infty \overline{b_n} r^n a_{n+p}, \ p < 0, \tag{5.8.12}$$

and so, from eq.(5.8.11) and eq.(5.8.12), the p-th Fourier coefficient of the function

$$\overline{\phi_r} f - ((I_d - P)\overline{\phi} f_r)_r$$

is

$$\sum_{n=0}^\infty \overline{b_n} r^n a_{n+p}, \ p \in \mathbb{N} \cup \{0\}. \tag{5.8.13}$$

Since (from Lemma 5.8.2) $\|B^n f\|_{L^1} = O(n+1)$ and $|b_n| = o(1)$, the series

$$\sum_{n=0}^\infty \overline{b_n} r^n B^n f$$

converges in L^1 for each fixed $r \in (0,1)$. Furthermore, since $f \in \mathcal{M}$ and $B\mathcal{M} \subset \mathcal{M}$, this sum belongs to $\mathcal{M}$. The p-th Fourier coefficient of this function is equal to

$$\int \Big\{ \sum_{n=0}^{\infty} \overline{b_n} r^n B^n f \Big\} \overline{\zeta}^p\, dm = \sum_{n=0}^{\infty} \overline{b_n} r^n \int B^n f \overline{\zeta}^p\, dm$$

$$= \sum_{n=0}^{\infty} \overline{b_n} r^n \int f \overline{\zeta}^{n+p}\, dm$$

$$= \sum_{n=0}^{\infty} \overline{b_n} r^n a_{n+p},\ p \in \mathbb{N} \cup \{0\}.$$

From eq.(5.8.13) observe these are precisely the Fourier coefficients of $\overline{\phi}_r f - ((I_d - P)\overline{\phi} f_r)_r$ and hence

$$\overline{\phi}_r f - ((I_d - P)\overline{\phi} f_r)_r = \sum_{n=0}^{\infty} \overline{b_n} r^n B^n f \in \mathcal{M}.$$

From eq.(5.8.10) we conclude $P(\overline{\phi} f)$ belongs to $\mathcal{M}$.

Since $\overline{\phi} f \in L^{\infty}$, $P(\overline{\phi} f) \in H^2 \cap \mathcal{M}$. Finally, note that $P(\overline{\phi} f) \neq 0$ (since otherwise, from the F. and M. Riesz theorem, $\overline{\phi} f \in \overline{H_0^1}$ and so $f \in \overline{h}/\overline{\phi}$ for some $h \in H_0^1$. But this would make f co-analytic and hence constant). Thus $\mathcal{M} \cap H^2 \neq (0)$. $\square$

LEMMA 5.8.14. $H^2 \cap \overline{IH_0^2}$ is dense in $H^1 \cap \overline{IH_0^1}$.

PROOF. Clearly $H^2 \cap \overline{IH_0^2} \subset H^1 \cap \overline{IH_0^1}$. We will show that if $g \in BMOA$ annihilates $H^2 \cap \overline{IH_0^2}$, then g annihilates $H^1 \cap \overline{IH_0^1}$. The lemma will then follow from the Hahn-Banach theorem. By Theorem 5.1.4,

$$H^2 \cap \overline{IH_0^2} = (IH^2)^{\perp_{H^2}}$$

and so by the Hahn-Banach theorem $g \in IH^2$. From here, one concludes that $g/I = G \in BMOA$ [13].

From Proposition 3.5.19, the pairing

$$<f,g> = \lim_{r \to 1^-} \int f_r \overline{g}\, dm$$

can be written alternatively as

$$(f,g) = \lim_{\rho \to \infty} \int f \overline{h_\rho(g)}\, dm.$$

If $f \in H^1 \cap \overline{IH_0^1}$, then $f\overline{I} = \overline{F}$ for some $F \in H_0^1$ and it is routine to show

$$(f,g) = (f\overline{I}, g\overline{I}) = (\overline{F}, G).$$

---

[13] If p is an analytic polynomial, then from the H^1-$BMOA$ duality (Theorem 3.5.13),

$$|\int p\overline{Ig}\, dm| = |\int (pI)\overline{g}\, dm| \leq c\|pI\|_{H^1}\|g\|_{BMOA} \leq c\|p\|_{H^1}\|g\|_{BMOA}.$$

Thus, the linear functional $p \to \int p\overline{Ig}\, dm$ can be extended to be continuous on H^1 and so $\overline{I}g \in BMOA$. See [82] for generalizations and further details.

But
$$<f,g> = (f,g) = (\overline{F},G) = <\overline{F},G> = \lim_{s\to 1}\int \overline{F}(s\zeta)\overline{G}(\zeta)\,dm(\zeta) = 0$$
since for each $s \in (0,1)$, $F(s\zeta)G(\zeta)$ are the boundary values of an H^1 function which vanishes the origin. □

PROOF OF THEOREM 5.8.1. From Proposition 5.8.5, $\mathcal{M} \cap H^2$ is a non-zero B-invariant subspace, and a routine argument shows it is closed in H^2 and hence, by Theorem 5.1.4,
$$\mathcal{M} \cap H^2 = H^2 \cap I\overline{H_0^2}$$
for some inner function I. By Lemma 5.8.14, $H^1 \cap I\overline{H_0^1} \subset \mathcal{M}$, being the closure of $H^2 \cap I\overline{H_0^2}$ in L^1.

For the reverse inequality, let $f \in \mathcal{M}$. For each $n \in \mathbb{N}$, define ϕ_n to be the outer function with boundary values equal to
$$|\phi_n| = e^{-f^*/n}$$
almost everywhere on $\mathbb{T}$. Applying the proof of Proposition 5.8.5, we see that $P(\overline{\phi_n}f) \in \mathcal{M} \cap H^2$. Note also that $\overline{\phi_n}f \to f$ almost everywhere since $\phi_n \to 1$ almost everywhere and, from the dominated convergence theorem,

(5.8.15) $$\overline{\phi_n}f \to f \text{ in } L^1.$$

By Kolmogorov's theorem (Theorem 3.4.1) and eq.(5.8.15), $P(\overline{\phi_n}f) \to f$ in L^s for fixed $s \in (0,1)$. But
$$P(\overline{\phi_n}f) \in H^2 \cap I\overline{H_0^2} \subset H^s \cap I\overline{H_0^s}$$
for each n and so $f \in H^s \cap I\overline{H_0^s}$. This means that $f = I\overline{h}$ for some $h \in H_0^s$. Since $|I| = 1$ almost everywhere and $f \in H^1$, then $h \in H_0^1$ and consequently $f \in H^1 \cap I\overline{H_0^1}$. Thus we have shown $\mathcal{M} = H^1 \cap I\overline{H_0^1}$ and the proof is complete. □

5.9. Cyclic vectors

DEFINITION 5.9.1. We say a vector $f \in H^p$ ($1 \leq p < \infty$) is *cyclic* for B if
$$\bigvee\{B^n f : n \in \mathbb{N} \cup \{0\}\} = H^p.$$

The problem of characterizing the cyclic vectors of B on H^2 was officially posed (although there were some earlier results in the literature [83]) by Don Sarason in 1965 at a conference in Lexington, Kentucky. The problem was solved by Douglas, Shapiro, and Shields [29] in a paper which served as the motivation for much of the work on the backward shift. The main theorem about cyclic vectors follows from the main theorems, Theorem 5.1.4 and Theorem 5.8.1, concerning the classification of the B-invariant subspaces of H^p. Before we state this theorem, we need a definition.

Recall from Chapter 3 the spaces
$$\mathfrak{N}(\mathbb{D}) = \{\,f/g : f,g \in H^\infty(\mathbb{D})\}, \quad \mathfrak{N}(\mathbb{D}_e) = \{\,f/g : f,g \in H^\infty(\mathbb{D}_e)\}$$
of functions of "bounded type". From the Nevanlinna factorization theory, every $F \in \mathfrak{N}(\mathbb{D}_e)$ can be factored as

(5.9.2) $$F = \frac{\phi_1}{\phi_2}G,$$

where ϕ_1, ϕ_2 are "inner" functions on $\mathbb{D}_e$ ($\phi_j(z) = \overline{I_j(1/\overline{z})}$, where I_j is an inner function in the usual sense) and G is an "outer" function on $\mathbb{D}_e$ ($G = \overline{O(1/\overline{z})}$, where O is an outer function in the usual sense).

DEFINITION 5.9.3. We will say that $f \in H^p$ has a *pseudocontinuation of bounded type* if f has a pseudocontinuation across $\mathbb{T}$ to a function belonging to $\mathfrak{N}(\mathbb{D}_e)$.

REMARK 5.9.4. Note that some care must be taken with the previous definition. For example, the function

$$f(z) := \sum_{n=1}^{\infty} \frac{1}{n^3} \frac{1}{z - (1+1/n)}$$

belongs to H^2 since

$$\left\| \frac{1}{z - (1+1/n)} \right\|_{H^2} \leq n$$

and it is not difficult to show that $f \in \mathfrak{M}(\mathbb{D}_e)$. Moreover f has a pseudocontinuation across $\mathbb{T}$ (since it has an analytic continuation across $\mathbb{T}\setminus\{1\}$). However this pseudocontinuation is not of bounded type since the poles $\{1 + 1/n : n \in \mathbb{N}\}$ are not the zeros of a Blaschke function on the exterior disk.

The characterization of the cyclic vectors for H^p $(1 \leq p < \infty)$ is the following.

THEOREM 5.9.5. *A vector $f \in H^p$ $(1 \leq p < \infty)$, is not cyclic for B if and only if f has a pseudocontinuation of bounded type.*

PROOF. Suppose that f is non-cyclic for B. From our main theorems, Theorem 5.1.4 and Theorem 5.8.1, f/I has a pseudocontinuation

$$T_{f/I} \in N^+(\mathbb{D}_e) \subset \mathfrak{N}(\mathbb{D}_e).$$

Since I has a pseudocontinuation T_I defined by $(T_I)(z) = 1/\overline{I(1/\overline{z})} \in \mathfrak{N}(\mathbb{D}_e)$, then f has a pseudocontinuation defined by $T_f = T_I T_{f/I} \in \mathfrak{N}(\mathbb{D}_e)$.

On the other hand, suppose that f has a pseudocontinuation $T_f \in \mathfrak{N}(\mathbb{D}_e)$. Then by eq.(5.9.2),

$$(T_f)(z) = \frac{\overline{I_1(1/\overline{z})}}{\overline{I_2(1/\overline{z})}} \overline{O(1/\overline{z})} \quad z \in \mathbb{D}_e,$$

were I_1, I_2 are inner functions and O is outer. At almost every point of $\mathbb{T}$ we have $f\overline{zI_2} = \overline{zI_1O} \in \overline{H_0^p}$.[14] Thus $f \in H^p \cap zI_2\overline{H_0^p}$ and so f belongs to some non-trivial B-invariant subspace. Hence f is not cyclic. □

REMARK 5.9.6. Although the above result completely characterizes cyclicity, it may be difficult to apply since determining whether or not a given function in H^p has a pseudocontinuation of bounded type does not seem to be an easy task. A precise "testable" condition for the existence of a pseudocontinuation of bounded type is still very much an open problem worthy of future study. We would like to mention several known results about cyclic vectors and refer the reader to the appropriate references for the proofs. This list is by no means complete but rather a sample of some of the results. The tacit assumption below is that $p \in (1, \infty)$.

---

[14]In this last step, we are using the fact that $\overline{zI_1O} \in \overline{N^+}$ and has L^p boundary values and hence belongs to $\overline{H^p}$.

1. The first set of results, found in the Douglas, Shapiro, Shields paper [**29**], discuss when a function $f \in H^p$ has an analytic continuation across a portion of the circle.
    (a) If $f \in H^p$ is analytic on $R\mathbb{D}$ where $R > 1$, then f is either cyclic or is a rational function (and hence not cyclic).
    (b) If $f \in H^p$ has an analytic continuation across all points of an arc $J \subset \mathbb{T}$ with the exception of an isolated branch point, then f is cyclic.
2. There is some information which can be obtained about cyclic vectors from the growth of the Taylor coefficients. For example, H. S. Shapiro [**83**], in perhaps one of the earliest published results about cyclic vectors, showed that if $f = \sum_n a_n z^n$ with $|a_n|^{1/n} \to 0$ and f is not a polynomial, then f is cyclic for H^2. Kriete [**52**] was able to obtain further information along these lines. He showed that if $f = \sum_n a_n z^n \in H^2$ is a non-cyclic vector and not a polynomial, then there is an $r > 0$ and $k \in \mathbb{N} \cup \{0\}$ such that

$$\sum_{n=k+j}^{\infty} |a_n|^2 \geq r^j \sum_{n=k}^{\infty} |a_n|^2 \ \forall j \in \mathbb{N}.$$

Unfortunately, this condition does not characterize the cyclic vectors. In fact, given $\{d_j : j \in \mathbb{N}\} \subset (0,1)$ with $d_j \to 0$, there is a bounded cyclic outer function $f = \sum_n a_n z^n$ such that

$$\sum_{n=j}^{\infty} |a_n|^2 \geq d_j \|f\|_{\infty}^2.$$

3. The next set of results relate cyclicity with the modulus of the function.
    (a) [**29**] If $f \in H^p$ is non-constant and

    $$\int \log |\Re f(\zeta)| \, dm(\zeta) = -\infty,$$

    then f is cyclic.
    (b) (Kriete [**52**]) If f is non-cyclic and $a > 0$. Then either

    $$\int \log \big| a - |f(\zeta)| \big| \, dm(\zeta) > -\infty,$$

    or $|f(\zeta)| = a$ almost everywhere.
    (c) (Kriete [**52**]) If f, g are non-cyclic, then either

    $$\int \log \big| |f(\zeta)| - |g(\zeta)| \big| \, dm(\zeta) > -\infty,$$

    or $|f(\zeta)| = |g(\zeta)|$ almost everywhere.
4. These next set of results [**29**] (a nice summary can be found in [**28**]) say that both the cyclic and non-cyclic vectors occupy a large portion of H^p. We let $\mathcal{N}$ denote the non-cyclic vectors and $\mathcal{C}$ denote the cyclic ones.
    (a) $H^p = \mathcal{C} + \mathcal{C}$ (every function in H^p is the sum of two cyclic vectors).
    (b) (Tumarkin [**100**]) $\mathcal{N}$ is a dense F_σ set of first category in H^p.
    (c) $\mathcal{C}$ is dense G_δ set in H^p.
    (d) $\mathcal{N} + \mathcal{N} \subset \mathcal{N}$; $\mathcal{N} + \mathcal{C} \subset \mathcal{C}$; $(\mathcal{N} \cdot \mathcal{N}) \cap H^p \subset \mathcal{N}$; $(\mathcal{N} \cdot \mathcal{C}) \cap H^p \subset \mathcal{C}$; if $f \in \mathcal{C}$ and $1/f \in H^p$, then $1/f \in \mathcal{C}$.

5. We say a sequence $\{n_k : k \in \mathbb{N}\} \subset \mathbb{N}$ is a *lacunary sequence* if
$$n_{k+1} \geq d \cdot n_k \ \forall\, k \in \mathbb{N}$$
for some $d > 1$. A function $f = \sum_k a_k z^{n_k}$ is called a lacunary function if $\{n_k : k \in \mathbb{N}\}$ is a lacunary sequence. In general, H^2 lacunary functions are non-continuable and hence cyclic.

   (a) [29] If $f = \sum_k a_k z^{n_k}$ is a lacunary function in H^2 ($\sum_k |a_k|^2 < \infty$) with an infinite number of $a_k \neq 0$, then f is cyclic.

   (b) In [1] Abakumov proved the same result is true if the lacunary sequence $\{n_k : k \in \mathbb{N}\}$ is replaced by a finite union of lacunary sequences.

   (c) From (1 a), if g is analytic on $R\mathbb{D}$ for some $R > 1$, then g is either cyclic or rational. Abakumov [1] proved the same result is true if the function g is replaced by $g + f$, where f is an H^2 lacunary function.

   (d) Other interesting results about pseudocontinuations and lacunary functions can be found in papers of Aleksandrov [6] [7].

We comment that in the Bergman space setting, pseudocontinuations do not completely describe non-cyclicity. From Theorem 5.4.5, a non-cyclic vector has a pseudocontinuation of bounded type across $\mathbb{T}$. However the converse is not true. Using the dual pairing (between the Bergman space L_a^p and the Dirichlet space D_q from eq.(5.4.2))

$$<f,g> = \lim_{r \to 1^-} \int f_r \bar{g}\, dm \quad f \in L_a^p, g \in D_q,$$

one can show that an inner function ϕ is non-cylic for the Bergman space L_a^p if and only ϕ is the inner factor for some non-zero function from D_q. It is certainly possible to create such an inner function which is not the divisor of some Dirichlet function [15]. This particular inner function ϕ has a pseudocontinuation of bounded type (all inner functions do) but *is* a cyclic vector for L_a^p. We also comment that in the Hardy space case, the description of the non-cyclic vectors is the same for all p (pseudocontinuations of bounded type). For the Bergman space, this is not the case. For example, one can create an inner function (in fact a Blaschke product) which is the inner divisor for some non-zero D_2 function but is not the inner divisor of any D_q function for $q > 2$ [16]. This inner function would be non-cyclic for L_a^2 but cyclic for $L_a^p, 1 < p < 2$.

For other spaces of analytic functions, the problem of characterizing the cyclic vectors is very much an open question. For example, in the Dirichlet space D_q, certainly any vector with a pseudocontinuation of bounded type is non-cyclic (use Theorem 5.9.5 and the fact that the Dirichlet norm dominates the H^q norm). However the converse fails miserably. In fact [9] there are non-cyclic vectors for the Dirichlet space which do not have a pseudocontinuation (to any contiguous domain) across *any* set of positive measure in $\mathbb{T}$.

---

[15]By [99] an inner function ϕ divides a non-zero $D_q, q > 2$ function if and only if $\int \log \mathrm{dist}(\zeta, \sigma(\phi))\, dm(\zeta) > -\infty$. So just choose a ϕ for which $\sigma(\phi) = \mathbb{T}$. For $q \leq 2$, the complete description of inner divisors of D_q functions is not known. However there are many partial results [19] and from this, one can create inner functions which do not divide non-zero D_q functions.

[16]Using [19], one can create a Blaschke product b for which $\sigma(b) = \mathbb{T}$ and such that b divides a D_2 function. This b could not be the inner divisor of any $D_q, q > 2$ function since $\int \log \mathrm{dist}(\zeta, \sigma(b))\, dm(\zeta) = -\infty$.

5.10. Duality

For the sake of completeness, we include the following result which seems to be in the folklore of H^p theory. We found this version in [20], p. 743.

THEOREM 5.10.1. *If $p \in (1, \infty)$, then $\ell \in (H^p \cap I\overline{H_0^p})^*$ if and only if there is a $g \in H^q \cap I\overline{H_0^q}$ such that*

$$(5.10.2) \qquad \ell(f) = \int f\overline{g}\, dm \quad f \in H^p \cap I\overline{H_0^p}.$$

Moreover $\|\ell\| \asymp \|g\|_q$.

PROOF. Clearly eq.(5.10.2) defines a element of $(H^p \cap I\overline{H_0^p})^*$. Conversely, suppose $\ell \in (H^p \cap I\overline{H_0^p})^*$. By the Hahn-Banach extension theorem, ℓ can be extended to H^p and so, by Theorem 3.5.6, there is a $G \in H^q$ such that $\|G\|_q \asymp \|\ell\|$ and

$$\ell(f) = \int f\overline{G}\, dm$$

on $H^p \cap I\overline{H_0^p}$. Using the Riesz projection operator (Theorem 3.3.5),

$$\overline{I}G = g_1 + \overline{g_2},$$

where $g_1 \in H^q$ and $g_2 \in H_0^q$ with $\|g_2\|_q \leq c_q \|G\|_q$. From this we observe

$$g := I\overline{g_2} = G - Ig_1 \in H^q \cap I\overline{H_0^q}$$

and

$$\ell(f) = \int f\overline{g}\, dm.$$

But since $\|g\|_q = \|g_2\|_q$ we have $\|g\|_q \leq c\|\ell\|$. To show $\|\ell\| \leq c'\|g\|_q$, we use some elementary functional analysis. Indeed,

$$\begin{aligned}\|g\|_q &\asymp \sup\left\{ \left|\int F\overline{g}\, dm\right| : F \in H^p, \|F\|_p \leq 1 \right\} \\ &\geq \sup\left\{ \left|\int F\overline{g}\, dm\right| : F \in H^p \cap I\overline{H_0^p}, \|F\|_p \leq 1 \right\} \\ &= \|\ell\|.\end{aligned}$$

Thus $\|g\|_q \asymp \|\ell\|$. $\qquad\square$

5.11. The commutant

We now make a few remarks about the commutant of B.

DEFINITION 5.11.1. *The commutant of B on H^p ($0 < p < \infty$), denoted by $\{B\}'$, is the set of all continuous linear operators A on H^p for which $AB = BA$.*

To classify such operators A, first notice that for any $p \in (0, \infty)$,

$$(5.11.2) \qquad \bigvee \left\{ \frac{1}{1 - \lambda z} : \lambda \in \mathbb{D} \right\} = H^p.$$

One can see this by observing that the closed linear span, in the sup-norm on $\mathbb{T}$, of

$$\left\{ \frac{1}{1 - \lambda z} : \lambda \in \mathbb{D} \right\}$$

contains all the polynomials. From here, recall that the polynomials are dense in H^p. [17]

Suppose $A \in \{B\}'$. Then

$$AB\left(\frac{1}{1-\lambda z}\right) = BA\left(\frac{1}{1-\lambda z}\right) \quad \forall \lambda \in \mathbb{D}.$$

Since $B(1-\lambda z)^{-1} = \lambda(1-\lambda z)^{-1}$,

$$\lambda A\left(\frac{1}{1-\lambda z}\right) = BA\left(\frac{1}{1-\lambda z}\right)$$

and so letting

$$\phi(\lambda, z) := A\left(\frac{1}{1-\lambda z}\right),$$

we get the functional equation

$$\lambda \phi(\lambda, z) = \frac{\phi(\lambda, z) - \phi(\lambda, 0)}{z},$$

where we understand

$$\phi(\lambda, 0) = A\left(\frac{1}{1-\lambda z}\right)\bigg|_{z=0}.$$

A little algebra yields

$$A\left(\frac{1}{1-\lambda z}\right) = \frac{\phi(\lambda, 0)}{1-\lambda z}.$$

If we let $\psi(\lambda) := \phi(\lambda, 0)$, we see that ψ is uniquely determined by the above equation (from eq.(5.11.2)) and is an analytic function on $\mathbb{D}$. Thus if $A \in \{B\}'$, there is a unique analytic function ψ on $\mathbb{D}$ so that

$$A\left(\frac{1}{1-\lambda z}\right) = \frac{\psi(\lambda)}{1-\lambda z}.$$

Also notice that

$$|\psi(\lambda)|^p \left\|\frac{1}{1-\lambda z}\right\|^p = \left\|A\left(\frac{1}{1-\lambda z}\right)\right\|^p \leq \|A\|^p \left\|\frac{1}{1-\lambda z}\right\|^p \quad \forall \lambda \in \mathbb{D},$$

and so ψ is a bounded analytic function on $\mathbb{D}$. Since ψ is bounded, this allows us to define, at least for bounded analytic functions f, the co-analytic Toeplitz operator $T_{\overline{\psi}}$ by

$$T_{\overline{\psi}}f = P(\overline{\psi}f),$$

where P is the Riesz projection operator. A routine computation shows that when f is one of the kernels $(1-\lambda z)^{-1}$, $\lambda \in \mathbb{D}$, then $Af = T_{\overline{\psi}}f$ and so

$$A = T_{\overline{\psi}}$$

on a dense set of H^p (see eq.(5.11.2)).

For $p \in (1, \infty)$, the Riesz projection P is continuous (Theorem 3.3.5) and so $T_{\overline{\psi}}$ extends to be a continuous operator on H^p. Thus for $p \in (1, \infty)$,

$$\{B\}' = \left\{ T_{\overline{\psi}} : \psi \in H^\infty \right\}.$$

---

[17] For each $f \in H^p$, $f_r \to f$ in the norm of H^p (Theorem 3.2.3) as $r \to 1^-$. Now approximate f_r uniformly on $\mathbb{D}^-$ with its power series expansion.

When $p = 1$, P and hence $T_{\overline{\psi}}$ do not extend to be continuous on H^1 for general H^∞ symbol ψ. A result of Stegenga [**92**] says that $T_{\overline{\psi}}$ extends to be continuous on H^1 if and only if $\Re\psi$ and $\Im\psi$ are "multipliers" of BMO, that is to say

$$(\Re\psi)BMO \subset BMO, \quad (\Im\psi)BMO \subset BMO.$$

So in this case

$$\{B\}' = \{\, T_{\overline{\psi}} : \Re\psi, \Im\psi \text{ are multipliers of } BMO \,\}.$$

Though we will not include a proof here, there is also a description by Sarason [**79**] (and simplified by Nikolskiĭ in [**65**], p. 179) of the commutant of B when restricted to one of its invariant subspaces $H^2 \cap I\overline{H_0^2}$. The result says

$$\{B|H^2 \cap I\overline{H_0^2}\}' = \{\, T_{\overline{\psi}}|H^2 \cap I\overline{H_0^2} : \psi \in H^\infty \,\}.$$

Curiously, both the Sarason and Nikolskiĭ proofs are Hilbert space in flavor and there does not seem to be an extension of this result to the general case when $p \in (1, \infty)$. Clearly

$$\{\, T_{\overline{\psi}}|H^p \cap I\overline{H_0^p} : \psi \in H^\infty \,\} \subset \{B|H^p \cap I\overline{H_0^p}\}'.$$

Equality certainly seems reasonable but we do not know of any proof of this in the literature.

We will take up the description of the symbols $\psi \in H^\infty$ for which $T_{\overline{\psi}}$ extends to be continuous on H^p ($0 < p < 1$), in Section 8 of Chapter 6.

5.12. Compactness of the inclusion operator

For $1 \leq p' < p < \infty$, the inclusion operator

$$i : H^p \cap I\overline{H_0^p} \to H^{p'} \cap I\overline{H_0^{p'}}$$

is continuous. This next result says slightly more.

THEOREM 5.12.1 (Aleksandrov [**6**]). *For $1 \leq p' < p < \infty$, the inclusion operator $i : H^p \cap I\overline{H_0^p} \to H^{p'} \cap I\overline{H_0^{p'}}$ is compact.*

PROOF. For $s > 0$ and $f \in L^s$, define

$$\|f\|_s^* := \inf \{\, \|f + g\|_s : g \in \overline{H_0^s} \,\}$$

Claim 1: Let ϕ be a unimodular function on $\mathbb{T}$ (not necessarily inner) and for fixed $1 \leq p' < p < \infty$, let $1/r = 1/p' - 1/p$. Then for any $f \in \phi\overline{H_0^p}$,

$$\|f\|_{p'}^* \leq \|\phi\|_r^* \|f\|_p.$$

Proof of Claim 1: Let $g \in \overline{H_0^r}$ and observe from Hölder's inequality that $g\overline{\phi}f \in \overline{H_0^{p'}}$ (since $\overline{\phi}f \in \overline{H_0^p}$) and so

$$\begin{aligned}\|f\|_{p'}^* &\leq \|f - g\overline{\phi}f\|_{p'} \\ &= \|(\phi + g)\overline{\phi}f\|_{p'} \\ &\leq \|\phi + g\|_r \|f\|_p.\end{aligned}$$

Now take an infimum over all $g \in \overline{H_0^r}$ to get the desired inequality. This proves Claim 1.

Claim 2: If I is an inner function and $r > 0$, then
$$\lim_{N \to \infty} \|\bar{z}^N I\|_r^* = 0.$$

Proof of Claim 2: Since $\bar{z}^N I \in L^\infty$, then $P(\bar{z}^N I) \in H^r$ and $(I-P)(\bar{z}^N I) \in \overline{H_0^r}$ for any $r > 0$. Thus
$$\|\bar{z}^N I\|_r^* \leq \|P(\bar{z}^N I) + (I-P)(\bar{z}^N I)\|_r^* \leq \|P(\bar{z}^N I)\|_r \quad \forall\, N \in \mathbb{N}.$$
If $r \leq 2$, then we can apply Hölder's inequality and Parseval's theorem to obtain
$$\begin{aligned}
\|P(\bar{z}^N I)\|_r &\leq C_r \|P(\bar{z}^N I)\|_2 \\
&= C_r \Big\| \sum_{k=N}^{\infty} \zeta^{k-N} \hat{I}(k) \Big\|_2 \\
&= C_r \Big\{ \sum_{k=N}^{\infty} |\hat{I}(k)|^2 \Big\}^{1/2}
\end{aligned}$$
which goes to zero as $N \to \infty$.

If $r > 2$, then
$$\begin{aligned}
\|P(\bar{z}^N I)\|_r^r &= \int |P(\bar{z}^N I)|^{r-1} |P(\bar{z}^N I)| \, dm \\
&\leq \Big\{ \int |P(\bar{z}^N I)|^{2(r-1)} \, dm \Big\}^{1/2} \Big\{ \int |P(\bar{z}^N I)|^2 \, dm \Big\}^{1/2} \\
&\leq C_r \|\bar{z}^N I\|_{2(r-1)}^{r-1} \|P(\bar{z}^N I)\|_2 \\
&\leq C_r \|P(\bar{z}^N I)\|_2
\end{aligned}$$
which goes to zero (by Parseval's theorem above) as $N \to \infty$. This proves Claim 2.

For each $n \in \mathbb{N}$, let $S_n : H^p \cap \overline{IH_0^p} \to H^{p'}$ be defined by
$$(S_n f)(z) := \sum_{k=0}^{n-1} \hat{f}(k) z^k.$$

To prove the compactness of the inclusion operator $i : H^p \cap \overline{IH_0^p} \to H^{p'} \cap \overline{IH_0^{p'}}$, we will first prove that the operator norm of $I_d - S_n : H^p \cap \overline{IH_0^p} \to H^{p'}$ goes to zero as $n \to \infty$. Since each S_n is a finite rank operator, this will say that $I_d : H^p \cap \overline{IH_0^p} \to H^{p'}$ is compact. From this follows that $i : H^p \cap \overline{IH_0^p} \to H^{p'} \cap \overline{IH_0^{p'}}$ is compact.

We assume for the moment that $1 < p' < p < \infty$ and make a remark concerning $p' = 1$ at the end. First note that for $f \in H^p \cap \overline{IH_0^p}$ and $n \in \mathbb{N}$, we have $\bar{z}^n f \in \bar{z}^n \overline{IH_0^p}$ and $\bar{z}^n S_n f \in \overline{H_0^{p'}}$. Thus

(5.12.2) $$\|\bar{z}^n(f - S_n f)\|_{p'}^* = \|\bar{z}^n f\|_{p'}^*.$$

By Claim 1, applied to the function $\bar{z}^n f$ and the unimodular function $\bar{z}^n I$, we obtain

(5.12.3) $$\|\bar{z}^n f\|_{p'}^* \leq \|\bar{z}^n I\|_r^* \|\bar{z}^n f\|_p \leq \|\bar{z}^n I\|_r^* \|f\|_p.$$

Since $\bar{z}^n(f - S_n f) \in H^{p'}$,
$$\|\bar{z}^n(f - S_n f) - g\|_{p'} \geq C_{p'} \Big\{ \|\bar{z}^n(f - S_n f)\|_{p'} + \|g\|_{p'} \Big\} \quad \forall\, g \in \overline{H_0^{p'}}.$$

This last inequality can be obtained using the Riesz projection, see eq.(3.3.8). Hence
(5.12.4) $$\|\bar{z}^n(f - S_n f)\|_{p'}^* \geq C_{p'} \|\bar{z}^n(f - S_n f)\|_{p'}.$$
Finally
$$\begin{aligned}
\|f - S_n f\|_{p'} &= \|\bar{z}^n(f - S_n f)\|_{p'} \\
&\leq C_{p'} \|\bar{z}^n(f - S_n f)\|_{p'}^* \quad \text{from eq.(5.12.4)} \\
&\leq C_{p'} \|\bar{z}^n f\|_{p'}^* \quad \text{from eq.(5.12.2)} \\
&\leq C_q \|\bar{z}^n I\|_r^* \|f\|_p \quad \text{from eq.(5.12.3)}.
\end{aligned}$$
From here, it follows from Claim 2 that $I_d - S_n : H^p \cap \overline{IH_0^p} \to H^{p'}$ goes to zero in operator norm.

When $p' = 1$, observe, for $f \in H^p \cap \overline{IH_0^p}$ and $1 < p'' < p$, that
$$\int |f - S_n f|\, dm \leq C_{p''} \|f - S_n f\|_{p''} \leq C_{p''} \|\bar{z}^n I\|_{r(p'')}^* \|f\|_p$$
which goes to zero as before. $\square$

REMARK 5.12.5.   1. It turns out that the above inclusion operator is compact for p and p' satisfying $0 < p' < p < \infty$ and $1 \leq p < \infty$. When $p > 1$, observe that Claim 1 can be extended to all $0 < p' < p$ and the rest of the proof follows as before.
2. When $p = 1$ some adjustments need to be made since in this case, the inclusion i cannot be approximated by the finite rank operators S_n, see [**6**].
3. When $0 < p' < p < 1$, the inclusion is not compact.

CHAPTER 6

The backward shift on H^p for $p \in (0,1)$

6.1. Introduction

Several facts make the problem of characterizing the B-invariant subspaces of H^p ($0 < p < 1$) especially challenging. On a function theory level, the Cauchy integral formula, a key tool used in Chapter 5, is no longer valid since there are H^p functions which are not reproduced by the Cauchy integral of their boundary values. On a functional analysis level, H^p is not locally convex and the Hahn-Banach separation theorem, another key tool used in Chapter 5, is also invalid. Before launching into the technical details of Aleksandrov's characterization of the B-invariant subspaces of H^p ($0 < p < 1$), we want to make some further remarks about the complexity of this space of functions.

REMARK 6.1.1. Throughout this chapter, we will leave the index p off the H^p "norm" $\|\cdot\|_p$ and just write $\|\cdot\|$.

Let $n_p := \max\{[1, 1/p) \cap \mathbb{N}\} = [1/p]$. From a simple change of variables, we obtain the following.

LEMMA 6.1.2. Let $p \in (0,1)$ be fixed. For each $j = 1, \cdots, n_p$ and $\zeta \in \mathbb{T}$,
$$\frac{1}{(1-\overline{\zeta}z)^j} \in H^p$$
and
$$\left\| \frac{1}{(1-\overline{\zeta}z)^j} \right\| = \left\| \frac{1}{(1-z)^j} \right\|.$$

Two important facts result from this lemma. First, H^p ($0 < p < 1$) contains more functions than H^1 since $(1-\overline{\zeta}z)^{-1} \notin H^1$ for $\zeta \in \mathbb{T}$. This is important since these functions are eigenvectors for B on H^p and as a result, play a key role in Aleksandrov's characterization. Secondly, opposite to the $p \geq 1$ case, $H^p \cap \overline{H_0^p} \neq (0)$ for $p < 1$. For example, if $w \in \mathbb{T}$, then
$$\frac{1}{(1-\overline{\zeta}w)^j} = \frac{\overline{w}^j}{(\overline{w}-\overline{\zeta})^j}$$
is the boundary function for
$$\left(\frac{z^j}{(z-\zeta)^j} \right)$$
which belongs to $\overline{H_0^p}$ for $j = 1, \cdots, n_p$. In fact, by the next several results of de Leeuw [27], $H^p \cap \overline{H_0^p}$ is quite a rich class of functions.

PROPOSITION 6.1.3. For any $n \in \mathbb{Z}$, $z^n \notin H^p \cap \overline{H_0^p}$.

PROOF. If not, argue that $z^{-n} \in H^p \cap \overline{H_0^p}$ for some $n \in \mathbb{N}$. But since $H^p \cap L^1 = H^1$, $z^{-n} \in H^1$, a contradiction. □

PROPOSITION 6.1.4. *If $\mu \in M_s(\mathbb{T})$ (singular measures) and $p \in (0,1)$, then*

$$(P\mu)(z) = \int_{\mathbb{T}} \frac{d\mu(\zeta)}{1 - \overline{\zeta}z} \in H^p \cap \overline{H_0^p}.$$

PROOF. By Theorem 3.4.1, $P\mu \in H^p$ for all $p \in (0,1)$. Similarly,

$$F_\mu(z) := \int \frac{d\mu(\zeta)}{1 - \overline{\zeta}z}, \quad z \in \mathbb{D}_e,$$

belongs to $H^p(\mathbb{D}_e)$ and vanishes at infinity. Thus by eq.(3.2.10), the function $G(z) := \overline{F_\mu(1/\overline{z})}, z \in \mathbb{D}$, belongs to $\in H_0^p$. The measure μ is singular and so by Fatou's jump theorem (Theorem 3.4.8), $\overline{(P\mu)(\zeta)} = G(\zeta)$ almost everywhere and so $P\mu \in \overline{H_0^p}$. □

PROPOSITION 6.1.5. *If $f \in H^p \cap \overline{H_0^p}$ and I is inner with $I(0) = 0$, then*

$$f \circ I \in H^p \cap \overline{H_0^p}.$$

PROOF. By definition, there is a $G \in H^p, H \in H_0^p$ such that $f = G$ and $\overline{f} = H$ almost everywhere on $\mathbb{T}$. By the Littlewood subordination principle [1], $G \circ I$ and $H \circ I$ both belong to H^p. Note also that since $I(0) = 0$, $H \circ I \in H_0^p$.

From the first line of the proof, the set $A := \{\zeta \in \mathbb{T} : G(\zeta) = \overline{H(\zeta)}\}$ has full measure in $\mathbb{T}$. Moreover, since I is inner, then $I(\zeta) \in A$ for almost every $\zeta \in \mathbb{T}$ and so

$$(G \circ I)(\zeta) = (f \circ I)(\zeta) \text{ and } (H \circ I)(\zeta) = (\overline{f} \circ I)(\zeta)$$

almost everywhere. Thus $f \circ I \in H^p \cap \overline{H_0^p}$. □

By the Riesz projection theorem (Theorem 3.3.5) we can decompose the space L^p $(1 < p < \infty)$ as

$$L^p = H^p \oplus \overline{H_0^p}.$$

For $p \in (0,1)$, the following result says the above decomposition is still possible, although, from our previous results, it is far from being direct.

PROPOSITION 6.1.6 (Aleksandrov [3] [4]). *Let $p \in (0,1)$. Then*

1. $H^p + \overline{H_0^p} = L^p$.
2.
$$\bigvee \left\{ \frac{1}{1 - \overline{\zeta}z} : \zeta \in \mathbb{T} \right\} = H^p \cap \overline{H_0^p}.$$

PROOF. To prove statement (1), we need to establish two claims.

Claim 1: Every $f \in L^p$ can be written as

(6.1.7) $$f = \sum_{k=1}^{\infty} P_k,$$

where P_k is a trigonometric polynomial, the convergence is in the L^p metric, and $\sum_k \|P_k\|^p < \infty$.

---

[1] The Littlewood subordination principle is the following: Let $g, h, w \in \mathfrak{H}(\mathbb{D})$ with $w(0) = 0$ and $|w| \leq 1$ (By the Schwartz lemma, $|w(z)| \leq |z|$). Suppose also that $g = h \circ w$. Then, for every $p \in (0, \infty]$, $M_p(r, g) \leq M_p(r, h)$ for all $r \in (0, 1)$. A proof of this can be found in [31], p. 10.

6.1. INTRODUCTION

To prove this, we first observe that the trigonometric polynomials are dense in L^p.[2] To finish, we choose a sequence $\{r_n : n \in \mathbb{N}\}$ of trigonometric polynomials with $\|f - r_n\|^p \leq 1/n^2$ and set

$$P_1 = r_1, \quad P_n = r_n - r_{n-1} \ (n = 2, 3, \cdots).$$

Then

$$\sum_{k=2}^{\infty} \|P_k\|^p \leq \sum_{k=2}^{\infty} \|r_k - r_{k-1}\|^p < \sum_{k=2}^{\infty} 2/(k-1)^2 < \infty$$

and $\sum_{k=1}^{n} P_k = r_n$, which converges (in L^p) to f as $n \to \infty$. This proves Claim 1.

Claim 2: Every trigonometric polynomial P_k above can be written as

(6.1.8) $$P_k = Q_k + R_k,$$

where $Q_k \in H^p$, $R_k \in \overline{H_0^p}$, $\|Q_k\| \leq C\|P_k\|$, and C is a constant (independent of the polynomial P_k).

To see this, define for each $\zeta \in \mathbb{T}$,

$$Q_k^{(\zeta)}(z) := z^n \frac{P_k(z)}{z^n - \zeta}, \quad R_k^{(\zeta)}(z) := z^{-n} \frac{P_k(z)}{z^{-n} - \overline{\zeta}}, \quad z \in \mathbb{T},$$

where $n = n_k$ is any integer greater than the degree of P_k. By Fubini's theorem along with Lemma 6.1.2, observe that

$$\begin{aligned}
\int \|Q_k^{(\zeta)}\|^p \, dm(\zeta) &= \int_{|\zeta|=1} \int_{|z|=1} \left| \frac{z^n P_k(z)}{z^n - \zeta} \right|^p \, dm(z) \, dm(\zeta) \\
&= \int_{|z|=1} \int_{|\zeta|=1} \frac{|P_k(z)|^p}{|1 - \overline{z}^n \zeta|^p} \, dm(\zeta) \, dm(z) \\
&= \int_{|z|=1} |P_k(z)|^p \left\{ \int_{|\zeta|=1} \frac{1}{|1 - \overline{z}^n \zeta|^p} \, dm(\zeta) \right\} dm(z) \\
&= \int_{|z|=1} |P_k(z)|^p \left\| \frac{1}{1 - \zeta} \right\|^p \, dm(z) \\
&= \left\| \frac{1}{1 - z} \right\|^p \|P_k\|^p.
\end{aligned}$$

Thus, there is a number $\zeta \in \mathbb{T}$ such that

(6.1.9) $$\|Q_k^{(\zeta)}\|^p \leq \left\| \frac{1}{1 - z} \right\|^p \|P_k\|^p.$$

For this specially chosen $\zeta \in \mathbb{T}$, define $Q_k := Q_k^{(\zeta)}$ and $R_k := R_k^{(\zeta)}$. A simple observation shows that $Q_k \in H^p$, $R_k \in \overline{H_0^p}$ (since $n > \deg(P_k)$), and $P_k = Q_k + R_k$. Thus we have shown eq.(6.1.8), which proves Claim 2.

Finally, to show statement (1) (i.e., $f = u + v$, where $u \in H^p$ and $v \in \overline{H_0^p}$), we let

$$u = \sum_k Q_k, \quad v = \sum_k R_k.$$

---

[2]Indeed if $f \in L^p$ and g_n is the characteristic function of the set $\{|f| \leq n\}$, then by the dominated convergence theorem, $g_n f \to f$ in L^p. Now for fixed n, approximate the bounded function $g_n f$ in L^p by trigonometric polynomials (approximate first in L^1 and then note that $\|\cdot\|_p \leq \|\cdot\|_1$).

Notice (from Claim 2 and the fact that $\sum_k \|P_k\|^p < \infty$) that both sums converge in L^p norm, hence $u \in H^p$ and $v \in \overline{H_0^p}$. Also notice that $u + v = \sum_k P_k = f$. The proof of statement (1) is now complete.

To prove statement (2), let $f \in H^p \cap \overline{H_0^p}$ and $\varepsilon > 0$. Since the analytic polynomials are dense in H^p, there are analytic polynomials P_1 and P_2 with
$$\|f - P_1\| < \varepsilon \text{ and } \|f - \overline{z}P_2(\overline{z})\| < \varepsilon.$$
Set
$$P(z) := P_1(z) - \overline{z}P_2(\overline{z}).$$
It is easy to check that if $n > \max(\deg(P_1), \deg(P_2))$ and $\zeta \in \mathbb{T}$, the function
$$r^{(\zeta)}(z) := P_1(z) - z^n \frac{P(z)}{z^n - \zeta}, \quad z \in \mathbb{T},$$
is a (finite) linear combination of Cauchy kernels
$$\left\{ \frac{1}{1 - \overline{\zeta}z} : \zeta \in \mathbb{T} \right\}.$$
Also, just as in the first part of the proof, the number $\zeta \in \mathbb{T}$ can be chosen so that
$$\begin{aligned}\|r^{(\zeta)} - P_1\|^p &\leq \left\|\frac{1}{1-z}\right\|^p \|P\|^p \\ &\leq C_p \left\{ \|P_1 - f\|^p + \|f - \overline{z}P_2(\overline{z})\|^p \right\} \\ &\leq C_p \varepsilon^p.\end{aligned}$$
Therefore
$$\|r^{(\zeta)} - f\|^p \leq \|r^{(\zeta)} - P_1\|^p + \|P_1 - f\|^p \leq C_p \varepsilon^p.$$
This completes the proof of statement (2). $\square$

Adding to the already complicated nature of $H^p \cap \overline{H_0^p}$ is the fact that it can not be separated from its compliment by continuous linear functionals on H^p. From Theorem 3.5.31, the dual of H^p can be identified with $P\Lambda_{1/p-1}$ via the pairing
$$\lim_{r \to 1^-} \int_{\mathbb{T}} f(r\zeta)\overline{g(\zeta)}\, dm(\zeta).$$
Thus if g annihilates $H^p \cap \overline{H_0^p}$, then g annihilates all of the Cauchy kernels $(1 - \overline{\xi}z)^{-1}, \xi \in \mathbb{T}$, and so

(6.1.10) $$0 = \lim_{r \to 1^-} \int_{\mathbb{T}} \frac{1}{1 - \overline{\xi}r\zeta}\overline{g(\zeta)}\, dm(\zeta) = \lim_{r \to 1^-} \overline{g(r\xi)} = \overline{g(\xi)} \quad \forall \xi \in \mathbb{T}.$$

Hence $(H^p \cap \overline{H_0^p})^\perp = (0)$. This makes characterizing the B-invariant subspaces of H^p by relating them to a certain S-invariant subspaces of the dual space, as in Chapter 5, unworkable for $p \in (0, 1)$.

To characterize the B-invariant subspaces of H^p, Aleksandrov takes a direct approach and discovers that the B-invariant subspaces $\mathcal{E} \subset H^p$ $(0 < p < 1)$ depend on three parameters

1. An inner function I
2. A closed set $F \subset \mathbb{T}$ with $\sigma(I) \cap \mathbb{T} \subset F$ [3]

---

[3] Recall that $\sigma(I) = \{ z \in \mathbb{D}^- : \liminf_{\lambda \to z} |I(\lambda)| = 0 \}$ is the spectrum of the inner function I.

3. A function $k : F \to [1, n_p] \cap \mathbb{N}$ with the additional property
$$k|\sigma(I) \cup (F \backslash F_0) = n_p,$$
where F_0 are the isolated points of F. Such a function k is called (I, p)-permissible.

DEFINITION 6.1.11. $\mathcal{E}^p(I, F, k)$ is the set of $f \in H^p$ with

(6.1.12) $$f \in H^p \cap I\overline{H_0^p}.$$

(6.1.13) f has an analytic continuation to a neighborhood of $\mathbb{T} \backslash F$.

(6.1.14) f has a pole of order not more than $k(\zeta)$ for $\zeta \in F_0 \backslash \sigma(I)$.

REMARK 6.1.15.   1. It is clear that $\mathcal{E}^p(I, F, k)$ is a B-invariant linear manifold of H^p. Furthermore, since $\mathcal{E}^p(I, F, k) \subset H^p \cap I\overline{H_0^p}$, every $f \in \mathcal{E}^p(I, F, k)$ has the property that f/I has a pseudocontinuation belonging to $H^p(\mathbb{D}_e)$ which vanishes at infinity (see Remark 5.2.1).

2. As mentioned in Remark 5.2.3,
$$\frac{1}{(1-\bar{\zeta}z)^j} \in H^p \cap I\overline{H_0^p}, \; \forall \zeta \in \mathbb{T}, \; j = 1, \cdots, n_p,$$
which is why the restrictions in eq.(6.1.13) and eq.(6.1.14) are needed in the description of $\mathcal{E}^p(I, F, k)$.

3. The manifold $\mathcal{E}^p(I, F, k)$ is finite dimensional only if I is a finite Blaschke product and F is a finite set.

4. Directly from the definition of $\mathcal{E}^p(I, F, k)$, it follows that
$$\mathcal{E}^p(I_1, F_1, k_1) \subset \mathcal{E}^p(I_2, F_2, k_2) \quad \Leftrightarrow \quad I_2/I_1 \in H^\infty, \; F_1 \subset F_2, \; k_1 \leq k_2|F_1.$$
Therefore $\mathcal{E}^p(I_1, F_1, k_1) = \mathcal{E}^p(I_2, F_2, k_2)$ if and only if $F_1 = F_2$, $k_1 = k_2$, and I_2/I_1 is a constant.

The main theorem of this chapter is due to Aleksandrov [3] and says that $\mathcal{E}^p(I, F, k)$ is closed and every B-invariant subspace of H^p ($0 < p < 1$) takes this form.

THEOREM 6.1.16 (Aleksandrov [3]). Let $p \in (0, 1)$.

1. $\mathcal{E}^p(I, F, k)$ is a proper B-invariant subspace of H^p.
2. For every non-trivial B-invariant subspace $\mathcal{E} \subset H^p$, there is a unique triple (I, F, k) such that $\mathcal{E} = \mathcal{E}^p(I, F, k)$.

For a given B-invariant subspace $\mathcal{E}$, the parameters I, F, and k will be determined as follows: Since the H^2 norm dominates the H^p norm, $\mathcal{E} \cap H^2$ is a closed B-invariant subspace of H^2 and so, by Theorem 5.1.4, is of the form $H^2 \cap I\overline{H_0^2}$ [4]. Set $I_{\mathcal{E}} = I$. The set $F = F_{\mathcal{E}}$ is defined by
$$F := \left\{ \zeta \in \mathbb{T} : \frac{1}{1 - \bar{\zeta}z} \in \mathcal{E} \right\}.$$

---

[4]The inner function I can be taken to be the constant function $I = 1$ here which yields $H^2 \cap \overline{H_0^2} = (0)$ (by the F. and M. Riesz theorem). This will correspond to the case when $\mathcal{E} \cap H^2 = (0)$.

The function $k = k_{\mathcal{E}} : F \to [1, n_p] \cap \mathbb{N}$ is defined by

$$k(\zeta) = \max \left\{ s \in \mathbb{N} : \frac{1}{(1 - \overline{\zeta}z)^s} \in \mathcal{E} \right\}$$

and can be shown to be (I, p)-permissible. It will turn out that if

$$e^p(I, F, k) := \left(H^2 \cap I\overline{H_0^2} \right) \bigvee \left\{ \frac{1}{(1 - \overline{\zeta}z)^s} : \zeta \in F, s = 1, \cdots, k(\zeta) \right\}$$

then

$$e^p(I, F, k) \subset \mathcal{E} \subset \mathcal{E}^p(I, F, k).$$

To finish, we need to solve the (difficult) approximation problem

$$e^p(I, F, k) = \mathcal{E}^p(I, F, k).$$

Our presentation of Aleksandrov's result will be in three parts. In the first part, called "The parameters", we will discuss how the parameters I, F, and k interact with each other and determine the B-invariant subspace $\mathcal{E}$. In the second part, called "A reduction", we focus on the approximation problem $e^p(I, F, k) = \mathcal{E}^p(I, F, k)$ and show how this problem can be reduced to showing $e^p(1, F, k) = \mathcal{E}^p(1, F, k)$. The third part, called "Rational approximation", will rephrase the rational approximation problem $e^p(1, F, k) = \mathcal{E}^p(1, F, k)$ for the Hardy space of the upper half-plane where the tools of distributions and the atomic decomposition from Chapter 4 can be employed. Special cases of this problem (for example when $k \equiv n_p$) were solved by Aleksandrov in an earlier paper [**2**] and those results will be needed and developed here. To understand these approximation results, the reader might want to review Chapter 4 on the Hardy spaces of the upper half-plane, especially Coifman's atomic decomposition theorem.

6.2. The parameters

6.2.1. The space $\mathcal{E}^p(I, F, k)$. We begin our analysis with a discussion of the basic properties of $\mathcal{E}^p(I, F, k)$. Clearly $\mathcal{E}^p(I, F, k)$ is a B-invariant linear manifold of H^p.

PROPOSITION 6.2.1. *The linear manifold $\mathcal{E}^p(I, F, k)$ is closed in H^p.*

PROOF. To prove this result, we will show that for

$$a \notin F \cup I^{-1}(\{\infty\}) \ [5]$$

there is a $C_a > 0$, independent of $f \in \mathcal{E}^p(I, F, k)$, such that

(6.2.2) $$|f(a)| \leq C_a \|f\|.$$

We will also show that for each $\zeta \in F_0 \setminus \sigma(I)$, there is a $C_{a,\zeta} > 0$ such that for all $f \in \mathcal{E}^p(I, F, k)$,

(6.2.3) $$|a - \zeta|^{k(\zeta)} |f(a)| \leq C_{a,\zeta} \|f\|.$$

Using the definition of $\mathcal{E}^p(I, F, k)$, it follows from eq.(6.2.2) and eq.(6.2.3) that $\mathcal{E}^p(I, F, k)$ is closed. [6]

---

[5] Note here that I has been extended via its pseudocontinuation $I(z) = 1/\overline{I(1/\overline{z})}$, $z \in \mathbb{D}_e$, to be a meromorphic function on $\mathbb{C}_\infty \setminus \mathbb{T}$ with values in $\mathbb{C}_\infty$.

[6] The constant C_a is bounded in a small neighborhood of a and the constant $C_{a,\zeta}$ remains bounded as $a \to \zeta$.

6.2. THE PARAMETERS

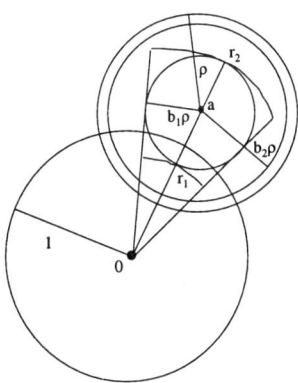

FIGURE 1. The "box" type region $R(a, b_1\rho)$ along with $B(a, b_1\rho)$ and $B(a, b_2\rho)$.

To prove eq.(6.2.2), we let
$$\rho = \rho_{F,I}(a) = \text{dist}\left(a,\ F \cup I^{-1}(\{\infty\})\right).$$
For $z \in \mathbb{C}$ and $r > 0$, let $B(z,r) = \{w \in \mathbb{C} : |w - z| < r\}$. Choose $0 < b_1 < b_2 < 1$ so that
$$B(a, b_1\rho) \subset R(a, b_1\rho) \subset B(a, b_2\rho),$$
where
$$R(a, b_1\rho) := \{\ se^{i\theta} : r_1 \le s \le r_2, \theta_1 \le \theta \le \theta_2\ \},$$
is the "box" type region which circumscribes $B(a, b_1\rho)$ and consists of part of a sector with vertex at the origin (see Figure 1).

Also define
$$(6.2.4) \qquad I_F^*(a) := \max\left\{\ 1,\ \max\{|I(z)| : z \in B(a, b_2\rho)\}\ \right\}.$$

If $f \in \mathcal{E}^p(I, F, k)$, then $|f|^p$ is subharmonic [7] in a neighborhood of $B(a, b_1\rho)$ and so, using the sub-mean value property of subharmonic functions, we obtain
$$(6.2.5) \qquad |f(a)|^p \le \frac{1}{|B(a, b_1\rho)|} \int_{B(a,b_1\rho)} |f|^p dx\, dy = \frac{C}{\rho^2} \int_{B(a,b_1\rho)} |f|^p dx\, dy.$$
and so
$$|f(a)|^p \le \frac{C}{\rho^2} \int_{R(a,b_1\rho)} |f|^p dx\, dy$$
which is bounded above by
$$(6.2.6) \quad \frac{C}{\rho^2} \int_{[r_1,1]} \int_{[\theta_1,\theta_2]} |f(se^{i\theta})|^p d\theta s ds + \frac{CI_F^*(a)^p}{\rho^2} \int_{[1,r_2]} \int_{[\theta_1,\theta_2]} |\frac{f}{I}(se^{i\theta})|^p d\theta s ds.$$
The first integral above is bounded above by
$$(6.2.7) \qquad \frac{C}{\rho^2}(1 - r_1^2)\|f\|^p.$$

---

[7] See [31], p. 7, or [39], p. 33, for the definition and the basic properties of subharmonic functions.

To estimate the second integral in eq.(6.2.6), we notice that
$$f \in \mathcal{E}^p(I, F, k) \subset H^p \cap I\overline{H_0^p}$$
and so, by Remark 5.2.3, f/I belongs to $H^p(\mathbb{D}_e)$. Thus for $s \in (1, r_2]$,
$$\int_{[\theta_1, \theta_2]} |\frac{f}{I}(se^{i\theta})|^p \, d\theta \leq \int_{\mathbb{T}} |f/I|^p d\theta = \|f\|^p.$$
Hence, the second integral in eq.(6.2.6) is bounded above by
$$(6.2.8) \qquad \frac{C}{\rho^2} I_F^*(a)^p (r_2^2 - 1) \|f\|^p.$$
Combining eq.(6.2.5) with eq.(6.2.7) and eq.(6.2.8) we get
$$\begin{aligned} |f(a)|^p &\leq \frac{C}{\rho^2}(1 - r_1^2)\|f\|^p + \frac{C}{\rho^2} I_F^*(a)^p (r_2^2 - 1)\|f\|^p \\ &\leq \frac{C}{\rho^2} I_F^*(a)^p (1 - r_1)\|f\|^p + \frac{C}{\rho^2} I_F^*(a)^p (r_2 - 1)\|f\|^p \\ &\leq \frac{C}{\rho^2} I_F^*(a)^p (r_2 - r_1)\|f\|^p \\ &\leq \frac{C}{\rho} I_F^*(a)^p \|f\|^p, \end{aligned}$$
which is the inequality in eq.(6.2.2).

To prove eq.(6.2.3), we use (as before) the fact that $|f|^p |z - \zeta|^{pk(\zeta)}$ is subharmonic to obtain
$$|f(a)||a - \zeta|^{k(\zeta)} \leq C \frac{I_F^{\zeta *}(a)}{\rho^{1/p}} \|f\|,$$
where
$$\rho = \rho_{F,I}^\zeta(a) = \text{dist}\left(a, \; F \cup I^{-1}(\{\infty\}) \setminus \{\zeta\} \right),$$
and $I_F^{\zeta *}(a)$ is appropriately (re)defined as in eq.(6.2.4). $\square$

6.2.2. The inner function $I = I_\mathcal{E}$. We begin our study of how the inner function I affects the properties of $\mathcal{E}^p(I, F, k)$ by first focusing our attention a bit more on $H^p \cap I\overline{H_0^p}$.

PROPOSITION 6.2.9. *Let $f \in H^p \cap I\overline{H_0^p}$. Then for almost every $\zeta \in \mathbb{T}$, there is a $g \in H^p \cap \overline{H_0^p}$ such that $f = g(I - \zeta)$.*

PROOF. Define
$$g^{(\zeta)}(z) := \frac{f(z)}{I(z) - \zeta}, \quad z \in \mathbb{T}$$
and observe that the proof of eq.(6.1.9) shows
$$\int_{\mathbb{T}} \|g^{(\zeta)}\|_{L^p}^p \, dm(\zeta) = \|\frac{1}{1-z}\|^p \|f\|^p < \infty.$$
Thus, for almost every $\zeta \in \mathbb{T}$, $g^{(\zeta)} \in L^p$. Furthermore, $\Re(1 - \overline{\zeta}I) > 0$ and so $(1 - \overline{\zeta}I)$ is outer (see Remark 3.2.5), from which $g^{(\zeta)} \in N^+ \cap L^p = H^p$.

Since $f \in H^p \cap I\overline{H_0^p}$, then $f = I\overline{h}$ almost everywhere on $\mathbb{T}$ for some $h \in H_0^p$. Thus
$$g^{(\zeta)} = \frac{f}{I - \zeta} = \frac{\overline{h}}{1 - \overline{I}\zeta}$$

and one argues as before to conclude $g^{(\zeta)} \in \overline{H_0^p}$. □

COROLLARY 6.2.10. *Let I be an inner function. Then the following statements are true.*
$$H^p \cap I\overline{H_0^p} = H^p \cap \overline{H_0^p} + I(H^p \cap \overline{H_0^p}),$$
$$H^p \cap I\overline{H_0^p} = \bigcup_{\zeta \in \mathbb{T}} (I - \zeta)(H^p \cap \overline{H_0^p}).$$

Let $\mathcal{E}$ be a B-invariant subspace of H^p with $\mathcal{E} \neq H^p$. Then, it is easy to check that $\mathcal{E} \cap H^2$ is a (closed) B-invariant subspace of H^2 which, by Theorem 5.1.4, is equal to $H^2 \cap I\overline{H_0^2}$. If $\mathcal{E} \cap H^2 = (0)$, we set the inner function I equal to the constant function one and note that $H^2 \cap \overline{H_0^2} = (0)$ by the F. and M. Riesz theorem.

THEOREM 6.2.11. *If $\mathcal{E}$ is a B-invariant subspace of H^p, then $\mathcal{E} \subset H^p \cap I\overline{H_0^p}$.*

COROLLARY 6.2.12. *A B-invariant subspace $\mathcal{E}$ of H^p is contained in $H^p \cap \overline{H_0^p}$ if and only if $\mathcal{E} \cap H^2 = \{0\}$.*

The proof of Theorem 6.2.11 will require several additional pieces of information. The first is an estimate on the growth of the Taylor coefficients of H^p functions and can be found in [**31**], p. 98.

LEMMA 6.2.13. *If $p \in (0,1)$ and $f = \sum_n a_n z^n \in H^p$, then*
$$|a_n| \leq c_p(1+n)^{1/p-1}\|f\| \quad \forall\, n \in \mathbb{N} \cup \{0\}.$$

LEMMA 6.2.14. *If $p \in (0,1)$ and $f \in H^p$, then*
$$\|B^n f\| \leq c_p\|f\|(1+n)^{2/p-1} \quad \forall\, n \in \mathbb{N} \cup \{0\}.$$

PROOF. If $f = \sum a_k z^k \in H^p$, a power series computation yields
$$(B^n f)(\zeta) = \frac{1}{\zeta^n}\left\{f(\zeta) - \sum_{k=0}^{n-1} a_k \zeta^k\right\},$$
for almost every $\zeta \in \mathbb{T}$ and so
$$\begin{aligned}
\|B^n f\|^p &\leq \|f\|^p + \sum_{k=0}^{n-1} |a_k|^p \\
&\leq c_p\|f\|^p\Big(1 + \sum_{k=0}^{n-1}(1+k)^{1-p}\Big) \text{ from Lemma 6.2.13}\\
&\leq c_p\|f\|^p(1 + n \cdot n^{1-p})\\
&\leq c_p\|f\|^p(n+1)^{2-p}.
\end{aligned}$$
□

Recall from Theorem 4.4.8 that for a function $f \in H^p$ ($0 < p < \infty$), the non-tangential maximal function
$$f^*(\zeta) = \sup_{z \in \Gamma_\alpha(\zeta)} |f(z)|$$
satisfies the inequality $\|f^*\|_{L^p} \leq c_{\alpha,p}\|f\|$. Here α (the opening of the vertex) will be fixed.

LEMMA 6.2.15. *Let* $f \in \mathcal{E}$, $h \in H^\infty$, *with* $hf^* \in L^2$. *Then* $P(\overline{h}f) \in \mathcal{E} \cap H^2$.

PROOF. The proof of this result is very similar to the proof of Proposition 5.8.5 but we record it here anyway. From Fatou's theorem (Theorem 3.2.3), $f_r \to f$ almost everywhere as $r \to 1^-$ and furthermore $|\overline{h}f_r - \overline{h}f| \leq 2|h|f^*$ almost everywhere on $\mathbb{T}$. By the dominated convergence theorem

$$\overline{h}f_r \to \overline{h}f \text{ in } L^2 \text{ as } r \to 1^-. \tag{6.2.16}$$

The continuity of the Riesz projections on L^2 (Theorem 3.3.5) says,

$$(I_d - P)\overline{h}f_r \to (I_d - P)\overline{h}f \text{ in } L^2 \text{ as } r \to 1^-$$

and so, using Lemma 5.8.4,

$$((I_d - P)\overline{h}f_r)_r \to (I_d - P)\overline{h}f \text{ in } L^2 \text{ as } r \to 1^-. \tag{6.2.17}$$

Note that $\overline{h_r}f \to \overline{h}f$ almost everywhere as $r \to 1^-$ and

$$|\overline{h_r}f - \overline{h}f| \leq 2\|h\|_\infty |f|$$

almost everywhere on $\mathbb{T}$. By the dominated convergence theorem,

$$\overline{h_r}f \to \overline{h}f \text{ in } L^p \text{ as } r \to 1^-. \tag{6.2.18}$$

From the identity

$$P(\overline{h}f) = \overline{h}f - (I_d - P)\overline{h}f$$

along with eq.(6.2.17) and eq.(6.2.18), one sees that

$$P(\overline{h}f) = \lim_{r \to 1^-} \left\{ \overline{h_r}f - ((I_d - P)\overline{h}f_r)_r \right\}, \tag{6.2.19}$$

where the limit is in L^p.

Let $h(z) = \sum_n b_n z^n$ and $f = \sum_n a_n z^n$. For fixed $r \in (0,1)$, the series

$$\sum_{n=0}^\infty \overline{b_n} r^n B^n f$$

converges in L^p since, by Lemma 6.2.14, $\|B^n f\| = O((1+n)^{2/p-1})$ and $|b_n| = o(1)$. Moreover, since $B\mathcal{E} \subset \mathcal{E}$, this series converges to a function belonging to $\mathcal{E}$.

Finally,

$$\sum_{n=0}^{\infty} \overline{b_n} r^n B^n f$$

$$= \sum_{n=0}^{\infty} \overline{b_n} r^n \overline{\zeta}^n \left\{ f - \sum_{j=0}^{n-1} a_j \zeta^j \right\}$$

$$= \sum_{n=0}^{\infty} \overline{b_n} r^n \overline{\zeta}^n f - \sum_{n=0}^{\infty} \overline{b_n} r^n \overline{\zeta}^n \sum_{j=0}^{n-1} a_j \zeta^j$$

$$= \sum_{n=0}^{\infty} \overline{b_n} r^n \overline{\zeta}^n f - \sum_{j=0}^{\infty} a_j \zeta^j \sum_{n=j+1}^{\infty} \overline{b_n} r^n \overline{\zeta}^n$$

$$= \overline{h_r} f - \sum_{j=0}^{\infty} a_j \sum_{n=j+1}^{\infty} \overline{b_n} r^n \zeta^{j-n}$$

$$= \overline{h_r} f - \sum_{l=-\infty}^{-1} r^{-l} \zeta^l \sum_{j=0}^{\infty} a_j \overline{b_{j-l}} r^j$$

$$= \overline{h_r} f - ((I_d - P)\overline{h} f_r)_r.$$

From this and eq.(6.2.19), we conclude $P(\overline{h}f) \in \mathcal{E} \cap H^2$. □

We are now ready to prove the main result of this subsection, Theorem 6.2.11.

PROOF OF THEOREM 6.2.11. Let $f \in \mathcal{E}$ and define the bounded outer function h with boundary modulus function[8]

$$|h(\zeta)| := \begin{cases} 1/f^*(\zeta) & \text{if } f^*(\zeta) \geq 1, \\ 1 & \text{if } f^*(\zeta) < 1. \end{cases}$$

Clearly $hf^* \in L^\infty$ and moreover,

(6.2.20) $$\overline{h}f = P(\overline{h}f) + (I_d - P)(\overline{h}f).$$

By Lemma 6.2.15, the first term above belongs to $\mathcal{E} \cap H^2 = H^2 \cap I\overline{H_0^2}$ and so $P(\overline{h}f) = I\overline{g}$ for some $g \in \overline{H_0^2}$. By the continuity of the Riesz projection (Theorem 3.3.5), the second term belongs to $\overline{H_0^2}$. Dividing by $\overline{h}$ in eq.(6.2.20) yields

(6.2.21) $$f = I \left\{ \overline{\frac{g}{h}} + \frac{I}{h} \overline{(I_d - P)(\overline{h}f)} \right\}.$$

Since h is an outer function,

$$\overline{\frac{g}{h}} + \frac{I}{h} \overline{(I_d - P)(\overline{h}f)} \in N^+$$

and is zero at $z = 0$. Furthermore, by eq.(6.2.21), the above function has L^p boundary values and so belongs to H_0^p. Thus $f \in H^p \cap I\overline{H_0^p}$ and the proof is complete. □

---

[8] The existence of such an outer function follows from the equality $\int \log|h(\zeta)| dm(\zeta) = -\int \log^+ f^*(\zeta) dm(\zeta)$ and the estimate

$$\int \log^+ f^*(\zeta) dm(\zeta) = \frac{1}{p} \int \log^+ f^*(\zeta)^p dm(\zeta) \leq \frac{1}{p} \int f^*(\zeta)^p dm(\zeta) \leq c_p \|f\|^p.$$

Now apply the Nevanlinna factorization theory, Theorem 3.2.4.

6.2.3. The closed set F. For a B-invariant subspace $\mathcal{E}$ of H^p ($0 < p < 1$), let
$$F = F_{\mathcal{E}} := \Big\{ \zeta \in \mathbb{T} : \frac{1}{1 - \overline{\zeta}z} \in \mathcal{E} \Big\}.$$

THEOREM 6.2.22. 1. F is closed.
2. $\sigma(I) \cap \mathbb{T} \subset F$ if $\mathcal{E} \cap H^2 = H^2 \cap \overline{IH_0^2}$.
3. Every $f \in \mathcal{E}$ has an analytic continuation to a neighborhood of $\mathbb{T} \setminus F$.

We will prove this theorem in several parts since the different parts depend on several technical lemmas. Part 1 follows from the simple observation (which is really just the L^p continuity of translations) that if $\zeta_n \to \zeta_0$ ($\zeta_n \in \mathbb{T}$), then for each $p \in (0, 1)$,
$$\frac{1}{1 - \overline{\zeta_n}z} \to \frac{1}{1 - \overline{\zeta_0}z} \quad \text{in } H^p.$$

We remark (and will use several times later) the following more general fact.

LEMMA 6.2.23. *Let $\{a_n\} \subset \mathbb{D}^-$ with $a_n \to 1$. Then for each $p \in (0, 1)$,*
$$\frac{1}{1 - \overline{a_n}\zeta} \to \frac{1}{1 - \zeta} \quad \text{in } L^p.$$

PROOF. If $a_n = r_n e^{it_n}$, then
$$\begin{aligned}
|1 - \overline{a_n}e^{i\theta}|^p &= \big\{ (1 - r_n)^2 + 2r_n \big(1 - \cos(\theta - t_n)\big) \big\}^{p/2} \\
&\geq C \big\{ (1 - r_n)^2 + r_n(\theta - t_n)^2 \big\}^{p/2} \\
&\geq C(\theta - t_n)^p.
\end{aligned}$$

Thus for any measurable set $E \subset \mathbb{T}$,
$$\int_E \frac{1}{|1 - \overline{a_n}e^{i\theta}|^p} d\theta \leq C \int_{E_n} \frac{1}{|\theta|^p} d\theta, \tag{6.2.24}$$
where E_n is the translate in the circle of E by t_n (note that $|E_n| = |E|$).

If $\varepsilon > 0$ is given, then
$$\int_{\mathbb{T}} \Big| \frac{1}{1 - \overline{a_n}e^{i\theta}} - \frac{1}{1 - e^{i\theta}} \Big|^p \frac{d\theta}{2\pi} = \int_{|\theta| > \varepsilon} + \int_{|\theta| < \varepsilon}.$$

The first integral goes to zero as $n \to \infty$. By eq.(6.2.24), the second integral is bounded by $C_p \varepsilon^{1-p}$ uniformly in n. $\square$

The proof of statement (2) of Theorem 6.2.22 will depend on the following technical lemma which proves slightly more than we need right now but will be very instrumental later.

LEMMA 6.2.25. *Let I be an inner function. Assume*
$$a \in \mathbb{C} \setminus \big\{ 1/\overline{z} : z \in \sigma(I) \big\}$$
and let $I(\eta) = \sum_{j=0}^{\infty} c_j(\eta - a)^j$ be the series expansion of I near a. [9]
1. *For $m \in \mathbb{N}$, the function*
$$h(z) := \frac{I(z) - \sum_{j=0}^{m-1} c_j(z - a)^j}{(z - a)^m}, \quad z \in \mathbb{D},$$
belongs to $H^\infty \cap \overline{IH_0^\infty}$.

---
[9] Recall from Remark 3.2.5 that I is analytic on the set $\mathbb{C} \setminus \{1/\overline{z} : z \in \sigma(I)\}$.

2. *The functions*
$$\frac{1}{(1-\bar{\zeta}z)^k} \quad \text{and} \quad \frac{I}{(1-\bar{\zeta}z)^k}$$
belong to the L^p $(0 < p < 1)$ closure of $H^\infty \cap \overline{IH_0^\infty}$, whenever
$$\zeta \in \sigma(I) \cap \mathbb{T} \text{ and } k = 1, \cdots, n_p.$$

PROOF. Clearly h belongs to $H^\infty(\mathbb{D})$. Moreover, for almost every $\zeta \in \mathbb{T}$,
$$h(\zeta) = I(\zeta)\overline{\left\{\frac{\zeta^m - I(\zeta)\sum_{j=0}^{m-1} \overline{c_j}\zeta^{m-j}(1-\bar{a}\zeta)^j}{(1-\bar{a}\zeta)^m}\right\}} \in \overline{IH_0^\infty}.$$

This proves statement (1).

We first prove statement (2) in the special case where $k = 1$. Without loss of generality, it can be assumed that $\zeta = 1$. Since $1 \in \sigma(I)$, I is not continuous (but bounded) at the point 1 and so there are sequences $\{a_n : n \in \mathbb{N}\}$ and $\{b_n : n \in \mathbb{N}\}$ in $\mathbb{D}$ such that $a_n \to 1$ and $b_n \to 1$ but
$$d_1 = \lim_{n\to\infty} I(a_n) \neq \lim_{n\to\infty} I(b_n) = d_2.$$

By part (1),
$$\frac{I - I(a_n)}{z - a_n}, \quad \frac{I - I(b_n)}{z - b_n} \in H^\infty \cap \overline{IH_0^\infty}.$$

A routine computation using Lemma 6.2.23 shows that
$$\frac{I - I(a_n)}{z - a_n} \to \frac{I - d_1}{z - 1}, \quad \frac{I - I(b_n)}{z - b_n} \to \frac{I - d_2}{z - 1}$$

in H^p norm and so
$$\frac{1}{d_2 - d_1}\left\{\frac{I - d_1}{z - 1} - \frac{I - d_2}{z - 1}\right\} = \frac{1}{z - 1}$$

belongs to the H^p closure of $H^\infty \cap \overline{IH_0^\infty}$. Also notice that
$$\frac{I}{z - 1} = \frac{I - d_1}{z - 1} + \frac{d_1}{z - 1}$$

belongs to the H^p closure of $H^\infty \cap \overline{IH_0^\infty}$.

We now proceed to case where $k \geq 1$. First notice that if $1 \in \sigma(I)$ and $k \in \mathbb{N} \cap [1, 1/p]$, then
$$(6.2.26) \qquad I = \prod_{1 \leq j \leq k} I_j$$

where I_j are inner functions with $1 \in \sigma(I_j)$. [10]

Exactly as before,
$$\frac{I_j - I_j(a_n)}{z - a_n} \to \frac{I_j - d_1}{z - 1} \quad \text{and} \quad \frac{I_j - I_j(b_n)}{z - b_n} \to \frac{I_j - d_2}{z - 1}$$

---

[10] We look at two cases. Suppose $I = S_\mu \prod_{w \in A} B_w$, where B_w is a single Blaschke factor with zero at w (with corresponding multiplicity) and 1 is not in the support of μ. Then, since $1 \in \sigma(I)$, $1 \in A^-$. Partition the sequence A into k disjoint subsequences $A_1, \cdots, A_k$ with $1 \in A_j^-$ and set $I_j = S_\mu^{1/k} \prod_{w \in A_j} B_w$. Then I_j is inner and has the property eq.(6.2.26). Suppose, $I = BS_\mu$, where B is a Blaschke product and 1 belongs to the support of μ. Then, let $I_1 = BS_\mu^{1/k}$, $I_j = S_\mu^{1/k}$, $2 \leq j \leq k$.

in L^{pk} (note $pk < 1$). Thus $1/(z-1)$ and $I_j/(z-1)$ belong to the L^{pk} closure of $H^\infty \cap I_j \overline{H_0^\infty}$. It follows from the generalized Hölder's inequality [11] that $I/(1-z)^k$ belongs to the L^p closure of $H^\infty \cap I\overline{H_0^\infty}$. A similar proof shows $1/(1-z)^k$ belongs to the L^p closure of $H^\infty \cap I\overline{H_0^\infty}$. This proves statement (2). □

PROOF OF THEOREM 6.2.22, PART 2. By Lemma 6.2.25(2), $(1-\bar{\zeta}z)^{-1}$ belongs to the L^p closure of $H^2 \cap I\overline{H_0^2} = \mathcal{E} \cap H^2$ whenever $\zeta \in \sigma(I) \cap \mathbb{T}$. Thus $(1-\bar{\zeta}z)^{-1} \in \mathcal{E}$ and so $\zeta \in F$. □

To prove statement (3) of Theorem 6.2.22, we will need several technical lemmas involving convolutions. If $q_1, q_2, r \in [1,\infty]$ with $1/q_1 + 1/q_2 \geq 1$ and $f \in L^{q_1}$, $g \in L^{q_2}$, we can define the convolution

$$(f * g)(w) = \int_\mathbb{T} f(w\bar{\zeta})g(\zeta)\, dm(\zeta), \quad w \in \mathbb{T}.$$

From the Hausdorff-Young inequality [**106**], p. 37,

$$\|f * g\|_{L^r} \leq \|f\|_{L^{q_1}} \|g\|_{L^{q_2}}, \quad \frac{1}{r} = \frac{1}{q_1} + \frac{1}{q_2} - 1.$$

For what follows, we let χ_δ ($\delta > 0$) denote the characteristic function on the arc

$$\{e^{it} \in \mathbb{T} : -\delta \leq t \leq \delta\}.$$

LEMMA 6.2.27. *Let $q_1, q_2, r \in [1,\infty]$ with $1/r = 1/q_1 + 1/q_2 - 1$. If $f \in L^1$, $\chi_\delta f \in L^{q_1}$, and $g \in L^{q_2} \cap L^\infty_{loc}(\mathbb{T}\setminus\{1\})$, then*

$$(f * g)\chi_\tau \in L^r \quad \forall \tau \in (0, \delta).$$

PROOF. Observe that

$$\begin{aligned} f * g &= (f\chi_\delta) * g + f(1-\chi_\delta) * g(1-\chi_{\delta_1}) + f(1-\chi_\delta) * g\chi_{\delta_1} \\ &= h_1 + h_2 + h_3, \end{aligned}$$

where $\delta_1 = \delta - \tau$. Two applications of the Hausdorff-Young inequality yield $h_1 \in L^r$ and $h_2 \in L^\infty$. To finish the proof, observe $h_3|[-\tau,\tau] = 0$. □

COROLLARY 6.2.28. *Let $h \in L^1$, $g \in L^q \cap L^\infty_{loc}(\mathbb{T}\setminus\{1\})$ for some $q \in (1,\infty)$ and $g, h \geq 0$. If $h\chi_\delta \leq h * g$, then $h\chi_\tau \in L^\infty$ for all $\tau \in (0,\delta)$.*

PROOF. In this proof, we will be using the notation for q_1, q_2, and r from Lemma 6.2.27. Choose $n \in \mathbb{N}$ so that $q \geq q_2 := n/(n-1)$. In the notation of Lemma 6.2.27, let $q_1 := 1$ (and so $r = n/(n-1)$) and $f = h$. By the Hausdorff-Young inequality, h and g satisfy the hypothesis of the lemma and so $(h*g)\chi_\tau \in L^{n/(n-1)}$ for $\tau \in (0,\delta)$. Since $h\chi_\delta \leq h * g$, one has $h\chi_\tau \in L^{n/(n-1)}$. Now take $q_2 = n/(n-1)$ and $q_1 = n/(n-1)$ and note that $r = n/(n-2)$ and that h and g satisfy the hypothesis of the lemma. Thus as before, $h\chi_\tau \in L^{n/(n-2)}$ for $\tau \in (0,\delta)$. Now take $q_2 = n/(n-1)$ and $q_1 = n/(n-2)$ and conclude that $h\chi_\tau \in L^{n/(n-3)}$. Repeat this procedure for a total of n times to get $h\chi_\tau \in L^{n/(n-n)} = L^\infty$ for $\tau \in (0,\delta)$. □

---

[11]If for each $1 \leq l \leq k$ $f_{l,j} \to f_l$ as $j \to \infty$ in L^{pk}, then $f_{1,j} \cdots f_{k,j} \to f_1 \cdots f_k$ in L^p as $j \to \infty$.

COROLLARY 6.2.29. *Let $f \in H^p$ and suppose*

(6.2.30) $$|f(a)|^p \leq C \int \frac{|f(\zeta)|^p}{|\zeta - a|^p} \, dm(\zeta)$$

for all $a \in \mathbb{D}$ with $|a - 1| < \delta$. Then $f\chi_\tau \in L^\infty$ whenever $|1 - e^{i\tau}| < \delta$.

PROOF. The function $g(e^{it}) := C|1 - e^{it}|^{-p}$ belongs to $L^q \cap L^\infty_{loc}(\mathbb{T}\setminus\{1\})$ for $q \in (1, 1/p)$. By the Hausdorff-Young inequality $|f|^p * g \in L^q$. Moreover

$$\chi_\delta |f|^p \leq |f|^p * g$$

almost everywhere.[12] Now apply Corollary 6.2.28. □

PROOF OF THEOREM 6.2.22, PART 3. Recall that

$$F = F_\mathcal{E} := \Big\{ \zeta \in \mathbb{T} : \frac{1}{1 - \bar{\zeta}z} \in \mathcal{E} \Big\}.$$

We will show that every $f \in \mathcal{E}$ has an analytic continuation to a neighborhood of $\mathbb{T}\setminus F$. To do this, we will first show that for every $\zeta \in \mathbb{T}\setminus F$

$$\lim_{z \to \zeta, z \in \mathbb{D}} f(z) \quad \text{exists}.$$

Without loss of generality, it can be assumed that $1 \notin F$ and that $\zeta = 1$. As $1 \notin F$, then $(1-z)^{-1} \notin \mathcal{E}$ and so

(6.2.31) $$\inf_{f \in \mathcal{E}} \Big\| f - \frac{1}{1-z} \Big\| = \delta > 0.$$

Moreover, for each $f \in \mathcal{E}$ and $a \in \mathbb{D}$

$$\frac{f - f(a)}{z - a} \in \mathcal{E}.^{13}$$

From Lemma 6.2.23, $(z-a)^{-1} \to (z-1)^{-1}$ in L^p norm as $a \to 1$ ($a \in \mathbb{D}$) and so using eq.(6.2.31),

$$\inf_{f \in \mathcal{E}} \Big\| f - \frac{1}{\zeta - a} \Big\|_{L^p}^p = \delta_1 > 0$$

for $a \in \mathbb{D}$ sufficiently close to 1. Thus, for fixed $f \in \mathcal{E}$,

$$\int \Big| \frac{1}{f(a)} \frac{f(\zeta) - f(a)}{\zeta - a} + \frac{1}{\zeta - a} \Big|^p \, dm(\zeta) > \delta_1$$

for $a \in \mathbb{D}$ near 1. This implies

$$|f(a)|^p \leq \frac{1}{\delta_1} \int \frac{|f(\zeta)|^p}{|\zeta - a|^p} \, dm(\zeta)$$

for $a \in \mathbb{D}$ close to 1. By Corollary 6.2.29, $f\chi_\tau \in L^\infty$ for sufficiently small τ.

Now suppose

$$\lim_{z \to 1, z \in \mathbb{D}} f(z)$$

---

[12]If $(|f|^p * g)(e^{it}) < \infty$, one can use the proof of Lemma 6.2.23 to show

$$\lim_{r \to 1^-} \int C|f|^p |re^{it} - \zeta|^{-p} dm \to (|f|^p * g)(e^{it}).$$

Now use eq.(6.2.30) and the fact that $f_r \to f$ almost everywhere.

[13]The operator $f \to (f - f(a))/(z - a)$ is continuous on H^p and a routine computation shows that, at least formally, $(I_d - aB)^{-1}Bf = (f - f(a))/(z - a)$. From Lemma 6.2.14, $\|B^n f\| = O((1+n)^{2/p-1})$, and so the series $\sum_{n=0}^\infty a^n B^{n+1} f$ converges in H^p. Since $B\mathcal{E} \subset \mathcal{E}$, the series converges in $\mathcal{E}$. Thus $(f - f(a))/(z - a) = (I_d - aB)^{-1}Bf = \sum_{n=0}^\infty a^n B^{n+1} f \in \mathcal{E}$.

does not exist. Then there are sequences $\{a_n : n \in \mathbb{N}\}$ and $\{b_n : n \in \mathbb{N}\} \subset \mathbb{D}$ such that $a_n \to 1$ and $b_n \to 1$ but

$$A = \lim f(a_n) \ne B = \lim f(b_n).\quad {}^{14}$$

Since $f\chi_\tau \in L^\infty$, Lemma 6.2.23 yields

$$\frac{f - f(a_n)}{z - a_n} \to \frac{f - A}{z - 1} \quad \text{and} \quad \frac{f - f(b_n)}{z - b_n} \to \frac{f - B}{z - 1} \quad \text{in } H^p.$$

But since

$$\frac{f - f(a_n)}{z - a_n}, \frac{f - f(b_n)}{z - b_n} \in \mathcal{E}$$

and $A \ne B$, then

$$\frac{1}{z - 1} = \frac{1}{B - A} \left\{ \frac{f - A}{z - 1} - \frac{f - B}{z - 1} \right\} \in \mathcal{E},$$

contradicting the fact that $1 \not\in F$. Thus the limit of $f(z)$ as $z \to \zeta$ exists for all $\zeta \in \mathbb{T} \setminus F$.

A little argument shows that f can be extended to be continuous on $\mathbb{D}^- \setminus F$. Furthermore, from Theorem 6.2.11, $\mathcal{E} \subset H^p \cap I\overline{H_0^p}$ and so $f \in \mathcal{E}$ has a meromorphic pseudocontinuation T_f on $\mathbb{D}_e$ given by

$$T_f(z) = T_I(z)\overline{h}(1/\overline{z}),$$

where $h \in H_0^p$. If J is an arc in $\mathbb{T} \setminus F$, then J does not intersect $\sigma(I)$ and so T_I is analytic on

$$G_1 := \{s\zeta : 1 < s < s_0, \zeta \in J\}.$$

Hence T_f is analytic on G_1 as well. By the above, T_f has continuous boundary values on ∂G_1 and so T_f is bounded on G_1. ^{15} As argued before, f is continuous on

$$G_2 := \{r\zeta : 0 < r \le 1, \zeta \in J\}$$

and so by Morera's theorem (Lemma 5.2.4 and Remark 5.2.5), T_f is an analytic continuation of f across J. □

6.2.4. The function $k_\mathcal{E}$. For a B-invariant subspace $\mathcal{E} \subset H^p$ we have defined an inner function $I = I_\mathcal{E}$ ($\mathcal{E} \cap H^2 = H^2 \cap I\overline{H_0^2}$) and a closed set

$$F = F_\mathcal{E} = \left\{ \zeta \in \mathbb{T} : (1 - \overline{\zeta}z)^{-1} \in \mathcal{E} \right\}.$$

Recall that F contains $\sigma(I) \cap \mathbb{T}$.

We now define a function $k = k_\mathcal{E}$ which determines the order of the pole of a function in $\mathcal{E}$ at the point ζ. More precisely, for $\zeta \in F$ define

$$k(\zeta) = k_\mathcal{E}(\zeta) := \max\left\{ s \in [1, n_p] \cap \mathbb{N} : \frac{1}{(1 - \overline{\zeta}z)^s} \in \mathcal{E} \right\}$$

and recall from the introduction that a general function $k : F \to [1, 1/p] \cap \mathbb{N}$ is (I, p)-*permissible* if

$$k(\zeta) = n_p \ \forall \, \zeta \in (\sigma(I) \cap \mathbb{T}) \cup (F \setminus F_0),$$

---

[14] Since f is bounded on the arc $\{e^{i\theta} : -\tau < \theta < \tau\}$, f is bounded on the sector formed from this arc. This is a "local" version of $N^+ \cap L^\infty = H^\infty$. From here it follows that such *finite* A and B exist.

[15] T_f is "H^p" on G_1 and has continuous boundary values. From here it follows that T_f is bounded on G_1 (again a local version of $N^+ \cap L^\infty = H^\infty$).

where F_0 are the isolated points of F.

LEMMA 6.2.32. *The function $k = k_\mathcal{E}$ as defined above is (I, p)-permissible.*

PROOF. By the definition of (I, p)-permissible, we must check that $k(\zeta) = n_p$ for all $\zeta \in (\sigma(I) \cap \mathbb{T}) \cup (F \backslash F_0)$.[16] By Lemma 6.2.25 (part 2), $(1 - \overline{\zeta}z)^{-j}$, $j = 1, \cdots, n_p$, belongs to the L^p closure of $H^\infty \cap I\overline{H_0^\infty}$ when $\zeta \in \sigma(I) \cap \mathbb{T}$. But

$$H^\infty \cap I\overline{H_0^\infty} \subset H^2 \cap I\overline{H_0^2} = \mathcal{E} \cap H^2$$

and so $(1 - \overline{\zeta}z)^{-j} \in \mathcal{E}$. Thus $k|\sigma(I) \cap \mathbb{T} = n_p$.

To finish, we will show that $k(\zeta) = n_p$ on $F \backslash F_0$ (the accumulation points of F). Since $\zeta \in F$, we know that $(1 - \overline{\zeta}z)^{-1} \in \mathcal{E}$ (by the definition of $F = F_\mathcal{E}$). We will show that if $s = 1, \cdots, n_p - 1$ then

$$(6.2.33) \qquad \frac{1}{(1 - \overline{\zeta}z)^l} \in \mathcal{E} \ \forall \, l = 1, \cdots, s \ \Rightarrow \ \frac{1}{(1 - \overline{\zeta}z)^{s+1}} \in \mathcal{E}.$$

Without loss of generality, we can assume $\zeta = 1$. For $w \in F \backslash \{1\}$,

$$\frac{1}{(1 - \overline{w}z)(1 - z)^s} - \frac{1}{(1 - \overline{w}z)(1 - w)^s}$$

$$= \frac{1}{1 - \overline{w}z} \left\{ \frac{(1 - w)^s - (1 - z)^s}{(1 - w)^s (1 - z)^s} \right\}$$

$$= \frac{1}{1 - \overline{w}z} \frac{z - w}{(1 - w)^s (1 - z)^s} \sum_{j=0}^{s-1} (1 - w)^j (1 - z)^{s-j-1}$$

$$= \frac{-w}{(1 - w)^s} \sum_{j=0}^{s-1} (1 - w)^j \frac{1}{(1 - z)^{j+1}} \in \mathcal{E},$$

by our inductive hypothesis in eq.(6.2.33). By the definition of $F = F_\mathcal{E}$, the function

$$\frac{1}{(1 - \overline{w}z)(1 - w)^s}$$

belongs to $\mathcal{E}$ and so

$$\frac{1}{(1 - \overline{w}z)(1 - z)^s} \in \mathcal{E}.$$

A similar proof to that of Lemma 6.2.23 shows

$$\frac{1}{(1 - \overline{w}z)(1 - z)^s} \to \frac{1}{(1 - z)^{s+1}} \quad \text{in } H^p \text{ as } w \to 1, w \in F.$$

Thus $(1 - z)^{-s-1} \in \mathcal{E}$ which shows eq.(6.2.33). □

By Remark 6.1.15, every $f \in \mathcal{E} \subset H^p \cap I\overline{H_0^p}$ has a pseudocontinuation of bounded type across $\mathbb{T}$ given by

$$T_f(z) = T_I(z)\overline{h}(1/\overline{z}),$$

where $h \in H_0^p$. If $\zeta \in F_0 \backslash \sigma(I)$, T_f will be analytic on $\{|z - \zeta| < \delta\} \cap \mathbb{D}_e$ for some $\delta > 0$. Theorem 6.2.22 says $f \in \mathcal{E}$ will have an analytic continuation across every arc of $\mathbb{T}$ which does not intersect F. Putting this all together, we conclude that f must have (at most) an isolated singularity about ζ. The final result of this subsection determines the order of this singularity.

---

[16]$k(\zeta)$ can be any other value in $[1, n_p] \cap \mathbb{N}$ on the set $F_0 \backslash \sigma(I)$.

LEMMA 6.2.34. *If $\zeta \in F_0 \setminus \sigma(I)$, then every $f \in \mathcal{E}$ has a pole of order at most $k(\zeta)$ at ζ.*

PROOF. Without loss of generality, we will assume $\zeta = 1$. Carefully following the proof of Proposition 6.2.1, one can show that for $a \in \mathbb{C}$ near 1 and $f \in \mathcal{E}$,

$$|f(a)| \leq \frac{C}{|1-a|^{1/p}} \|f\|.$$

Thus $(1-a)^{1/p} f(a)$ is bounded near 1 and so f can not have pole of order higher than n_p at 1. We now show slightly more, namely that f has a pole of order at most $s = k(1)$ at $\zeta = 1$.

Suppose that the lemma were not true. Then, by a Laurent series argument, there is an $f_1 \in \mathcal{E}$ with

$$f_1(z) = \frac{1}{(1-z)^{s+r}} + \cdots + \frac{b_{-1}}{1-z} + h_1,$$

where $r \in \mathbb{N}, s+r \leq n_p < 1/p$, for some $h_1 \in H^p$ with h_1 bounded near 1 (recall $s := k(1)$). Using the identity

$$B \frac{1}{(1-z)^j} = \sum_{l=1}^{j} \frac{1}{(1-z)^l}, \quad j = 1, 2, \cdots,$$

we see that

$$Bf_1 = \frac{1}{(1-z)^{s+r}} + \frac{1 + b_{-(s+r-1)}}{(1-z)^{s+r-1}} + \cdots + Bh_1$$

Thus

$$f_2 := Bf_1 - f_1 = \frac{1}{(1-z)^{s+r-1}} + \cdots + h_2 \in \mathcal{E},$$

where $h_2 = Bh_1 - h_1 \in H^p$ and is bounded near 1. Applying this process an appropriate number of times (each time lowering the order of the pole) we see that

$$(6.2.35) \qquad H := \frac{1}{(1-z)^{s+1}} + \sum_{j=1}^{s} \frac{c_j}{(1-z)^j} + G \in \mathcal{E},$$

where $G \in H^p$ and is bounded near 1. By the definition of $k(1)$ (i.e., $(1-z)^{-j} \in \mathcal{E}$ for $j = 1, \cdots, s = k(1)$) notice

$$\sum_{j=1}^{s} \frac{c_j}{(1-z)^j} \in \mathcal{E}$$

and so by eq.(6.2.35),

$$g := \frac{1}{(1-z)^{s+1}} + G \in \mathcal{E}.$$

As argued earlier,

$$\frac{g - g(a)}{z - a} \in \mathcal{E}, \quad \forall a \in \mathbb{D}$$

and a computation shows

$$(6.2.36) \qquad \frac{g - g(a)}{z - a} = \frac{G - G(a)}{z - a} + \sum_{j=0}^{s} \frac{1}{(1-z)^{j+1}} \frac{1}{(1-a)^{s-j+1}}.$$

But again, the definition of $k(1)$ implies
$$\sum_{j=0}^{s-1} \frac{1}{(1-z)^{j+1}} \frac{1}{(1-a)^{s-j+1}} \in \mathcal{E},$$
and so from eq.(6.2.36)
$$\frac{G - G(a)}{z - a} + \frac{1}{(1-z)^{s+1}} \frac{1}{1-a} \in \mathcal{E} \ \forall \, a \in \mathbb{D}.$$
Now since G is bounded near 1 and
$$\frac{1}{\zeta - a} \to \frac{1}{\zeta - 1} \ \text{in} \ L^p \ \text{as} \ a \to 1,$$
then
$$(1-a)\frac{G - G(a)}{z - a} + \frac{1}{(1-z)^{s+1}} \to \frac{1}{(1-z)^{s+1}} \in \mathcal{E} \ \text{as} \ a \to 1.$$
This contradicts the fact that $k(1) = s$. $\square$

6.3. A reduction

If $\mathcal{E}$ is a B-invariant subspace of H^p, then by Theorem 6.2.11, Theorem 6.2.22, and Lemma 6.2.34,
$$\mathcal{E} \subset \mathcal{E}^p(I, F, k),$$
where $I = I_\mathcal{E}$, $F = F_\mathcal{E}$, and $k = k_\mathcal{E}$ are defined as before. Let
$$e^p(I, F, k) := \left(H^2 \cap I\overline{H_0^2} \right) \bigvee \left\{ \frac{1}{(1 - \bar{\zeta}z)^j} : j = 1, \cdots, k(\zeta), \zeta \in F \right\}.$$
By the definitions of I, F, and k,
$$e^p(I, F, k) \subset \mathcal{E}.$$
To prove the main result of this chapter (Theorem 6.1.16), we will show that
$$(6.3.1) \qquad e^p(I, F, k) = \mathcal{E}^p(I, F, k).$$
This next theorem is a key reduction of this approximation problem to the case where the inner function I is equal to the constant function one. The proof will require more advanced tools which will be introduced in later sections.

THEOREM 6.3.2. *Let F be a closed subset of $\mathbb{T}$ and $k : F \to [1, 1/p) \cap \mathbb{N}$ a $(1,p)$-permissible function. Then $e^p(1, F, k) = \mathcal{E}^p(1, F, k)$.*

Assuming this result for now, we can finish the proof of the main result of this chapter.

COROLLARY 6.3.3. $e^p(I, F, k) = \mathcal{E}^p(I, F, k)$.

PROOF. Given $f \in \mathcal{E}^p(I, F, k)$ we must show that $f \in e^p(I, F, k)$.

Claim 1: There exists a $\zeta_0 \in \mathbb{T}$ with $\zeta_0 \notin I(F_0 \backslash \sigma(I))$ such that
$$\frac{f}{I - \zeta_0} \in H^p \cap \overline{H_0^p}.$$

Proof of Claim 1: By Proposition 6.2.9, the set
$$\left\{ \zeta \in \mathbb{T} : \frac{f}{I - \zeta} \in H^p \cap \overline{H_0^p} \right\}$$

has full measure in $\mathbb{T}$. Also notice that $F_0\backslash\sigma(I)$ is a countable set and so $I(F_0\backslash\sigma(I))$ is also countable. The claim now follows.

Claim 2: $\tilde{F} := F \cup I^{-1}(\{\zeta_0\})$ [17] is closed.

Proof of Claim 2: Since I is analytic near $\mathbb{T}\backslash\sigma(I)$, the set $I^{-1}(\{\zeta_0\})$ will be a sequence of points (assuming of course that I is not constant) which have no accumulation point in $\mathbb{T}\backslash\sigma(I)$. Since $\sigma(I) \cap \mathbb{T} \subset F$, the set $I^{-1}(\{\zeta_0\}) \cup F$ will contain all of its accumulation points and hence be closed.

Claim 3: The function $\tilde{k} : \tilde{F} \to [1, 1/p] \cap \mathbb{N}$ defined by

$$\tilde{k}(\zeta) = \begin{cases} k(\zeta) & \text{if } \zeta \in F \\ m_I(\zeta) & \text{if } \zeta \in \tilde{F}\backslash F \end{cases},$$

where $m_I(\zeta)$ is the multiplicity of the zero of $I - \zeta_0$ at the point ζ, is $(1, p)$-permissible (i.e. $\tilde{k} = n_p$ on the accumulation points of $\tilde{F}$)

It is important to remark here that I is analytic near $\zeta \in \tilde{F}\backslash F$ and so, using the definition of an inner function, one can show that $I'(\zeta) \neq 0$. Thus $m_I(\zeta) = 1$ for all $\zeta \in \tilde{F}\backslash F$. Also notice that $\zeta_0 \in \mathbb{T}$ was chosen in Claim 1 so that $F_0 \cap I^{-1}(\{\zeta_0\}) = \emptyset$ and so there is no ambiguity in the definition of $\tilde{k}$.

Proof of Claim 3: First note that $\tilde{k} \leq n_p$ since $\tilde{k}|F = k|F \leq n_p$ and $\tilde{k}|\tilde{F}\backslash F = 1$. As argued in Claim 2, the accumulation points of $I^{-1}(\{\zeta_0\})$ lie in $\sigma(I) \cap \mathbb{T}$. Thus the accumulation points of $\tilde{F}$ lie in the set of accumulation points for F together with $\sigma(I) \cap \mathbb{T}$. But $\tilde{k} = k = n_p$ on this set.

Claim 4: $g := f/(I - \zeta_0) \in \mathcal{E}^p(1, \tilde{F}, \tilde{k})$.

Proof of Claim 4: Since $f \in \mathcal{E}^p(I, F, k)$, f has an analytic continuation across $\mathbb{T}\backslash F$. But since $(I - \zeta_0)^{-1}$ has an analytic continuation across

$$\mathbb{T}\backslash\{\sigma(I) \cup I^{-1}(\{\zeta_0\})\},$$

g has an analytic continuation across $\mathbb{T}\backslash\tilde{F}$. From Claim 1, $g \in H^p \cap \overline{H_0^p}$ and thus we just need to check the order of the poles of g at the isolated points of $\tilde{F}$. This can be verified using the fact that ζ_0 was chosen in Claim 1 so that $F_0 \cap I^{-1}(\{\zeta_0\}) = \emptyset$.

Claim 5: $f \in e^p(I, F, k)$.

Proof of Claim 5: We are assuming the conclusion of Theorem 6.3.2 and hence $g = f/(I - \zeta_0) \in e^p(1, \tilde{F}, \tilde{k})$, meaning

$$g \in \bigvee \left\{ \frac{1}{(1 - \bar{\zeta}z)^j} : j = 1, \cdots, \tilde{k}(\zeta), \zeta \in \tilde{F} \right\}.\text{[18]}$$

Since $I - \zeta_0$ is bounded,

$$f \in \bigvee \left\{ \frac{I - \zeta_0}{(1 - \bar{\zeta}z)^j} : j = 1, \cdots, \tilde{k}(\zeta), \zeta \in \tilde{F} \right\}.$$

---

[17]Note that $I^{-1}(\{\zeta_0\}) = \{w \in \mathbb{T}\backslash\sigma(I) : I(w) = \zeta_0\}$ will be a sequence of points since I is analytic near $\mathbb{T}\backslash\sigma(I)$.

[18]Here we are taking $I = 1$ and so $H^2 \cap \overline{H_0^2} = (0)$.

6.3. A REDUCTION

To finish, we will show
$$\frac{I - \zeta_0}{(1 - \bar{\zeta}z)^j} \in e^p(I, F, k), \quad \forall j = 1, \cdots, \tilde{k}(\zeta), \zeta \in \tilde{F}.$$

We examine the several cases individually.

Case 1: $\zeta \in \sigma(I) \cap \mathbb{T}$.

In this case $\tilde{k}(\zeta) = k(\zeta) = n_p$ and by Lemma 6.2.25 (part 2), the functions
$$\frac{I}{(1 - \bar{\zeta}z)^j}, \quad \frac{1}{(1 - \bar{\zeta}z)^j}, \quad j = 1, \cdots, n_p$$

belong to the H^p closure of $H^\infty \cap I\overline{H_0^\infty}$ which (by definition) is contained in $e^p(I, F, k)$.

Case 2: $\zeta \in F \setminus \sigma(I)$.

Note that $\tilde{k}(\zeta) = k(\zeta)$ and by definition,
$$\frac{1}{(1 - \bar{\zeta}z)^j} \in e^p(I, F, k), \quad \forall j = 1, \cdots, k(\zeta).$$

To show that
$$\frac{I}{(1 - \bar{\zeta}z)^j} \in e^p(I, F, k),$$

we proceed as follows. Fix $1 \leq j \leq k(\zeta)$. I is analytic near ζ, and so by Lemma 6.2.25 (part 1),
$$\frac{1}{(z - \zeta)^j} \left\{ I - \sum_{l=0}^{j-1} \frac{I^l(\zeta)}{l!}(z - \zeta)^l \right\}$$

belongs to $H^\infty \cap I\overline{H_0^\infty} \subset e^p(I, F, k)$. But note that
$$\frac{1}{(z - \zeta)^j} \sum_{l=0}^{j-1} \frac{I^l(\zeta)}{l!}(z - \zeta)^l \in e^p(I, F, k)$$

and so $I/(z - \zeta)^j \in e^p(I, F, k)$.

Case 3: $\zeta \in \tilde{F} \setminus F$.

In this case $\tilde{k}(\zeta) = m_I(\zeta) = 1$ (see our remarks following Claim 3). Clearly $(I - \zeta_0)/(1 - \bar{\zeta}z) \in H^2$ Moreover, for almost every $z \in \mathbb{T}$,
$$\frac{I - \zeta_0}{1 - \bar{\zeta}z} = I \left\{ \bar{z} \frac{1 - \zeta_0 \bar{I}}{\bar{z} - \bar{\zeta}} \right\} \in I\overline{H_0^2}$$

and so
$$\frac{I - \zeta_0}{1 - \bar{\zeta}z} \in H^2 \cap I\overline{H_0^2} \subset e^p(I, F, k).$$

□

6.4. Rational approximation

6.4.1. An outline.
Recall from the last section that

$$e^p(1, F, k) = \mathcal{E}^p(1, F, k) \Rightarrow e^p(I, F, k) = \mathcal{E}^p(I, F, k),$$

leaving us to show that the rational functions in $\mathcal{E}^p(1, F, k)$ are dense in $\mathcal{E}^p(1, F, k)$. This is a highly non-trivial result and will be the content of most of the rest of this book. To prove this, we will prove an analogous statement for the Hardy spaces of the upper half plane where the theory of distributions and atomic decompositions play a critical role. Before getting involved in this approximation scheme, we would like to review some material from Chapter 4 on the Hardy spaces of the upper half plane. Since we will only highlight the material here, the reader is encouraged to (re)read Chapter 4 for further details.

For $p \in (0, 1)$, recall the Hardy class $\mathcal{H}^p = \mathcal{H}^p(\mathbb{C}_+)$ of analytic functions f on the upper half-plane $\mathbb{C}_+$ such that

$$\|f\| := \sup_{y > 0} \left\{ \int_{\mathbb{R}} |f(x + iy)|^p dx \right\}^{1/p}$$

is finite. Functions $f \in \mathcal{H}^p$ have non-tangential boundary values $f(x)$ for almost every $x \in \mathbb{R}$ and the boundary function belongs to L^p. Furthermore, the map which takes $f \in \mathcal{H}^p$ to its boundary function $f(x) \in L^p$ is an isometry whose range is denoted by $\mathcal{H}^p(\mathbb{R})$. Also recall the related F-space $\mathcal{H}^p(\mathbb{C}\setminus\mathbb{R})$ of analytic functions f on $\mathbb{C}\setminus\mathbb{R}$ such that

$$\sup_{y \neq 0} \left\{ \int_{\mathbb{R}} |f(x + iy)|^p dx \right\}^{1/p}$$

is finite. Notice that every $f \in \mathcal{H}^p(\mathbb{C}\setminus\mathbb{R})$ can be written as

$$f(z) = \begin{cases} f_+(z) & z \in \mathbb{C}_+, \\ f_-(z) & z \in \mathbb{C}_- \end{cases}$$

where $f_+, \overline{f_-(\bar{z})} \in \mathcal{H}^p$.

DEFINITION 6.4.1. Define $\mathcal{H}^p \cap \overline{\mathcal{H}^p}$ to be the space

$$\left\{ f \in \mathcal{H}^p(\mathbb{R}) : f(x) = \overline{g}(x) \text{ a.e. for some } g \in \mathcal{H}^p(\mathbb{R}) \right\}.$$

This space is the analog of $H^p \cap \overline{H_0^p}$ for the upper half-plane. It is important to realize that by $\mathcal{H}^p \cap \overline{\mathcal{H}^p}$ we do *not* mean the space of functions $f \in \mathcal{H}^p$ such that $f(z) = \overline{g}(z)$ for all $z \in \mathbb{C}_+$, as the notation suggests. Such a function f would have to be constant (since it would be both analytic and conjugate analytic). However, the only constant function belonging to $\mathcal{H}^p$ is the zero function.

EXAMPLE 6.4.2. From Example 4.6.1, the rational functions of the form

$$f(z) = \sum_{j=1}^{N} \frac{c_j}{z - a_j}, \quad c_j \in \mathbb{C}, a_j \in \mathbb{R},$$

with c_j and a_j chosen so that f belongs to $\mathcal{H}^p(\mathbb{C}\setminus\mathbb{R})$, have boundary functions belonging to $\mathcal{H}^p \cap \overline{\mathcal{H}^p}$. This follows from the fact that $\overline{f}(x)$ is the boundary function for $\overline{f}(\bar{z})$ which again is a rational function with simple poles on $\mathbb{R}$ and belongs to $\mathcal{H}^p$.

Observe that $\mathcal{H}^p \cap \overline{\mathcal{H}^p}$ is a closed subspace of L^p and moreover, we can identify this space with a certain subspace of $\mathcal{H}^p(\mathbb{C}\backslash\mathbb{R})$ in the following way. If $f \in \mathcal{H}^p \cap \overline{\mathcal{H}^p}$, then $f(x) = \overline{g}(x)$ almost everywhere for some $g \in \mathcal{H}^p(\mathbb{R})$. This suggests forming the function

$$(6.4.3) \qquad F(z) := \begin{cases} f(z) = F_+(z) & z \in \mathbb{C}_+, \\ \overline{g(\overline{z})} = F_-(z) & z \in \mathbb{C}_-. \end{cases}$$

A few moments thought will reveal that $F \in \mathcal{H}^p(\mathbb{C}\backslash\mathbb{R})$ and

$$F_+(x) := \lim_{y \to 0^+} F_+(x + iy) = f(x) = \overline{g}(x) = \lim_{y \to 0^-} F_-(x + iy) := F_-(x)$$

for almost every $x \in \mathbb{R}$. That is to say, every $f \in \mathcal{H}^p \cap \overline{\mathcal{H}^p}$ corresponds, via eq.(6.4.3), to a function $F \in \mathcal{H}^p(\mathbb{C}\backslash\mathbb{R})$ with $F_+(x) = F_-(x)$ almost everywhere (the boundary values from above and below the real axis are almost always the same).

DEFINITION 6.4.4. Let $\mathcal{H}^p(\mathbb{C}\backslash\mathbb{R})_*$ denote the functions

$$F(z) = \begin{cases} F_+(z) & z \in \mathbb{C}_+, \\ F_-(z) & z \in \mathbb{C}_- \end{cases} \in \mathcal{H}^p(\mathbb{C}\backslash\mathbb{R})$$

such that $F_+(x) = F_-(x)$ almost everywhere.

The above analysis says that every $f \in \mathcal{H}^p \cap \overline{\mathcal{H}^p}$ corresponds, via eq.(6.4.3), to a function $F \in \mathcal{H}^p(\mathbb{C}\backslash\mathbb{R})_*$. Now suppose $F \in \mathcal{H}^p(\mathbb{C}\backslash\mathbb{R})_*$. Then

$$(6.4.5) \qquad F(z) = \begin{cases} F_+(z) = f(z) & z \in \mathbb{C}_+, \\ F_-(z) = \overline{g(\overline{z})} & z \in \mathbb{C}_-, \end{cases}$$

where $f, g \in \mathcal{H}^p$ and by hypothesis, $f(x) = \overline{g}(x)$ almost everywhere. From this follows that $f \in \mathcal{H}^p \cap \overline{\mathcal{H}^p}$. Thus every $F \in \mathcal{H}^p(\mathbb{C}\backslash\mathbb{R})_*$ corresponds, via eq.(6.4.5), to a function in $\mathcal{H}^p \cap \overline{\mathcal{H}^p}$. Note also that the above correspondence between $\mathcal{H}^p(\mathbb{C}\backslash\mathbb{R})_*$ and $\mathcal{H}^p \cap \overline{\mathcal{H}^p}$, by means of eq.(6.4.3) and eq.(6.4.5), is one to one. Moreover, it is easy to check from the definition that the norms on both of these spaces are comparable.

REMARK 6.4.6. An important distinction needs to be made here concerning limits. For $F \in \mathcal{H}^p(\mathbb{C}\backslash\mathbb{R})_* \ (\simeq \mathcal{H}^p \cap \overline{\mathcal{H}^p})$, there are two limits to consider. The first is

$$\lim_{y \to 0^+} \left\{ F_+(x + iy) - F_-(x - iy) \right\}$$

which is $F_+(x) - F_-(x)$ and is, almost everywhere, the zero function. The second limit

$$\ell = \lim_{y \to 0^+} \left\{ F_+(\cdot + iy) - F_-(\cdot - iy) \right\}$$

is in the sense of distributions (see Theorem 4.8.13)

$$\ell(\psi) = \lim_{y \to 0^+} \int_{\mathbb{R}} \left\{ F_+(x + iy) - F_-(x - iy) \right\} \psi(x) dx,$$

where $\psi \in \mathcal{S}$, the class of rapidly decreasing infinitely differentiable functions. This last limit is not zero unless F is the zero function. Indeed, from Theorem 4.8.13, the linear functionals

$$\ell_1(\psi) := \lim_{y \to 0^+} \int F_+(x + iy)\psi(x) dx, \quad \ell_2(\psi) := \lim_{y \to 0^+} \int F_-(\overline{x + iy})\psi(x) dx$$

are tempered distributions (since $F_+(z), F_-(\bar{z}) \in \mathcal{H}_h^p$). If they are equal, then $F_+(z) = F_-(\bar{z})$ for all $z \in \mathbb{C}_+$. But this means that F_+ is both analytic and anti-analytic and thus constant. Hence F_+ and F_- are zero.

We are now ready to define the analog of $\mathcal{E}^p(1, F, k)$ for the plane.

DEFINITION 6.4.7. For a closed set $F \subset \mathbb{R}$ (with the set of isolated points denoted by F_0) and a permissible function $k : F \to [1, n_p] \cap \mathbb{N}$, namely, $k|F \backslash F_0 = n_p$, define $E^p(F, k)$ to be the functions $f \in \mathcal{H}^p(\mathbb{C}\backslash\mathbb{R})$ with

(6.4.8) $$f \in \mathcal{H}^p(\mathbb{C}\backslash\mathbb{R})_* \ (\simeq \mathcal{H}^p \cap \overline{\mathcal{H}^p}).$$

(6.4.9) $\qquad f$ has an analytic continuation across $\mathbb{R}\backslash F$.

(6.4.10) $\qquad f$ has a pole of order no more than $k(t)$ and each point $t \in F_0$.

The main rational approximation result which will be discussed in the next few sections is the following.

THEOREM 6.4.11. *The rational functions in $E^p(F, k)$ are dense in $E^p(F, k)$.*

A corollary to this result is our much desired approximation result for the disk.

COROLLARY 6.4.12. $e^p(1, F, k) = \mathcal{E}^p(1, F, k)$ [19].

Before proving the corollary (assuming Theorem 6.4.11), we need to make a remark about the relationship between $\mathcal{H}^p$ and H^p. The map

$$\phi(z) := i\frac{1+z}{1-z}$$

is a conformal map from $\mathbb{D}$ to $\mathbb{C}_+$ and the composition operator $T : \mathcal{H}^p(\mathbb{C}_+) \to H^p(\mathbb{D})$ defined by $Tf = f \circ \phi$ is an isomorphism between $\mathcal{H}^p$ and a subspace of H^p. Since $\mathcal{H}^p$ does not contain the constant functions, this map is not onto. However, the following result does identify the range of T.

PROPOSITION 6.4.13. *An analytic function f on $\mathbb{C}_+$ belongs to $\mathcal{H}^p(\mathbb{C}_+)$ if and only if $f \circ \phi \cdot (\phi')^{1/p}$ belongs to $H^p(\mathbb{D})$.*

A computation reveals $\phi' = 2i/(1-z)^2$ and so the theorem says $g \in H^p$ belongs to the range of T if and only if $g \cdot (1-z)^{-2/p}$ also belongs to H^p. But since $(1-z)^{-2/p}$ is an outer function, this is equivalent to saying that the boundary function $g(\zeta)(1-\zeta)^{-2/p}$ belongs to L^p. [20] A direct consequence of this result is the following.

LEMMA 6.4.14. *If $p \in (0,1)$ and $g \in H^p$ and analytic in a neighborhood of $\zeta = 1$ with*

$$g^{(k)}(1) = 0 \ \forall k = 0, \cdots, n_p + 1,$$

then $g \circ \phi^{-1} \in \mathcal{H}^p$.

---

[19] Here $F \subset \mathbb{T}$ and k is a $(1, p)$-permissible function on F.

[20] i.e., if $g \in H^p$, then $g \cdot (1-z)^{-2/p} \in N^+$ (since $(1-z)^{-2/p}$ is outer). Since $N^+ \cap L^p = H^p$, then $g \cdot (1-z)^{-2/p} \in H^p$ if and only if $g(\zeta)(1-\zeta)^{-2/p} \in L^p$.

6.4. RATIONAL APPROXIMATION

PROOF. To show
$$g(\zeta)/(1-\zeta)^{2/p} \in L^p,$$
it suffices to prove the integral
$$\int_J \frac{|g(\zeta)|^p}{|1-\zeta|^2}|d\zeta|$$
is finite, where J is an arc of the circle which contains the point 1. By our hypothesis, we will assume that g is analytic in a neighborhood of J. From our hypothesis about g and a Taylor series argument,
$$|g(\zeta)| \leq C|1-\zeta|^{n_p+2}$$
for ζ near the point 1, and so
$$\int_J \frac{|g(\zeta)|^p}{|1-\zeta|^2}|d\zeta| \leq C \int_J |1-\zeta|^{pn_p+2p-2}|d\zeta|.$$
But $1/p - 1 \leq n_p \leq 1/p$ and so $pn_p + 2p - 2 > -1$. Hence, the right-hand side of the above integral inequality is finite. □

PROOF OF COROLLARY 6.4.12 ASSUMING THEOREM 6.4.11. We begin with a few general remarks. First, if $f \in \mathcal{E}^p(1, F, k)$ is a rational function, then $f \in e^p(1, F, k)$. Indeed, let $\{\zeta_1, \cdots, \zeta_n\} \subset F$ be the poles of f. If we let P_j denote the principal part of the Laurent expansion of f about ζ_j, then by definition, P_j belongs to $e^p(1, F, k)$. Moreover,
$$g := f - \sum_{j=1}^n P_j$$
belongs to $\mathcal{E}^p(1, F, k)$ and is also an entire function. But recall that
$$g \in \mathcal{E}^p(1, F, k) \subset H^p \cap \overline{H_0^p}$$
and so g must vanish at infinity. By Liouville's theorem, g is the zero function and so $f \in e^p(1, F, k)$.

Secondly, if F is a finite set, it follows from the previous paragraph that $\mathcal{E}^p(1, F, k)$ must be equal to $e^p(1, F, k)$. Thus, we can assume that F is an infinite set.

Finally, if $F = \mathbb{T}$, we can apply Proposition 6.1.6 to get,
$$\mathcal{E}^p(1, F, k) = \mathcal{E}^p(1, \mathbb{T}, n_p) = H^p \cap \overline{H_0^p} = \bigvee \left\{ \frac{1}{1-\bar{\zeta}z} : \zeta \in \mathbb{T} \right\} = e^p(1, \mathbb{T}, n_p).$$

We are now ready to start the proof in earnest. By assuming that $F \neq \mathbb{T}$, we can rotate, if necessary (and this does not change anything) and assume that F does not contain the point $\zeta = 1$. Define $\tilde{F} = \phi(F)$ and the function $\tilde{k} : \tilde{F} \to \mathbb{N} \cap [1, n_p]$ by
$$\tilde{k}(x) := k(\zeta), \quad \zeta \in F, \quad \phi(\zeta) = x.$$
Let $g \in \mathcal{E}^p(1, F, k)$ be analytic near $\zeta = 1$ with
$$g^{(j)}(1) = 0, \quad \forall\, j = 0, \cdots, n_p + 1$$
and note, from Lemma 6.4.14, that $T^{-1}g = g \circ \phi^{-1} \in \mathcal{H}^p$. By the definition of $\mathcal{E}^p(1, F, k)$, $g(\zeta) = \bar{h}(\zeta)$ almost everywhere for some function $h \in H_0^p$. Since
$$|g(\zeta)| \leq C|1-\zeta|^{n_p+2}, \quad \zeta \in \mathbb{T} \text{ near } 1,$$

the same is true for h and by the proof of Lemma 6.4.14, $h \circ \phi^{-1} \in \mathcal{H}^p$. But $g \circ \phi^{-1} = \overline{h} \circ \phi^{-1}$ almost everywhere on $\mathbb{R}$ and so $T^{-1}g = g \circ \phi^{-1} \in \mathcal{H}^p \cap \overline{\mathcal{H}^p}$. From the fact that ϕ is a rational function, and the definitions of the spaces $\mathcal{E}^p(1, F, k)$ and $E^p(\tilde{F}, \tilde{k})$, one can show that $T^{-1}g$ belongs to $E^p(\tilde{F}, \tilde{k})$.

Since we are assuming Theorem 6.4.11, $T^{-1}g$ can be approximated by a sequence of rational functions $\{r_n : n \in \mathbb{N}\} \subset E^p(\tilde{F}, \tilde{k})$ and so $g = \lim_{n\to\infty} T(r_n)$. A few moments thought, using the fact that ϕ is a rational function, shows that $T(r_n)$ is a rational function from $\mathcal{E}^p(1, F, k)$ which, by our remark above, belongs to $e^p(1, F, k)$. Thus

$$\{ g \in \mathcal{E}^p(1, F, k) : g^{(j)}(1) = 0, j = 0, \cdots, n_p + 1 \} \subset e^p(1, F, k).$$

To approximate a general $f \in \mathcal{E}^p(1, F, k)$ with rational functions from $\mathcal{E}^p(1, F, k)$, we first find a rational function

$$R(z) := \sum_{l=0}^{n_p+1} \frac{c_l}{1 - \overline{\zeta_l} z}, \quad \zeta_l \in F$$

(which will certainly belong to $e^p(1, F, k)$) with the property that $g := f - R$ satisfies the conditions $g^{(j)}(1) = 0$ for all $j = 0, \cdots, n_p + 1$. To see how to do this, let A be the following $(n_p + 2) \times (n_p + 2)$ matrix

$$A = \begin{pmatrix} \frac{1}{1-\overline{\zeta_0}} & \frac{1}{1-\overline{\zeta_1}} & \cdots & \frac{1}{1-\overline{\zeta_{n_p+1}}} \\ \frac{\overline{\zeta_0}}{(1-\overline{\zeta_0})^2} & \frac{\overline{\zeta_1}}{(1-\overline{\zeta_1})^2} & \cdots & \frac{\overline{\zeta_{n_p+1}}}{(1-\overline{\zeta_{n_p+1}})^2} \\ \vdots & \vdots & \vdots & \vdots \\ \frac{(n_p+1)!\overline{\zeta_0}^{n_p+1}}{(1-\overline{\zeta_0})^{n_p+2}} & \frac{(n_p+1)!\overline{\zeta_1}^{n_p+1}}{(1-\overline{\zeta_1})^{n_p+2}} & \cdots & \frac{(n_p+1)!\overline{\zeta_{n_p+1}}^{n_p+1}}{(1-\overline{\zeta_{n_p+1}})^{n_p+2}} \end{pmatrix}$$

and note that the determinant of A is equal to

$$V \prod_{j=0}^{n_p+1} j! \prod_{j=0}^{n_p+1} (1 - \overline{\zeta_j}),$$

where V is the Vandermonde matrix for the points $\overline{\zeta_j}/(1 - \overline{\zeta_j}), j = 0, \cdots, n_p + 1$, and hence is non-zero. If Y is the column vector

$$Y = (f^{(j)}(1))_{j=0}^{n_p+1},$$

we can solve the matrix equation $AC = Y$ for the column vector $C = (c_j)_{j=0}^{n_p+1}$. If we now define

$$R(z) := \sum_{l=0}^{n_p+1} \frac{c_l}{1 - \overline{\zeta_l} z},$$

then R certainly belongs to $e^p(1, F, k)$ and for each $j = 0, \cdots, n_p + 1$,

$$R^{(j)}(1) = \sum_{l=0}^{n_p+1} \frac{j! c_l \overline{\zeta_l}^j}{(1 - \overline{\zeta_l})^{j+1}} = (AC)_j = f^{(j)}(1).$$

Thus $g := f - R$ has the desired property $g^{(j)}(1) = 0$ for all $j = 0, \cdots, n_p + 1$ and so, by the above argument, $g \in e^p(1, F, k)$. Since $R \in e^p(1, F, k)$,

$$f = g + R \in e^p(1, F, k).$$

$\square$

6.4.2. A preliminary approximation problem. To get started on Aleksandrov's approximation scheme, we will solve several preliminary rational approximation problems which can be found in [**2**]. For simplicity in notation, we let $L^p = L^p(\mathbb{R})$.

Let $p \in (0,1)$ and let X^p denote the L^p closure of the set of $f \in L^p$ which can be written in the form

(6.4.15) $$f = \sum_{j=1}^{N} \frac{c_j}{x - a_j}, \quad a_j \in \mathbb{R}, c_j \in \mathbb{C}.$$

From Example 4.6.1, such functions of the form eq.(6.4.15) are the boundary functions for $\mathcal{H}^p$ functions. Moreover, $\overline{f}$ is the boundary function for $\overline{f(\overline{z})} \in \mathcal{H}^p$ and so $f \in \mathcal{H}^p \cap \overline{\mathcal{H}^p}$. Our first rational approximation result is the following.

THEOREM 6.4.16. *For* $p \in (0,1)$, $X^p = \mathcal{H}^p \cap \overline{\mathcal{H}^p}$.

6.4.2.1. *Some topology and linear algebra.* The proof of Theorem 6.4.16 depends on some rather technical results involving the Hilbert transform. Before getting into this, we need to make some preliminary comments about the functional analysis tools that will appear often in this section. Recall that a topological vector space (TVS) is a (complex) vector space X with a topology τ such that the basic vector space operations are continuous with respect to τ. The following standard fact can be found in [**78**], p. 16.

PROPOSITION 6.4.17. *Let X be a TVS and E an n-dimensional linear manifold of X. Then*

1. *Every linear bijection of $\mathbb{C}^n$ onto E is also a homeomorphism (continuous with continuous inverse).*
2. *E is closed.*

This yields the following result.

COROLLARY 6.4.18. *Let X be a TVS and E be an n-dimensional subspace.*

1. *Any two (TVS) topologies on E are equivalent.*
2. *If there exists a closed subspace $Y \subset X$ such that*
   $$X = E + Y \quad \text{and} \quad E \cap Y = (0),$$
   then there is a continuous projection $P: X \to E$.
3. *If X^* (the topological dual of X) separates points, there is a continuous projection $P: X \to E$.*

PROOF. The proof of statement (1) follows from elementary linear algebra and Proposition 6.4.17(1).

To prove statement (2), note that every $f \in X$ can be written uniquely as $f = e + g$, where $e \in E$ and $g \in Y$ and so the map $P: X \to E$ defined by $Pf = e$ is a well defined linear projection. Since E is finite dimensional, we can use (1) to endow E with an equivalent topology it inherits from being linearly homeomorphic to X/Y (with the quotient topology); specifically, $A \subset E$ is open if and only if $[A]$ is open in X/Y ($[h]$ will be the coset of h in X/Y). The map $f \to [f]$ is continuous from $X \to X/Y$ and since $[f] = [e]$, the map $f \to e$, that is P, is also continuous.

The construction of the linear projection in statement (3) is really just basic linear algebra but, for the sake of completeness, we include it anyway. We first claim that if the vectors $e_1, \cdots, e_n$ form a basis for E, there are functionals $\lambda_1, \cdots, \lambda_n$ from X^* such that the matrix

$$A := \begin{pmatrix} \lambda_1(e_1) & \cdots & \lambda_n(e_1) \\ \vdots & \vdots & \vdots \\ \lambda_1(e_n) & \cdots & \lambda_n(e_n) \end{pmatrix}$$

is invertible. We will inductively build up this matrix. Since X^* separates points, there is a $\lambda_1 \in X^*$ so that $\lambda_1(e_1) \neq 0$ (λ_1 separates e_1 from the zero vector). So trivially, the matrix $A_1 = \lambda_1(e_1)$ is invertible. Suppose that for some $j = 1, \cdots, n-1$, we are able to choose $\lambda_1, \cdots, \lambda_j \in X^*$ so that the matrix

$$A_j := \begin{pmatrix} \lambda_1(e_1) & \cdots & \lambda_j(e_1) \\ \vdots & \vdots & \vdots \\ \lambda_1(e_j) & \cdots & \lambda_j(e_j) \end{pmatrix}$$

is invertible. Then, for any $\lambda \in X^*$, form the matrix

$$B(\lambda) := \begin{pmatrix} \lambda_1(e_1) & \cdots & \lambda_j(e_1) & \lambda(e_1) \\ \vdots & \vdots & \cdots & \cdots \\ \lambda_1(e_j) & \cdots & \lambda_j(e_j) & \lambda(e_j) \\ \lambda_1(e_{j+1}) & \cdots & \lambda_j(e_{j+1}) & \lambda(e_{j+1}) \end{pmatrix}.$$

We claim that $\lambda \in X^*$ can be chosen so that $B(\lambda)$ is invertible and from here, one can build up the matrix A. To see this claim, we suppose the contrary. Then $\det B(\lambda) = 0$ for all λ. By expanding along the last column, observe that

$$0 = \det B(\lambda) = C_1 \lambda(e_1) + \cdots + C_j \lambda(e_j) \pm (\det A_j)\lambda(e_{j+1}) \; \forall \, \lambda \in X^*.$$

Since X^* separates points,

$$C_1 e_1 + \cdots + C_j e_j \pm (\det A_j) e_{j+1} = 0$$

which, since the e_j's form a basis, makes all the coefficients equal to zero. This produces a contradiction since we assuming A_j is invertible.

Since A is invertible, we can find vectors $v_j = (a_{i,j})_{i=1}^n$ such that $Av_j = (\delta_{i,j})_{i=1}^n$. Letting $\Lambda_j = \sum_{i=1}^n a_{i,j} \lambda_i$ we see that $\Lambda_j(e_k) = \delta_{j,k}$. Hence the operator defined by

$$P(f) = \sum_{j=1}^n \Lambda_j(f) e_j$$

is the desired continuous projection onto E. $\square$

6.4.2.2. *Vanishing moments and the Hilbert transform.*

LEMMA 6.4.19. *Let f be a measurable function on $\mathbb{R}$ such that there exists a $C > 0$ and an $n \in \mathbb{N}$ such that*

$$\text{meas}\left\{ x : |f(x)|(1+|x|^n) > \alpha \right\} \leq C/\alpha$$

for all $\alpha > 0$. Then $f \in L^p$ for all $p \in (1/(n+1), 1)$.

PROOF. We begin the proof with the claim that if f satisfies the hypothesis of the lemma, then

(6.4.20) $$\text{meas}\left\{ x : |f(x)|^p > \alpha \right\} \leq \begin{cases} \frac{C+2}{\alpha^{1/p(n+1)}} & \text{if } \alpha \in (0,1), \\ \frac{C}{\alpha^{1/p}} & \text{if } \alpha \in [1, \infty). \end{cases}$$

To see this, observe that if $\alpha \geq 1$, then
$$\{\, x : |f(x)|^p > \alpha \,\} = \{\, x : |f(x)| > \alpha^{1/p} \,\} \subset \{\, x : |f(x)|(1+|x|^n) > \alpha^{1/p} \,\}.$$
Now use the hypothesis of the lemma to obtain
$$\operatorname{meas}\{\, x : |f(x)|^p > \alpha \,\} \leq C/\alpha^{1/p}.$$
If $\alpha \in (0,1)$, we will show that
$$\{\, x : |f(x)|^p > \alpha \,\} = B \cup C,$$
where
$$\operatorname{meas}(B) \leq C\alpha^{-1/p(n+1)}, \qquad \operatorname{meas}(C) \leq 2\alpha^{-1/p(n+1)}.$$
From here, eq.(6.4.20) follows. Let $t = \alpha^{-1/p(n+1)}$ and
$$B := \{\, |x| \geq t : |f(x)|^p > \alpha \,\}, \quad C := \{\, |x| < t : |f(x)|^p > \alpha \,\}.$$
Clearly $\operatorname{meas}(C) \leq 2t = 2\alpha^{-1/p(n+1)}$. To estimate $\operatorname{meas}(B)$, note that if $|x| \geq t$, then $(1+|x|^n) \geq \alpha^{-n/p(n+1)} + 1$. From here, it follows that
$$\begin{aligned}
B &= \{|x| \geq t : |f(x)|^p > \alpha\} \\
&= \{|x| \geq t : (1+|x|^n)^p |f(x)|^p > \alpha(1+|x|^n)^p\} \\
&\subset \{|x| \geq t : (1+|x|^n)^p |f(x)|^p > \alpha(\alpha^{-n/p(n+1)} + 1)^p\} \\
&\subset \{|x| \geq t : (1+|x|^n)^p |f(x)|^p > \alpha^{1/(n+1)}\} \\
&\subset \{x \in \mathbb{R} : (1+|x|^n)|f(x)| > \alpha^{1/p(n+1)}\}.
\end{aligned}$$

Now apply the hypothesis of the lemma to obtain $\operatorname{meas}(B) \leq C\alpha^{-1/p(n+1)}$.

To finish the proof of the lemma, we apply the basic distributional equality
$$\|f\|_{L^p}^p = \int_0^\infty \operatorname{meas}\{\, x : |f(x)|^p > \alpha \,\}\, d\alpha = \int_0^1 + \int_1^\infty.$$
By applying the estimate in eq.(6.4.20), one can show the above two integrals are finite. $\square$

DEFINITION 6.4.21. For $n \in \mathbb{N}$ and $d\mu = (1+|x|^n)dx$, let
$$L^1_{0,n} := \Big\{\, f \in L^1(d\mu) : \int x^k f(x) dx = 0 \;\; k = 0,1,\cdots n-1 \,\Big\}.$$

Note that $L^1_{0,n}$ is a closed subspace of $L^1(d\mu)$. Moreover, for $f \in L^1(d\mu)$ and $p \in (1/(n+1), 1)$,
$$\begin{aligned}
\int |f|^p dx &= \int |f|^p (1+|x|^n)^p \frac{1}{(1+|x|^n)^p} dx \\
&\leq \Big\{ \int |f|(1+|x|^n) dx \Big\}^p \Big\{ \int \frac{dx}{(1+|x|^n)^{p/(1-p)}} \Big\}^{1-p}
\end{aligned}$$
which is finite since $np/(1-p) > 1$. Thus we have
$$(6.4.22) \qquad L^1_{0,n} \subset L^1(d\mu) \subset L^p, \quad p \in (1/(n+1), 1).$$

LEMMA 6.4.23. $L^1_{0,n}$ is dense in L^p for $p \in (1/(n+1), 1)$.

PROOF. Let M denote the closure of $L^1_{0,n}$ in L^p. Suppose L^p/M is finite dimensional but non-zero. Then there is a non-zero linear functional Λ on L^p/M. Since L^p/M is finite dimensional, Λ will be necessarily continuous (its kernel will be finite dimensional and hence closed, see Proposition 6.4.17). If we define the linear functional λ on L^p by
$$\lambda(f) = \Lambda([f]),$$
where $[f]$ is the coset of f in L^p/M, then λ is continuous and non-zero, contradicting Day's theorem [25] $((L^p)^* = (0))$. Thus if L^p/M is finite dimensional, then it is zero and hence $M = L^p$.

To prove L^p/M is finite dimensional, it suffices to prove the linear manifold $L^1(d\mu)/M$ is finite dimensional. To see this reduction, observe that $L^1(d\mu)$ is dense in L^p (since $L^1(d\mu)$ contains all of the characteristic functions of bounded intervals) and so $L^1(d\mu)/M$ is dense in L^p/M. So if $L^1(d\mu)/M$ is finite dimensional, it is automatically closed (Proposition 6.4.17(2)), and so its closure, L^p/M, is also finite dimensional.

To prove that $L^1(d\mu)/M$ is finite dimensional, choose any n disjoint intervals $I_j = [a_j, b_j]$, $j = 0, \cdots, n-1$ and note that the $n \times n$ matrix
$$A = \begin{pmatrix} b_0 - a_0 & b_1 - a_1 & \cdots & b_{n-1} - a_{n-1} \\ (b_0^2 - a_0^2)/2 & (b_1^2 - a_1^2)/2 & \cdots & (b_{n-1}^2 - a_{n-1}^2)/2 \\ \vdots & \vdots & \vdots & \vdots \\ (b_0^n - a_0^n)/n & (b_1^n - a_1^n)/n & \cdots & (b_{n-1}^n - a_{n-1}^n)/n \end{pmatrix}$$
is non-singular.[21] For a function $f \in L^1(d\mu)$, let Y be the column vector
$$Y = \Big(\int f(x) x^j dx \Big)_{j=0}^{n-1},$$
and let $C = (c_j)_{j=0}^{n-1}$ be the column vector solution to the system $AC = Y$. Define
$$h = \sum_{j=0}^{n-1} c_j \chi_{I_j}$$
and observe that
$$\begin{aligned} \int (f-h) x^k dx &= \int f x^k dx - \sum_{j=0}^{n-1} c_j \int_{a_j}^{b_j} x^k dx \\ &= \int f x^k dx - \sum_{j=0}^{n-1} c_j \frac{b_j^{k+1} - a_j^{k+1}}{k+1} \\ &= \int f x^k dx - (AC)_k \\ &= 0. \end{aligned}$$

---

[21] It suffices to show that the only real vector $v = (c_0, \cdots, c_{n-1})$ solution to $A^T v = 0$ is the zero vector (note that A is a real matrix). Observe that the i-th entry of the product $A^T v$ is equal to
$$\int_{a_i}^{b_i} (c_0 + \cdots + c_{n-1} x^{n-1}) dx = 0.$$
This means that the polynomial $c_0 + \cdots + c_{n-1} x^{n-1}$ has a zero in the interior of $[a_i, b_i]$, $i = 0, \cdots, n-1$. From this follows that $c_0 = \cdots = c_{n-1} = 0$.

Hence $f - h \in L^1_{0,n}$ and so

$$[f] = [h] = \sum_{j=0}^{n-1} c_j [\chi_{I_j}].$$

Thus the cosets $[\chi_{I_j}]$, $j = 0, \cdots, n-1$, form a basis for $L^1(d\mu)/M$. $\square$

LEMMA 6.4.24. *Let S be the linear manifold of functions in $L^1_{0,n}$ which are finite linear combinations of characteristic functions of bounded intervals. Then S is dense in $L^1_{0,n}$.*

PROOF. Choose a set of bounded intervals $I_j = [a_j, b_j]$, $j = 0, \cdots, n-1$, as in the proof of Lemma 6.4.23 so that the matrix A (as defined in that proof) is invertible. If $m := \sum_{j=0}^{n-1} d_j \chi_{I_j}$ belongs to $L^1_{0,n}$, then as seen earlier, the column vector $D = (d_j)_{j=0}^{n-1}$ is a solution to the system $AD = 0$. But this can only happen when $D = 0$. Thus if

$$E := \bigvee \{ \chi_{I_j} : j = 0, \cdots, n-1 \},$$

then $E \cap L^1_{0,n} = (0)$. It is also true that

$$L^1(d\mu) = E + L^1_{0,n}.$$

To see this, we note from the proof of Lemma 6.4.23, that given $f \in L^1(d\mu)$, we can find a function $h \in E$ such that $f - h \in L^1_{0,n}$.

Now use the fact that E is finite dimensional to show the operator $P(f) = g$ (where $f = e + g$, $e \in E, g \in L^1_{0,n}$) is a continuous projection onto $L^1_{0,n}$, Corollary 6.4.18(2). Now let f be an arbitrary element of $L^1_{0,n}$. From basic measure theory, we can find a sequence of functions $\{r_k : k \in \mathbb{N}\}$, each of which is a finite linear combination of characteristic functions of bounded intervals, such that $r_k \to f$ in $L^1(d\mu)$. Note that r_k need not belong to $L^1_{0,n}$. Since $r_k = e_k + g_k$, and e_k is a finite linear combination of characteristic functions of bounded intervals, then $P(r_k) = g_k$ is also of that type and $P(r_k) \to P(f) = f$ in $L^1(d\mu)$. $\square$

Recall the Hilbert transform

$$(Hf)(x) := P.V. \frac{1}{\pi} \int \frac{f(t)}{x-t} \, dt$$

and its basic properties from Chapter 4.

PROPOSITION 6.4.25. *The Hilbert transform H acts continuously from $L^1_{0,n}$ to L^p for $p \in (1/(n+1), 1)$.*

PROOF. First we will show that $H L^1_{0,n} \subset L^p$. Indeed for $\alpha > 0$

$$\text{meas} \{ x : (1 + |x|^n)|(Hf)(x)| > \alpha \} = A \cup B,$$

where

$$A := \{ |x| \leq 1 : (1 + |x|^n)|(Hf)(x)| > \alpha \}$$
$$B := \{ |x| > 1 : (1 + |x|^n)|(Hf)(x)| > \alpha \}.$$

Observe that

$$A = \{ |x| \leq 1 : |(Hf)(x)| > \alpha/(1 + |x|^n) \} \subset \{ |x| \leq 1 : |(Hf)(x)| > \alpha/2 \}.$$

Using, Theorem 4.5.1 (the Hilbert transform is weak $(1,1)$), we obtain
$$\text{meas}(A) \leq \text{meas}\, \{\, |x| \leq 1 : |(Hf)(x)| > \alpha/2 \,\} \leq \frac{C}{\alpha}\|f\|_{L^1(dx)} \leq \frac{C}{\alpha}\|f\|_{L^1(d\mu)}.$$

To estimate meas(B), we first observe that since
$$\int x^k f(x)\,dx = 0, \quad k = 0, 1, \cdots, n-1,$$
a straightforward computation shows that
$$(Ht^n f)(x) = x^n (Hf)(x).$$

Thus
$$\begin{aligned}
B &= \{\, |x| > 1 : (1 + |x|^n)|(Hf)(x)| > \alpha \,\} \\
&= \left\{\, |x| > 1 : |(Hf)(t^n f)(x)| > \frac{\alpha |x|^n}{1 + |x|^n} \,\right\} \\
&\subset \{\, |x| > 1 : |(Hf)(t^n f)(x)| > \alpha/2 \,\}
\end{aligned}$$
since $|x|^n/(1+|x|^n) \geq 1/2$ for $|x| > 1$.

But since $x^n f \in L^1$, we can use the weak $(1,1)$ property of the Hilbert transform again to show
$$\text{meas}(B) \leq \text{meas}\, \{\, |x| > 1 : |(Ht^n f)(x)| > \alpha/2 \,\} \leq \frac{C}{\alpha}\|x^n f\|_{L^1(dx)} \leq \frac{C}{\alpha}\|f\|_{L^1(d\mu)}.$$

Combining this with the estimate of meas(A), we obtain the inequality
$$(6.4.26) \qquad \text{meas}\, \{\, x : (1 + |x|^n)|(Hf)(x)| > \alpha \,\} \leq \frac{C}{\alpha}\|f\|_{L^1(d\mu)}.$$

Now apply Lemma 6.4.19 to the function Hf to obtain
$$Hf \in L^p, \quad p \in (1/(n+1), 1).$$

Thus $HL^1_{0,n} \subset L^p$.

To show that H acts continuously between these two spaces, we will use the closed graph theorem. Let $\{f_k : k \in \mathbb{N}\}$ be a sequence in $L^1_{0,n}$ which converges to zero in the norm of $L^1(d\mu)$ with $Hf_k \to g$ in L^p. Since $f_k \to 0$ in $L^1(d\mu)$, we apply eq.(6.4.26) to show $Hf_k \to 0$ in measure. But $Hf_k \to g$ in L^p, and so by Chebychev's inequality, $Hf_k \to g$ in measure. But this means that g is the zero function. By the closed graph theorem, H acts continuously from $L^1_{0,n}$ to L^p. $\square$

Recall that X^p is the closure in L^p of functions (in L^p) of the form
$$f(x) = \sum_{j=1}^{N} \frac{c_j}{x - a_j}, \quad a_j \in \mathbb{R},\ c_j \in \mathbb{C}.$$

Define $\mathcal{P}$ to be the canonical projection from L^p onto L^p/X^p. Also recall from Lemma 6.4.23 that $L^1_{0,n}$ is dense in L^p.

PROPOSITION 6.4.27. *For $n \in \mathbb{N}$ and $p \in (1/(n+1), 1)$, the operator*
$$\mathcal{P} \circ H : L^1_{0,n} \to L^p/X^p$$
can be extended continuously to L^p.

PROOF. The proof of this result will be done by establishing the claims below.

Claim 1: There is a rational function g with simple poles on $\mathbb{R}$ (g not necessarily in L^p) such that $H\chi_{[0,1]} - g \in L^p$.

Proof of Claim 1: A simple computation shows that
$$(H\chi_{[0,1]})(x) = \frac{1}{\pi}\log\left|\frac{x-1}{x}\right|$$
which is p-th power integrable on every compact interval in $\mathbb{R}$. Now choose a rational function g with simple poles on $\mathbb{R}$ such that $H\chi_{[0,1]} - g$ dies off sufficiently fast at infinity that it belongs to L^p [22].

Claim 2: Given any bounded interval $[a,b]$, there is a rational function $g_{a,b}$ with simple poles on $\mathbb{R}$ such that
$$\|H\chi_{[a,b]} - g_{a,b}\|_{L^p} \leq A_p \|\chi_{[a,b]}\|_{L^p} = A_p (b-a)^{1/p}.$$

Proof of Claim 2: If $\phi(x) = (x-a)/(b-a)$ and $C_\phi f = f \circ \phi$, then $\chi_{[a,b]} = C_\phi \chi_{[0,1]}$. From Claim 1, there is a rational g with simple poles in $\mathbb{R}$ such that
$$\|H\chi_{[0,1]} - g\|_{L^p} = c < \infty.$$

Let $g_{a,b} = C_\phi g$ and notice that $g_{a,b}$ is also rational with simple poles in $\mathbb{R}$. Moreover
$$H\chi_{[a,b]} - g_{a,b} = C_\phi(H\chi_{[0,1]} - g)$$
and so, by a change of variables,
$$\|H\chi_{[a,b]} - g_{a,b}\|_{L^p} = c(b-a)^{1/p} = c\|\chi_{[a,b]}\|_{L^p}.$$

Claim 3: If f is a finite linear combination of characteristic functions of non-overlapping bounded intervals with
$$\int x^k f(x)dx = 0 \;\; \forall\, k = 0, 1, \cdots, n-1,$$
then
$$\|(\mathcal{P} \circ H)f\|_{L^p/X^p} \leq A_p \|f\|_{L^p}.$$

Proof of Claim 3: Write $f = \sum_{j=1}^{L} c_j \chi_{I_j}$, where the intervals do not overlap. For each $j = 1, \cdots, L$, choose a rational function g_{I_j} as in Claim 2 such that
$$\|c_j H\chi_{I_j} - g_{I_j}\|_{L^p} \leq A_p \|\chi_{I_j}\|_{L^p}.$$

Observe that Hf and $Hf - \sum_{j=1}^{L} g_{I_j}$ belong to L^p (since $f \in L^1_{0,n}$, see Proposition 6.4.25) and so $\sum_{j=1}^{L} g_{I_j}$ also belongs to L^p, and hence to X^p (since the sum is

---

[22] For large $|x|$, $|(H\chi_{[0,1]})(x)| = O(|x|^{-1})$. Now use eq.(4.6.3) with $a_1 = 0$.

an L^p rational function). By the definition of the quotient metric,

$$\|(\mathcal{P} \circ H)f\|_{L^p/X^p}^p \leq \left\| Hf - \sum_{j=1}^{L} g_{I_j} \right\|_{L^p}^p$$

$$\leq \sum_{j=1}^{L} \| c_j H\chi_{I_j} - g_{I_j} \|_{L^p}^p$$

$$\leq A_p \sum_{j=1}^{L} |c_j|^p \|\chi_{I_j}\|_{L^p}^p$$

$$= A_p \|f\|_{L^p}^p.$$

The last equality follows since the intervals do not overlap.

Claim 4: There is a $C > 0$ so that for all $f \in L^1_{0,n}$,

$$\|(\mathcal{P} \circ H)f\|_{L^p/X^p} \leq C\|f\|_{L^p}.$$

Proof of Claim 4: Let $f \in L^1_{0,n}$ and let $\{f_k : k \in \mathbb{N}\}$ be a sequence of functions which satisfy the hypothesis of Claim 3 with $f_k \to f$ in the norm of $L^1_{0,n}$, i.e., in $L^1(d\mu)$. Such a sequence exists by Lemma 6.4.24. By Claim 3,

(6.4.28) $$\|(\mathcal{P} \circ H)f\|_{L^p/X^p}^p \leq \|(\mathcal{P} \circ H)(f - f_k)\|_{L^p/X^p}^p + A_p \|f_k\|_{L^p}^p.$$

Let $\varepsilon > 0$ be given. From Proposition 6.4.25 the Hilbert transform acts continuously from $L^1_{0,n}$ to L^p and so for large enough k

$$\|(\mathcal{P} \circ H)(f - f_k)\|_{L^p/X^p}^p \leq \varepsilon.$$

To control the second term in eq.(6.4.28), note that

$$\|f_k\|_{L^p}^p \leq \|f_k - f\|_{L^p}^p + \|f\|_{L^p}^p.$$

From eq.(6.4.22), the inclusion operator $i : L^1_{0,n} \to L^p$ is continuous. Thus, since $f_k \to f$ in $L^1_{0,n}$, $\|f_k - f\|_{L^p}^p \leq \varepsilon$ for large enough k. From this we observe, from eq.(6.4.28), that

$$\|(\mathcal{P} \circ H)f\|_{L^p/X^p}^p \leq \varepsilon + A_p \varepsilon + A'_p \|f\|_{L^p}^p$$

for large enough k. The claim now follows.

To finish the proof of the proposition, we apply, via Lemma 6.4.23, the density of $L^1_{0,n}$ in L^p.

□

6.4.2.3. *An F. and M. Riesz theorem.* From Remark 4.5.4, a function $f \in L^1$ belongs to $\mathcal{H}^1(\mathbb{R})$ (i.e., is the boundary function for a $\mathcal{H}^1$ function), if and only if $H(\Re f) = \Im f$. An important corollary to Proposition 6.4.27 is that this result, in a sense, this "F. and M. Riesz theorem" can be extended to $\mathcal{H}^p$ for $p < 1$.

COROLLARY 6.4.29. *Let $p \in (0,1)$ and $f \in L^p$. Then $f \in \mathcal{H}^p(\mathbb{R})$ if and only if*

$$(\mathcal{P} \circ H)(\Re f) = \Im f + X^p.$$ [23]

Before proceeding, to the proof, we need the following lemma.

LEMMA 6.4.30. *For $p \in (1/(n+1), 1)$, $\mathcal{H}^p(\mathbb{R}) \cap L^1_{0,n}$ is dense in $\mathcal{H}^p(\mathbb{R})$.*

---

[23] Here we mean the extension of the operator $\mathcal{P} \circ H : L^1_{0,n} \to L^p/X^p$ to L^p.

PROOF. Let $f \in \mathcal{H}^p(\mathbb{R}) \cap \mathcal{H}^1(\mathbb{R})$ and set
$$f_\varepsilon := \frac{f}{(1 - i\varepsilon x)^n}.$$
Notice that $f_\varepsilon \in \mathcal{H}^p(\mathbb{R}) \cap \mathcal{H}^1(\mathbb{R})$ and, by the dominated convergence theorem, $f_\varepsilon \to f$ in $\mathcal{H}^p$ as $\varepsilon \to 0$. Moreover
$$x^k f_\varepsilon \in \mathcal{H}^1(\mathbb{R}), \quad k = 0, \cdots, n$$
and so (since the 0-th moment of an $\mathcal{H}^1(\mathbb{R})$ function vanishes, Corollary 4.3.6), $f_\varepsilon \in L^1_{0,n}$. Thus $\mathcal{H}^p(\mathbb{R}) \cap L^1_{0,n}$ is dense in $\mathcal{H}^1(\mathbb{R}) \cap \mathcal{H}^p(\mathbb{R})$. To finish, notice that $\mathcal{H}^1(\mathbb{R}) \cap \mathcal{H}^p(\mathbb{R})$ is dense in $\mathcal{H}^p(\mathbb{R})$ (see [**38**], p. 237 or the outline of this argument in Remark 4.2.3). □

PROOF OF COROLLARY 6.4.29. Suppose $f \in \mathcal{H}^p(\mathbb{R})$. For $g \in \mathcal{H}^p(\mathbb{R}) \cap L^1_{0,n}$ we have $H(\Re g) = \Im g$ and so certainly
$$(\mathcal{P} \circ H)(\Re g) = \Im g + X^p.$$
Now use Lemma 6.4.30 to approximate f in L^p-norm by such g. The continuity of the extension of the operator $\mathcal{P} \circ H : L^1_{0,n} \to L^p / X^p$ to L^p implies
$$(\mathcal{P} \circ H)(\Re f) = \Im f + X^p.$$

Conversely, suppose that $f \in L^p$ with $(\mathcal{P} \circ H)(\Re f) = \Im f + X^p$. Using the density of $L^1_{0,n}$ in L^p (Lemma 6.4.23), there is a sequence $\{f_k : k \in \mathbb{N}\}$ of real-valued functions in $L^1_{0,n}$ with compact support [24] such that
$$\Re f = \sum_k f_k \text{ in } L^p \text{ and } \sum_k \|f_k\|_{L^p}^p < \infty.$$
By the definition of the quotient metric on L^p/X^p and Proposition 6.4.27, there are $g_k \in X^p$ such that
$$\|Hf_k - g_k\|_{L^p} \leq A_p \|f_k\|_{L^p}$$
and we can even assume that g_k is real valued for all k [25]. The series $\sum_k Hf_k - g_k$ converges in L^p and hence $\sum_k (Hf_k - g_k) + X^p \in L^p/X^p$. Moreover
$$\begin{aligned}\sum_k (Hf_k - g_k) + X^p &= \sum_k (\mathcal{P} \circ H) f_k \\ &= (\mathcal{P} \circ H)\Big(\sum_k f_k\Big) \text{ by the continuity of } \mathcal{P} \circ H \\ &= (\mathcal{P} \circ H)(\Re f) \\ &= \Im f + X^p \text{ by assumption.}\end{aligned}$$
From this observe that
$$\sum_k (Hf_k - g_k) - \Im f \in X^p \subset \mathcal{H}^p(\mathbb{R}).$$

---

[24]Here we are using the facts that (i) if $g \in L^1_{0,n}$, then $\Re g \in L^1_{0,n}$; (ii) the functions in $L^1_{0,n}$ which are finite linear combinations of characteristic functions of bounded intervals are dense in $L^1_{0,n}$; (iii) the inclusion operator $i : L^1_{0,n} \to L^p$ is continuous.

[25]Indeed if $g \in X^p$, then $g = \lim_j r_j$, where r_j are rational functions in L^p with simple poles on $\mathbb{R}$. If $R_j := \frac{1}{2}(r_j + \overline{r_j})$, then $R_j \in X^p$ and $\Re g = \lim_j R_j$. Thus $\Re g \in X^p$ belongs to X^p whenever $g \in X^p$. Since Hf_k is real valued and $\|Hf_k - \Re(g_k)\|_{L^p} = \|\Re(Hf_k - g_k)\|_{L^p} \leq \|Hf_k - g_k\|_{L^p}$, we may choose g_k to be real-valued.

For each k, notice that
$$f_k + iHf_k \in \mathcal{H}^p(\mathbb{R})^{26} \text{ and } g_k \in X^p \subset \mathcal{H}^p \cap \overline{\mathcal{H}^p}.$$
Hence
$$f = \sum_k \left\{ f_k + i(Hf_k - g_k) \right\} - i \left\{ \sum_k (Hf_k - g_k) - \Im f \right\}$$
belongs to $\mathcal{H}^p(\mathbb{R})$. $\square$

6.4.2.4. Rational approximation.

COROLLARY 6.4.31. *Let $f \in \mathcal{H}^p(\mathbb{R})$ $(0 < p < 1)$ and real-valued. Then $f \in X^p$.*

PROOF. From the previous corollary
$$X^p = (\mathcal{P} \circ H)(\Re(if)) = if + X^p$$
and so $f \in X^p$. $\square$

At long last, we are finally ready to prove Theorem 6.4.16 ($X^p = \mathcal{H}^p \cap \overline{\mathcal{H}^p}$).

PROOF OF THEOREM 6.4.16. Note that $X^p \subset \mathcal{H}^p \cap \overline{\mathcal{H}^p}$. Now let $f \in \mathcal{H}^p \cap \overline{\mathcal{H}^p}$, i.e., $f, \overline{f} \in \mathcal{H}^p(\mathbb{R})$. Define the functions
$$g_1 := f + \overline{f}, \quad g_2 := i(f - \overline{f})$$
and observe that $g_1, g_2 \in \mathcal{H}^p(\mathbb{R})$ and are real-valued. From the previous corollary, g_1 and g_2 both belong to X^p. However, $ig_1 + g_2 = 2if$ and so $f \in X^p$. $\square$

6.4.3. Preliminaries for $1/p \notin \mathbb{N}$.

At this point in the main approximation theorem ($E^p(F, k)$ is the closure of the rational functions from this space), the proof will be broken down into two cases:
$$1/p \notin \mathbb{N} \text{ and } 1/p \in \mathbb{N}.$$

6.4.3.1. An approximation result.

To begin to deal with the case $1/p \notin \mathbb{N}$ recall, from Chapter 4, the spaces
$$\mathcal{H}^p = \left\{ f \in \mathfrak{H}(\mathbb{C}_+) : \sup_{y>0} \int |f(x+iy)|^p dx < \infty \right\}$$
$$\mathcal{H}^p(\mathbb{R}) = \left\{ g \in L^p : g(x) = \lim_{y \to 0^+} f(x+iy) \text{ a.e.}, f \in \mathcal{H}^p \right\}$$
$$\mathcal{H}^p_h = \left\{ u \in \text{Har}(\mathbb{C}_+) : u^* \in L^p \right\}$$
$$H^p(\mathbb{R}) = \left\{ \ell \in \mathcal{S}' : \ell(\psi) = \lim_{y \to 0^+} \int u(x+iy)\psi(x) dx, u \in \mathcal{H}^p_h \right\}$$
$$\mathcal{H}^p(\mathbb{C}\backslash\mathbb{R}) = \left\{ f \in \mathfrak{H}(\mathbb{C}\backslash\mathbb{R}) : \sup_{y \neq 0} \int |f(x+iy)|^p dx < \infty \right\}.$$

---

[26]The function $f_k + iHf_k$ is the boundary function of the $\mathcal{H}^p$ function
$$\int f_k(t) P_z(t)\, dt + i \int f_k(t) Q_z(t)\, dt = 1/i\pi \int f_k(t)(t-z)^{-1} dt,$$
see Theorem 4.3.1 and Remark 4.5.4. Also note that $1/i\pi \int f_k(t)(t-z)^{-1}\, dt$ belongs to $\mathcal{H}^p$ by Example 4.6.4 (since $f_k \in L^1_{0,n}$ and has compact support).

Also recall from Chapter 4, section 11, that to every $F \in \mathcal{H}^p(\mathbb{C}\setminus\mathbb{R})$, there is a (unique) distribution $\ell_F \in H^p(\mathbb{R})$ defined by

$$(6.4.32) \qquad \ell_F(\psi) := \lim_{y \to 0^+} \int \{ F(x+iy) - F(x-iy) \} \psi(x) dx$$

and

$$\|\ell_F\|_{H^p(\mathbb{R})} \asymp \|F\|_{\mathcal{H}^p(\mathbb{C}\setminus\mathbb{R})}.$$

Conversely, if $\ell \in H^p(\mathbb{R})$, the function

$$(6.4.33) \qquad F_\ell(z) := \frac{1}{2\pi i} \ell \left(\frac{1}{\cdot - z} \right)$$

belongs to $\mathcal{H}^p(\mathbb{C}\setminus\mathbb{R})$ and

$$\|F_\ell\|_{\mathcal{H}^p(\mathbb{C}\setminus\mathbb{R})} \asymp \|\ell\|_{H^p(\mathbb{R})}.$$

Moreover

$$\ell_{F_\ell} = \ell.$$

In our discussion below, we will abuse notation from time to time and equate functions in $\mathcal{H}^p(\mathbb{C}\setminus\mathbb{R})$ with their corresponding distributions in $H^p(\mathbb{R})$ via eq.(6.4.32) and eq.(6.4.33). To avoid the subscripts, we will let $\|\cdot\|$ denote the norm in $H^p(\mathbb{R})$ or $H^p(\mathbb{C}\setminus\mathbb{R})$.

From Definition 6.4.1 and Definition 6.4.4, recall the spaces

$$\mathcal{H}^p \cap \overline{\mathcal{H}^p} = \{ f \in \mathcal{H}^p(\mathbb{R}) : \overline{f} \in \mathcal{H}^p(\mathbb{R}) \}$$

$$\mathcal{H}^p(\mathbb{C}\setminus\mathbb{R})_* = \{ f \in \mathcal{H}^p(\mathbb{C}\setminus\mathbb{R}) : \lim_{y \to 0^+} f(x+iy) = \lim_{y \to 0^-} f(x+iy) \text{ a.e.} \}$$

and note that we can equate these spaces via eq.(6.4.3) and eq.(6.4.5).

DEFINITION 6.4.34. *For a closed set $E \subset \mathbb{R}$, we let $\mathcal{H}_E^p$ denote the functions in $\mathcal{H}^p(\mathbb{C}\setminus\mathbb{R})$ which are analytic across $\mathbb{R}\setminus E$.*

From Theorem 6.4.16, every $f \in \mathcal{H}^p(\mathbb{C}\setminus\mathbb{R})$ can be approximated by rational functions from $\mathcal{H}^p(\mathbb{C}\setminus\mathbb{R})$. This next result is a refinement of Theorem 6.4.16.

THEOREM 6.4.35. *Let $p \in (0,1)$ ($1/p \notin \mathbb{N}$) and F be a closed subset of $\mathbb{R}$. Then any $f \in \mathcal{H}_F^p \cap \mathcal{H}^p(\mathbb{C}\setminus\mathbb{R})_*$ is contained in the closure of the set of rational functions belonging to $\mathcal{H}^p(\mathbb{C}\setminus\mathbb{R})$ with poles in F.*

The proof of this result will require several steps. One important one being this next proposition.

PROPOSITION 6.4.36. *Suppose F is a closed subset of $\mathbb{R}$ and $p \in (0,1)$ ($1/p \notin \mathbb{N}$). Let $\{J_\gamma, \gamma \in \mathbb{N}\}$ be the (disjoint) complementary intervals to F. Then every function $f \in \mathcal{H}^p(\mathbb{C}\setminus\mathbb{R})$ can be written in the form*

$$f = \sum_\gamma f_\gamma + f_F,$$

where the above series converges in $\mathcal{H}^p(\mathbb{C}\setminus\mathbb{R})$, and

$$(6.4.37) \qquad f_\gamma \in \mathcal{H}_{J_\gamma^-}^p, \quad f_F \in \mathcal{H}_F^p, \quad \sum_\gamma \|f_\gamma\|_p^p + \|f_F\|_p^p \leq c_p \|f\|_p.$$

REMARK 6.4.38. We also remind the reader of the fact from Corollary 4.11.5 that $\ell \in H^p(\mathbb{R})$ has support (as a distribution) in a closed set E if and only if the corresponding function $F_\ell \in \mathcal{H}^p(\mathbb{C}\backslash\mathbb{R})$ is analytic across $\mathbb{R}\backslash E$, i.e., $F_\ell \in \mathcal{H}_E^p$. In several constructions below, we will not differentiate between the distributions and the functions and often say "$\ell \in \mathcal{H}_E^p$".

Recall from Example 4.6.6 that for fixed $p \in (0,1)$ ($1/p \notin \mathbb{N}$) and $f \in L^1([a,b])$, the function

$$(6.4.39) \qquad (Y_{[a,b]}f)(z) := \frac{1}{2\pi i}\int \frac{f(t)}{t-z}dt - \sum_{1 \leq j \leq n_p} \frac{c_j(f,a)}{(z-a)^j},$$

where

$$c_j(f,a) = -\frac{1}{2\pi i}\int (t-a)^{j-1}f(t)\,dt,$$

belongs to $\mathcal{H}^p$. A similar argument shows that in fact $Y_{[a,b]}f \in \mathcal{H}^p(\mathbb{C}\backslash\mathbb{R})$.

LEMMA 6.4.40. *If $p \in (0,1)$ ($1/p \notin \mathbb{N}$), then the map $f \to Y_{[a,b]}f$ is a continuous linear transformation from $L^1([a,b])$ to $\mathcal{H}^p(\mathbb{C}\backslash\mathbb{R})$ with*

$$\|Y_{[a,b]}f\| \leq C(b-a)^{1/p-1}\|f\|_{L^1},$$

where $C > 0$ does not depend on a,b or f.

PROOF. From the above, $f \to Y_{[a,b]}f$ is a linear operator from $L^1([a,b])$ to $\mathcal{H}^p(\mathbb{C}\backslash\mathbb{R})$. To prove it is continuous, we use the closed graph theorem. Suppose

$$f_n \to 0 \text{ in } L^1, \quad Y_{[a,b]}f_n \to g \text{ in } \mathcal{H}^p(\mathbb{C}\backslash\mathbb{R}).$$

Then

$$Y_{[a,b]}f_n \to 0, \quad Y_{[a,b]}f_n \to g \text{ pointwise in } \mathbb{C}\backslash\mathbb{R}.$$

Hence $g = 0$ and so, by the closed graph theorem, $f \to Y_{[a,b]}f$ is continuous.

To get the desired norm estimate, we define $\phi(x) := a+(b-a)x$ and $C_\phi g = g \circ \phi$. A computation reveals that

$$Y_{[a,b]}f = C_\phi^{-1} Y_{[0,1]} C_\phi f$$

and so

$$\begin{aligned}
\|Y_{[a,b]}f\| &= \|C_\phi^{-1}Y_{[0,1]}C_\phi f\| \\
&= (b-a)^{1/p}\|Y_{[0,1]}C_\phi f\| \quad \text{change of variables} \\
&\leq C(b-a)^{1/p}\|C_\phi f\|_{L^1} \quad Y_{[0,1]} \text{ is continuous} \\
&\leq C(b-a)^{1/p-1}\|f\|_{L^1} \quad \text{change of variables.}
\end{aligned}$$

$\square$

Since the proof of Proposition 6.4.36 will involve the atomic decomposition, it will be useful to view the functions $Y_{[a,b]}f \in \mathcal{H}^p(\mathbb{C}\backslash\mathbb{R})$ as their boundary distributions $\ell_{Y_{[a,b]}f} \in H^p(\mathbb{R})$ which turn out to be

$$\ell_{Y_{[a,b]}f} = f - \sum_{1 \leq j \leq [1/p]} (-1)^j 2\pi i \frac{c_j(f,a)}{(j-1)!}\delta_a^{(j-1)}.$$

The avoid all the subscripts, we will equate the function $Y_{[a,b]}f$ and the distribution $\ell_{Y_{[a,b]}f}$.

6.4. RATIONAL APPROXIMATION

From Theorem 4.9.3, a distribution $\ell \in H^p(\mathbb{R})$ has an atomic decomposition

$$\ell = \sum_{k=1}^{\infty} \lambda_k b_k,{}^{27}$$

where b_k is an atom, i.e., a function with support on an interval Δ_{b_k} with

$$|b_k| \leq |\Delta_{b_k}|^{-1/p} \text{ and } \int x^j b_k(x)dx = 0 \ \forall \ j = 0, 1, \cdots, [1/p] - 1.$$

Moreover,

$$\|\ell\|^p \asymp \sum_k |\lambda_k|^p.$$

PROOF OF PROPOSITION 6.4.36. We will begin by showing that for each atom b, there are $g_\gamma \in \mathcal{H}^p_{J_\gamma^-}$ and $g_F \in \mathcal{H}^p_F$ with

(6.4.41) $$b = \sum_\gamma g_\gamma + g_F$$

$$\sum_\gamma \|g_\gamma\|^p + \|g_F\|^p \leq C_p.$$

To this end, we have two cases to consider. If $\Delta_b \subset J_\gamma$ for some $\gamma \in \mathbb{N}$, then $b \in \mathcal{H}^p_{J_\gamma^-}$ and so we can take g_γ to be the atom b and $g_F = 0$, i.e., just a single term in the sum in eq.(6.4.41). Otherwise we are left to deal with the case where $\Delta_b \cap F \neq \emptyset$. For this case, define for each γ

(6.4.42) $$g_\gamma = Y_{J_\gamma \cap \Delta_b}(b\chi_{J_\gamma}),$$

where we understand that the a in the definition of "$Y_{[a,b]}f$" (from eq.(6.4.39)) is taken to be one of the endpoints of J_γ (possibly taking the right-hand endpoint of J_γ to be the a instead of the left). From Lemma 6.4.40,

$$\|g_\gamma\|^p \leq C_p \frac{|\Delta_b \cap J_\gamma|}{|\Delta_b|}.$$

Furthermore,

$$\sum_\gamma \|g_\gamma\|^p \leq \frac{C_p}{|\Delta_b|} \sum_\gamma |\Delta_b \cap J_\gamma| \leq C_p$$

and so $\sum_\gamma g_\gamma \in H^p(\mathbb{R})$. The distribution

$$g_F := b - \sum_\gamma g_\gamma$$

also belongs to $H^p(\mathbb{R})$ and $\|g_F\|^p \leq \|b\|^p + C_p \leq C_p$. Recall from eq.(4.9.2) that the $H^p(\mathbb{R})$ norms of atoms have a universal upper bound depending only on p.

The support of the distribution g_γ is contained in J_γ^- and so $g_\gamma \in \mathcal{H}^p_{J_\gamma^-}$. Furthermore, in terms of distributions,

$$g_F = b - \sum_\gamma \left\{ b\chi_{J_\gamma} - \sum_{j=0}^{[1/p]-1} C_{j,\gamma} \delta^{(j)}_{a_\gamma} \right\}{}^{28}.$$

---

[27] i.e., $\ell(\psi) = \sum_k \lambda_k \int b_k(x)\psi(x)dx$.

[28] Here a_γ is one of the endpoints of J_γ chosen as in eq.(6.4.42).

Thus, if $\psi \in C_0^\infty(J_{\gamma_0})$, then

$$g_F(\psi) = \int \left(b\chi_{J_{\gamma_0}} - \sum_\gamma b\chi_{J_\gamma \cap J_{\gamma_0}} \right) \psi(x) dx = 0.$$

Thus the support of g_F is contained in $\mathbb{R}\setminus J_\gamma$ for all $\gamma \in \mathbb{N}$ which means the support of g_F is contained in F, whence $g_F \in \mathcal{H}_F^p$. Thus we have shown the result for atoms b.

For a general $\ell \in H^p(\mathbb{R})$, with atomic decomposition $\ell = \sum_k \lambda_k b_k$, we decompose each atom

$$b_k = \sum_\gamma g_\gamma^{b_k} + g_F^{b_k}$$

as in eq.(6.4.41) and observe that, at least formally,

$$(6.4.43) \quad \ell = \sum_k \lambda_k b_k = \sum_k \lambda_k \left\{ \sum_\gamma g_\gamma^{b_k} + g_F^{b_k} \right\} = \sum_\gamma \left\{ \sum_k \lambda_k g_\gamma^{b_k} \right\} + \sum_k \lambda_k g_F^{b_k}.$$

Letting

$$g_\gamma := \sum_k \lambda_k g_\gamma^{b_k} \quad \text{and} \quad g_F := \sum_k \lambda_k g_F^{b_k},$$

we obtain the desired decomposition. To finish the proof, we will show that the sums which define the distributions g_γ and g_F converge in $H^p(\mathbb{R})$ and that reversing the order of summation in eq.(6.4.43) is justified.

Since

$$\sum_\gamma \|g_\gamma^{b_k}\|^p + \|g_F^{b_k}\|^p \leq C_p \quad \text{and} \quad \sum_k |\lambda_k|^p < \infty, \quad ^{29}$$

both the series $\sum_k \lambda_k g_\gamma^{b_k}$ and $\sum_k \lambda_k g_F^{b_k}$ converge in $H^p(\mathbb{R})$ for each fixed γ. Furthermore,

$$\infty > \sum_k \sum_\gamma |\lambda_k|^p \|g_\gamma^{b_k}\|^p = \sum_\gamma \sum_k |\lambda_k|^p \|g_\gamma^{b_k}\|^p$$
$$\geq C_p \sum_\gamma \|\sum_k \lambda_k g_\gamma^{b_k}\|^p \geq C_p \|\sum_\gamma \sum_k \lambda_k g_\gamma^{b_k}\|^p.$$

Thus

$$\sum_\gamma g_\gamma = \sum_\gamma \sum_k \lambda_k g_\gamma^{b_k}$$

converges in $H^p(\mathbb{R})$. To finish the proof, we justify the change of summation in

$$\sum_k \sum_\gamma \lambda_k g_\gamma^{b_k} = \sum_\gamma \sum_k \lambda_k g_\gamma^{b_k}.$$

---

[29] It is important to note that the constant C_p is independent of the atom b_k.

Indeed for each $K \in \mathbb{N}$,

$$\begin{aligned}
\|\sum_\gamma \sum_{k=1}^\infty \lambda_k g_\gamma^{b_k} - \sum_{k=1}^K \sum_\gamma \lambda_k g_\gamma^{b_k}\|^p &= \|\sum_\gamma \sum_{k=1}^\infty \lambda_k g_\gamma^{b_k} - \sum_\gamma \sum_{k=1}^K \lambda_k g_\gamma^{b_k}\|^p \\
&\leq \sum_\gamma \|\sum_{k=K+1}^\infty \lambda_k g_\gamma^{b_k}\|^p \\
&\leq \sum_\gamma \sum_{k=K+1}^\infty |\lambda_k|^p \|g_\gamma^{b_k}\|^p \\
&= \sum_{k=K+1}^\infty |\lambda_k|^p \sum_\gamma \|g_\gamma^{b_k}\|^p \\
&\leq C_p \sum_{k=K+1}^\infty |\lambda_k|^p
\end{aligned}$$

which goes to zero as $K \to \infty$ since the series $\sum_k |\lambda_k|^p$ converges. $\square$

At this point in the exposition, we give some examples of this decomposition theorem for rational functions in $\mathcal{H}^p(\mathbb{C}\backslash\mathbb{R})$. These examples will not only give the reader somewhat of a feel for this important result but will serve as valuable results in our construction later on.

EXAMPLE 6.4.44. As we saw from Example 4.11.3, if $p \in (1/2, 1)$,

$$f(z) = \sum_{j=1}^N \frac{c_j}{z - a_j}, \quad a_j \in \mathbb{R}, c_j \in \mathbb{C}$$

belongs to $\mathcal{H}^p(\mathbb{C}\backslash\mathbb{R})$ if and only if $N \geq 2$ and $\sum_j c_j = 0$ [30]. So suppose $F = [0, 1]$ and f as above belongs to $\mathcal{H}^p(\mathbb{C}\backslash\mathbb{R})$. Then $J_1 = (-\infty, 0)$ and $J_2 = (1, \infty)$ are the complimentary intervals of F. We write

$$f = \sum_{a_j \in J_1} \frac{c_j}{z - a_j} + \sum_{a_j \in F} \frac{c_j}{z - a_j} + \sum_{a_j \in J_2} \frac{c_j}{z - a_j}$$

and note that each summand need not belong to $\mathcal{H}^p(\mathbb{C}\backslash\mathbb{R})$ since, for example, $\sum_{a_j \in F} c_j$ may not be zero.

Let

$$d_1 = -\sum_{a_j \in J_1} c_j, \quad d_2 = -\sum_{a_j \in F} c_j + d_1.$$

Then, letting

$$f_{J_1} := \sum_{a_j \in J_1} \frac{c_j}{z - a_j} + \frac{d_1}{z}$$

$$f_F := -\frac{d_1}{z} + \sum_{a_j \in F} \frac{c_j}{z - a_j} + \frac{d_2}{z - 1}$$

$$f_{J_2} := -\frac{d_2}{z - 1} + \sum_{a_j \in J_2} \frac{c_j}{z - a_j},$$

---

[30]This conditions says that for large $|x|$, $|f(x)| = O(|x|^{-2})$ and so $f \in L^p$ for $p \in (1/2, 1)$.

we observe $f = f_{J_1} + f_F + f_{J_2}$. By construction, $\sum_{a_j \in J_1} c_j + d_1 = 0$ and so $f_{J_1} \in \mathcal{H}^1_{J_1^-}$. Also $-d_1 + \sum_{a_j \in F} c_j + d_2 = 0$ and so $f_F \in \mathcal{H}^p_F$. Finally, by our above computations and the fact that $\sum_j c_j = 0$, we conclude $-d_2 + \sum_{a_j \in J_2} c_j = 0$. Hence $f_{J_2} \in \mathcal{H}^p_{J_2^-}$. Note also that all the functions in the decomposition are rational functions.

This example suggests the more general principle that if f is a rational function in $\mathcal{H}^p(\mathbb{C}\setminus\mathbb{R})$ ($p \in (0,1)$, $1/p \notin \mathbb{N}$), then f can be written in the form

$$f = \sum_\gamma f_\gamma + f_F,$$

where the above sum is *finite* and f_γ and f_F are rational functions from $\mathcal{H}^p_{J_\gamma^-}$ ($\mathcal{H}^p_F$ respectively). As it turns out, this is indeed true and the proof follows along the same lines as the example above but gets more technical when $p < 1/2$. In this case, one needs to insert higher order poles at the endpoints of the intervals J_γ.

PROPOSITION 6.4.45. *Let $p \in (0,1)$ ($1/p \notin \mathbb{N}$), F a closed set in $\mathbb{R}$ with disjoint complimentary open intervals $\{J_\gamma : \gamma \in \mathbb{N}\}$, and*

$$r = \sum_{j=1}^N \frac{c_j}{z - a_j} \in \mathcal{H}^p(\mathbb{C}\setminus\mathbb{R}).$$

Then

$$r = \sum_\gamma r_\gamma + r_F,$$

where the above sum is finite, r_γ is a rational function from $\mathcal{H}^p_{J_\gamma^-}$, and r_F is a rational function from $\mathcal{H}^p_F$.

PROOF. From Example 4.11.3, we write a general rational function f with poles on the real axis in its distributional form as

(6.4.46) $$\ell_f = \ell = \sum_{j=1}^N \sum_{k=0}^{n_p-1} c_{j,k} \delta_{a_j}^{(k)}$$

and note that

$$\ell \in H^p(\mathbb{R}) \Leftrightarrow \ell(x^r) = 0 \;\; \forall\, r = 0, \cdots, n_p - 1.$$

For a distribution ℓ of the form eq.(6.4.46) (not necessarily in $H^p(\mathbb{R})$), we make the following claim: Given $a \in \mathbb{R}$, there exists unique constants $b_0, \cdots, b_{n_p-1}$ such that

(6.4.47) $$S := \ell - \sum_{s=0}^{n_p-1} b_s \delta_a^{(s)} \in H^p(\mathbb{R}).$$

To prove this claim, we will produce constants $b_0, \cdots, b_{n_p-1}$ so that $S(x^r) = 0$ for all $r = 0, \cdots, n_p - 1$. Since

$$\delta_a^{(s)}(x^r) = (-1)^s a^{r-s} \frac{r!}{(r-s)!},$$

then

$$S(x^r) = \ell(x^r) - \sum_{s=0}^r (-1)^s b_s a^{r-s} \frac{r!}{(r-s)!}.$$

6.4. RATIONAL APPROXIMATION

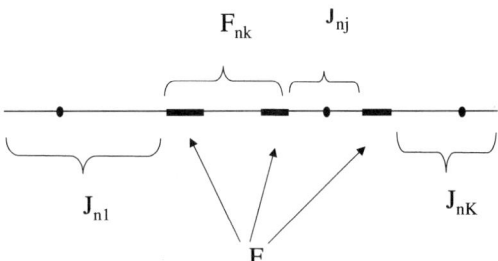

FIGURE 2. The closed set F along with the complimentary intervals which contain the poles of the rational function.

To show that $b_0, \cdots, b_{n_p-1}$, can be chosen so that $S(x^r) = 0$ for $r = 0, \cdots, n_p - 1$, we need to solve the matrix equation

$$\begin{pmatrix} 1 & 0 & 0 & 0 & \cdots & 0 \\ a & -1 & 0 & 0 & \cdots & 0 \\ a^2 & -2a & 1 & 0 & \cdots & 0 \\ \vdots & \vdots & \vdots & \vdots & \cdots & 0 \\ a^{n_p-1} & \cdots & \cdots & \cdots & \vdots & \pm 1 \end{pmatrix} \begin{pmatrix} b_0 \\ \vdots \\ \vdots \\ \vdots \\ b_{n_p-1} \end{pmatrix} = \begin{pmatrix} \ell(1) \\ \ell(x) \\ \ell(x^2) \\ \vdots \\ \ell(x^{n_p-1}) \end{pmatrix}$$

which of course can be done since the matrix is invertible. Thus we have shown the claim.

To prove the proposition, we let

$$\ell = \sum_{j=1}^{N} c_j \delta_{a_j} \in H^p(\mathbb{R})$$

be our distributional realization of a rational function from $\mathcal{H}^p(\mathbb{C}\setminus\mathbb{R})$ with simple poles $a_1, \cdots, a_N$. Let

$$J_{n_1}, \cdots, J_{n_K}$$

be the complimentary intervals of F which contain the points a_j. Without loss of generality (the proof can easily be adjusted otherwise), we will assume that $J_{n_1} = (-\infty, e_{n_1})$ and $J_{n_K} = (d_{n_K}, \infty)$. We denote the bounded intervals by $J_{n_k} = (d_{n_k}, e_{n_k})$. We also let F_{n_k} be the portion of F which lies between J_{n_k} and $J_{n_{k+1}}$ ($F \subset [e_{n_k}, d_{n_{k+1}}]$, see Figure 2).

For any distribution of the form eq.(6.4.46) and $a \in \mathbb{R}$, we let

$$L(\ell, a) = \sum_{s=0}^{n_p-1} b_s \delta_a^{(s)}$$

be the distribution obtained in eq.(6.4.47) such that $\ell - L(\ell, a) \in H^p(\mathbb{R})$.

For our distribution $\ell = \sum_{j=1}^{N} c_j \delta_{a_j}$, define the distributions

$$\ell_{J_{n_k}} = \sum_{a_j \in J_{n_k}} c_j \delta_{a_j}, \quad \ell_{F_{n_k}} = \sum_{a_j \in F_{n_k}} c_j \delta_{a_j}$$

and observe that

$$\ell = \sum_{k=1}^{K} \left\{ \ell_{J_{n_k}} + \ell_{F_{n_k}} \right\}.$$

Define inductively the following distributions

$$L_{J_{n_1}} := \ell_{J_{n_1}} - L(\ell_{J_{n_1}}, e_{n_1})$$

$$L_{F_{n_1}} := L(\ell_{J_{n_1}}, e_{n_1}) + \ell_{F_{n_1}} - L(L(\ell_{J_{n_1}}, e_{n_1}) + \ell_{F_{n_1}}, d_{n_2})$$

$$L_{J_{n_2}} := L(L(\ell_{J_{n_1}}, e_{n_1}) + \ell_{F_{n_1}}, d_{n_2}) + \ell_{J_{n_2}} - L(L(L(\ell_{J_{n_1}}, e_{n_1}) + \ell_{F_{n_1}}, d_{n_2}), e_{n_2})$$

and so on (at the last step, do not subtract any term off). By our definitions, $L_{J_{n_k}} \in \mathcal{H}^p_{J_{n_k}^-}$ and $L_{F_{n_k}} \in \mathcal{H}^p_{F_{n_k}}$. Moreover, they represent "rational functions". Finally

$$\sum_{k=1}^{K} L_{J_{n_k}} + \sum_{k=1}^{K} L_{F_{n_k}} = \ell.$$

$\square$

COROLLARY 6.4.48. *Let p, F, and $\{J_\gamma : \gamma \in \mathbb{N}\}$ be as in Proposition 6.4.36. Suppose that for every $\gamma \in \mathbb{N}$ there is a bounded operator $A_\gamma : \mathcal{H}^p_{J_\gamma^-} \to \mathcal{H}^p(\mathbb{C}\backslash\mathbb{R})$ and there is also a bounded operator $A_F : \mathcal{H}^p_F \to \mathcal{H}^p(\mathbb{C}\backslash\mathbb{R})$. Moreover, suppose that the operators $A_\gamma, \gamma \in \mathbb{N}$ are uniformly bounded* [31] *and $A_\gamma f = A_F f$ for every $f \in \mathcal{H}^p_{F \cap J_\gamma^-}$. Then there is a unique bounded operator $A : \mathcal{H}^p(\mathbb{C}\backslash\mathbb{R}) :\to \mathcal{H}^p(\mathbb{C}\backslash\mathbb{R})$ such that $A|\mathcal{H}^p_{J_\gamma^-} = A_\gamma$ for all $\gamma \in \mathbb{N}$ and $A|\mathcal{H}^p_F = A_F$.*

PROOF. By Proposition 6.4.36, a given $f \in \mathcal{H}^p(\mathbb{C}\backslash\mathbb{R})$ can be decomposed as $f = \sum_\gamma f_\gamma + f_F$ and the conditions in eq.(6.4.37) hold. The desired operator is then

$$Af := \sum_\gamma A_\gamma f_\gamma + A_F f_F.$$

The operators $A_\gamma, \gamma \in \mathbb{N}$ are uniformly bounded, and so A is bounded. Furthermore, A satisfies the conditions $A|\mathcal{H}^p_{J_\gamma^-} = A_\gamma$ for all $\gamma \in \mathbb{N}$ and $A|\mathcal{H}^p_F = A_F$. We need to show that A is well-defined; that is to say, the definition above does not depend on the particular choice of expansion $f = \sum_\gamma f_\gamma + f_F$.

To this end, suppose

$$\sum_\gamma f_\gamma + f_F = 0$$

with

$$f_\gamma \in \mathcal{H}^p_{J_\gamma^-}, \quad f_F \in \mathcal{H}^p_F, \quad \sum_\gamma \|f_\gamma\|^p < \infty.$$

Observe that if $\gamma \neq \gamma_0$, then $f_\gamma \in \mathcal{H}^p_{\mathbb{R}\backslash J_{\gamma_0}}$ and $f_F \in \mathcal{H}^p_{\mathbb{R}\backslash J_{\gamma_0}}$. Therefore

$$f_{\gamma_0} = -f_F - \sum_{\gamma \in \mathbb{N}\backslash\{\gamma_0\}} f_\gamma \in \mathcal{H}^p_{\mathbb{R}\backslash J_{\gamma_0}}.$$

---

[31] i.e., $\|A_\gamma\| \leq C$ for all γ.

As a result,
$$f_{\gamma_0} \in \mathcal{H}^p_{\mathbb{R}\setminus J_{\gamma_0}} \cap \mathcal{H}^p_{J^-_{\gamma_0}} \subset \mathcal{H}^p_{F\cap J^-_\gamma}.$$
Thus, by our hypothesis,
$$\sum_\gamma A_\gamma f_\gamma + A_F f_F = \sum_\gamma A_F f_\gamma + A_F f_F = A_F \left(\sum_\gamma f_\gamma + f_F\right) = 0.$$
Hence A is well defined and has the desired properties. □

COROLLARY 6.4.49. *Suppose $p \in (0,1)$ ($1/p \notin \mathbb{N}$) and F is a closed subset of $\mathbb{R}$. Then there is a continuous projection from $\mathcal{H}^p(\mathbb{C}\setminus\mathbb{R})$ onto $\mathcal{H}^p_F$.*

PROOF. Applying Corollary 6.4.48, it suffices to define the operators
$$A_F : \mathcal{H}^p_F \to \mathcal{H}^p_F \quad \text{and} \quad A_\gamma : \mathcal{H}^p_{J^-_\gamma} \to \mathcal{H}^p_{F\cap J^-_\gamma}$$
and show they are uniformly bounded with $A_F f = A_\gamma f$ whenever $f \in \mathcal{H}^p_{F\cap J^-_\gamma}$.

Define $A_F f = f$ for $f \in \mathcal{H}^p_F$. The definition of the operators A_γ separates into two distinct cases, $|J_\gamma| = \infty$ and $|J_\gamma| < \infty$. If $|J_\gamma| = \infty$, then $J^-_\gamma \cap F$ is a single point and so $\mathcal{H}^p_{F\cap J^-_\gamma} = (0)$.[32] In this case we define $A_\gamma : \mathcal{H}^p_{J^-_\gamma} \to \mathcal{H}^p_{F\cap J^-_\gamma}$ by $A_\gamma = 0$. If $|J_\gamma| < \infty$, we proceed as follows: Notice that $\mathcal{H}^p_{\{0,1\}}$ is a non-zero finite dimensional subspace (consisting only of rational functions with poles at $z = 0$ and $z = 1$).[33] Since the linear functionals on $\mathcal{H}^p_{[0,1]}$ separate points (indeed for each $a \in \mathbb{C}\setminus\mathbb{R}$, the linear functional $f \to f(a)$ is continuous), there is a continuous projection $P : \mathcal{H}^p_{[0,1]} \to \mathcal{H}^p_{\{0,1\}}$ (Corollary 6.4.18). The affine function ϕ_γ with $\phi_\gamma(J_\gamma) = (0,1)$ induces the bounded composition operator
$$C_{\phi_\gamma} : \mathcal{H}^p_{[0,1]} \to \mathcal{H}^p_{J^-_\gamma}$$
with uniformly bounded norm in γ. Hence the operator
$$A_\gamma := C_{\phi_\gamma} P C_{\phi_\gamma}^{-1}$$
is a projection operator from $\mathcal{H}^p_{J^-_\gamma}$ onto $\mathcal{H}^p_{F\cap J^-_\gamma}$ with uniformly bounded norm in γ. Since A_F and A_γ are continuous projection operators (which also satisfy $A_F f = A_\gamma f$ whenever $f \in \mathcal{H}^p_{F\cap J^-_\gamma}$), we can apply Corollary 6.4.48, to conclude the operator $A : \mathcal{H}^p(\mathbb{C}\setminus\mathbb{R}) \to \mathcal{H}^p_F$ is the desired continuous projection operator. □

We are finally ready for the proof of Theorem 6.4.35.

PROOF OF THEOREM 6.4.35. Let $f \in \mathcal{H}^p(\mathbb{C}\setminus\mathbb{R})_* \cap \mathcal{H}^p_F$. By Theorem 6.4.16, f can be approximated by a sequence of rational functions $\{r_j : j \in \mathbb{N}\}$ from $\mathcal{H}^p(\mathbb{C}\setminus\mathbb{R})$ whose poles are on the real axis. By the construction of the operator A in Corollary 6.4.49 and an application of Proposition 6.4.45, one can check that for each $j \in \mathbb{N}$, Ar_j is a rational function with poles in F. Moreover
$$f = Af = \lim_{j\to\infty} Ar_j \quad \text{in } \mathcal{H}^p(\mathbb{C}\setminus\mathbb{R}).$$
□

---

[32] $\mathcal{H}^p_{\{a\}}$ would be the space of constant multiples of the function $(z-a)^{-1}$. But $(x-a)^{-1}$ does not belong to L^p for any $p \in (0,1)$.

[33] By eq.(4.8.14), a distribution $\ell \in H^p(\mathbb{R})$ has order less than $[1/p]-1$. Since this distribution also has finite support in $\{0,1\}$, then by eq.(4.8.12), it can be written as a finite linear combination of point masses (and their derivatives) at the points $x = 0$ and $x = 1$.

6.4.3.2. *An operator.* In this section, we will define a certain operator which will help us control the order and location of the poles of our functions in Aleksandrov's rational approximation argument. The argument is quite technical and we begin with a few elementary facts.

LEMMA 6.4.50. *Let $\alpha > 0$ and $\phi \in C_0^\infty$ be given. Then there is a constant $C_{\phi,\alpha} > 0$ such that for any interval $I = [a, b]$, there is a polynomial p_I with $\deg(p_I) \leq [\alpha]$ such that*
$$|\phi - p_I| \leq C_{\phi,\alpha}|I|^\alpha \quad on\ I.^{34}$$

PROOF. If $[\alpha] = 0$, set $p_I = \phi(a)$ and notice that for $x \in I$,
$$\begin{aligned}|\phi(x) - p_I| &= \frac{|\phi(x) - \phi(a)|}{|x - a|^\alpha}|x - a|^\alpha \\ &\leq \sup_{x,y \in \mathbb{R}} \frac{|\phi(x) - \phi(y)|}{|x - y|^\alpha}|I|^\alpha \\ &\leq C_{\phi,\alpha}|I|^\alpha.\end{aligned}$$

For $[\alpha] \geq 1$, define $f := \phi - x^{[\alpha]}\phi^{([\alpha])}(a)/[\alpha]!$ and let
$$q := \sum_{k=0}^{[\alpha]-1} \frac{f^{(k)}(a)}{k!}(x-a)^k.$$

By Taylor's theorem,
$$(6.4.51) \quad |f(x) - q(x)| \leq \sup_{c \in I}|f^{([\alpha])}(c)|\frac{|x-a|^{[\alpha]}}{[\alpha]!} \leq \sup_{c \in I}|f^{([\alpha])}(c)|\frac{|I|^{[\alpha]}}{[\alpha]!} \quad \forall\, x \in I.$$

However, notice that $f^{([\alpha])}(c) = \phi^{([\alpha])}(c) - \phi^{([\alpha])}(a)$ and so
$$\begin{aligned}|f^{([\alpha])}(c)| &= \frac{|\phi^{([\alpha])}(c) - \phi^{([\alpha])}(a)|}{|c-a|^{[\alpha]-\alpha}}|c-a|^{[\alpha]-\alpha} \\ &\leq |I|^{[\alpha]-\alpha}\sup_{x,y \in \mathbb{R}}\frac{|\phi^{([\alpha])}(x) - \phi^{([\alpha])}(y)|}{|x-y|^{[\alpha]-\alpha}} \\ &= |I|^{[\alpha]-\alpha}C_{\phi,\alpha}.\end{aligned}$$

Combining this with eq.(6.4.51) we get
$$\left|\left\{\phi - x^{[\alpha]}\phi^{([\alpha])}(a)/[\alpha]!\right\} - q\right| = |f - q| \leq C_{\phi,\alpha}|I|^{[\alpha]} \quad on\ I.$$

Now define
$$p_I := x^{[\alpha]}\phi^{([\alpha])}(a)/[\alpha]! + q.$$
□

COROLLARY 6.4.52. *Let $\alpha > 0$ and $\phi \in C_0^\infty(a, b)$ be given. Then there is a constant $C_{\phi,\alpha} > 0$ such that for any interval $I = [a, b]$,*
$$|\phi| \leq C_{\phi,\alpha}|I|^\alpha \quad on\ I.$$

PROOF. In the proof of Lemma 6.4.50, observe that the polynomial p_I is equal to zero. □

---

[34]This result is reminiscent of the equivalent definition of the Lipschitz/Zygmund classes due to Campanato [18] and Meyers [62], see Remark 4.8.15.

6.4. RATIONAL APPROXIMATION

DEFINITION 6.4.53. For a fixed compact interval $I \subset \mathbb{R}$ and $a \in \mathbb{R}$, define for each function $f \in L^1(I)$ the complex numbers

$$c_j(f,a) := \frac{1}{j!} \int_I (a-t)^j f(t)\, dt, \quad j = 0, 1, \cdots, [1/p] - 1.$$

Also define the distribution

$$Y_a f := f - \sum_{j=0}^{[1/p]-1} c_j(f,a) \delta_a^{(j)}.$$

By the proof of Lemma 6.4.40, the mapping $f \to Y_a f$ is a continuous linear operator from $L^1(I)$ to $H^p(\mathbb{R})$ whenever $p \in (0,1)$ ($1/p \notin \mathbb{N}$). The main result of this subsection is the following extension of this operator.

PROPOSITION 6.4.54. Let $p \in (0,1)$ ($1/p \notin \mathbb{N}$), $\phi \in C_0^\infty$, and $a \in \mathbb{R}$ be fixed. Then for each $\ell \in H^p(\mathbb{R})$, the distribution $X_{a,\phi}\ell$ defined by

$$X_{a,\phi}\ell := \phi\ell - \sum_{j=0}^{[1/p]-1} \ell\left(\frac{1}{j!}\phi(\cdot)(a-\cdot)^j\right) \delta_a^{(j)}$$

belongs to $H^p(\mathbb{R})$. Moreover the mapping $\ell \to X_{a,\phi}\ell$ is continuous from $H^p(\mathbb{R})$ to itself.

PROOF. For an atom b, notice that $X_{a,\phi}b = Y_a(\phi b)$. Here the interval I in the definition of Y_a will be fixed in the remark below. We will show that

(6.4.55) $$\|Y_a(\phi b)\| \leq C_{a,\phi}.$$

Once this is verified, we can conclude that if $\ell \in H^p(\mathbb{R})$ has atomic decomposition $\ell = \sum_k \lambda_k b_k$ (with $\|\ell\|^p \asymp \sum_k |\lambda_k|^p$), then

(6.4.56) $$\left\| \sum_k \lambda_k Y_a(\phi b_k) \right\|^p \leq C_{a,\phi} \sum_k |\lambda_k|^p \leq C_{a,\phi}\|\ell\|^p$$

and so the distribution $\sum_k \lambda_k Y_a(\phi b_k)$ belongs to $H^p(\mathbb{R})$ (since each $Y_a(\phi b_k)$ belongs to $H^p(\mathbb{R})$ and the series converges in the norm of $H^p(\mathbb{R})$). Furthermore,

$$\begin{aligned}
\sum_k \lambda_k Y_a(\phi b_k) &= \sum_k \lambda_k \phi b_k - \sum_k \lambda_k \sum_{j=0}^{[1/p]-1} c_j(\phi b_k, a)\delta_a^{(j)} \\
&= \phi\ell - \sum_{j=0}^{[1/p]-1} \left\{ \sum_k \lambda_k c_j(\phi b_k, a) \right\} \delta_a^{(j)} \\
&= \phi\ell - \sum_{j=0}^{[1/p]-1} \ell\left(\left(\frac{1}{j!}\phi(\cdot)(a-\cdot)^j\right)\right) \delta_a^{(j)} \\
&= X_{a,\phi}\ell.
\end{aligned}$$

Finally, from eq.(6.4.56), $\|X_{a,\phi}\ell\| \leq C_{a,\phi}\|\ell\|$ and so the mapping $\ell \to X_{a,\phi}\ell$ is continuous.

The rest of the proof will be dedicated to proving the estimate in eq.(6.4.55). Before starting, however, we need to make a few remarks.

REMARK 6.4.57.   1. Throughout this proof, J will be a compact interval which contains the support of ϕ and I be a compact interval which has the same center as J but twice as long as I. This will be our fixed interval in the definition of Y_a above.

2. By the definition of Y_a on atoms, $Y_a(\phi b) = 0$ whenever the support of b does not intersect the support of ϕ.

3. Without loss of generality, we can assume that the atoms $\{b_k : k \in \mathbb{N}\}$ in the atomic decomposition of ℓ are such that

$$\int b_k(t) t^j \, dt = 0 \ \forall \, j = 0, \cdots, 2([1/p] - 1).$$

This follows from the Coifman theory, see [95], Chapter 3 (also see Remark 4.9.5).

4. From Corollary 6.4.52,

$$|\phi| \leq C_\phi |I|^{1/p-1}. \tag{6.4.58}$$

5. From Lemma 6.4.50, there is a constant $C_\phi > 0$ such that for each Δ_b (the interval which supports the atom b), there is a polynomial P of degree less than $1/p - 1$ (note that $1/p \notin \mathbb{N}$) such that

$$|\phi - P| \leq C_\phi |\Delta_b|^{1/p-1} \text{ on } \Delta_b. \tag{6.4.59}$$

We are now ready to prove eq.(6.4.55).

Case 1: $|\Delta_b| \geq |I|$. In this case

$$\begin{aligned}
\|Y_a(\phi b)\| &\leq C_{a,\phi} \int_{\Delta_b \cap I} |\phi||b| \, dt \quad \text{from the continuity of } f \to Y_a f \\
&\leq C_{a,\phi} |\Delta_b| |I|^{1/p-1} |\Delta_b|^{-1/p} \quad \text{from eq.(6.4.58)} \\
&\leq C_{a,\phi} \quad \text{since } |\Delta_b| \geq |I|.
\end{aligned}$$

Case 2: $|\Delta_b| \leq |I|$. In this case, we chose our polynomial P (which depends on Δ_b) as in eq.(6.4.59). We write

$$\phi b = b(\phi - P) + Pb.$$

Note that by our construction of the interval I, our tacit assumption that $\Delta_b \cap J \neq \emptyset$, and our assumption that $|\Delta_b| \leq |I|$, we have $\Delta_b \subset I$. [35]

$$\begin{aligned}
\|Y_a(b(\phi - P))\| &\leq C_{a,\phi} \int_{\Delta_b} |b||\phi - P| \, dt \quad \text{from the continuity of } f \to Y_a f \\
&\leq C_{a,\phi} |\Delta_b| |\Delta_b|^{-1/p} |\Delta_b|^{1/p-1} \quad \text{from eq.(6.4.59)} \\
&\leq C_{a,\phi}
\end{aligned}$$

To estimate $\|Y_a(Pb)\|$, we note that

$$\int b(t) t^j \, dt = 0 \ \forall \, j = 0, \cdots, 2([1/p] - 1)$$

---

[35] The condition $\Delta_b \subset I$ is needed to insure that $b(\phi - P)$ is a function on I so that $Y_a(b(\phi - P))$ makes sense. Recall that the interval in the definition of Y_a is I.

and P is a polynomial of degree less than $1/p - 1$. From this and the definition of Y_a, we observe that $Y_a(Pb) = Pb$. Moreover, by remark (3) above,

$$Pb/\|P|\Delta_b\|_\infty$$

is an atom (the appropriate moments vanish) and so has a uniform upper bound on its $H^p(\mathbb{R})$ norm which is independent of b P. From remarks (4) and (5) above along with the hypothesis that $|\Delta_b| \leq |I|$, we get

$$\|P|\Delta_b\|_\infty \leq C_\phi |I|^{1/p-1} = C_\phi$$

(I is dictated by ϕ) and it follows, from the fact that the atom $Pb/\|P|\Delta_b\|_\infty$ has a uniform upper bound on its $H^p(\mathbb{R})$ norm, that

$$\|Pb\| = \left\| \|P|\Delta_b\|_\infty \frac{Pb}{\|P|\Delta_b\|_\infty} \right\| \leq C_\phi.$$

Thus $\|Y_a(Pb)\| = \|Pb\| \leq C_\phi$ and so

$$\|Y_a(\phi b)\|^p \leq \|Y_a(b(\phi - P))\|^p + \|Y_a(Pb)\|^p \leq C_{a,\phi}.$$

Combining the two cases, $|\Delta_b| \geq |I|$ and $|\Delta_b| \leq |I|$, we have shown eq.(6.4.55) and have thus completed the proof. $\square$

Next, we define a sequence of operators which control the singularities of functions in $\mathcal{H}^p(\mathbb{C}\backslash\mathbb{R})_*$ on the real line.

DEFINITION 6.4.60. Let $\phi \in C_0^\infty$ with $0 \leq \phi \leq 1$ and with $\phi|_{[-1,1]} = 1$. Define, for $a \in \mathbb{R}$ and $t > 0$, the functions

$$\psi_{a,t}(x) = \phi(\frac{x-a}{t})$$

and the operators

$$\mathcal{X}_{a,t} = X_{a,\psi_{a,t}}.$$

REMARK 6.4.61. The operator $X_{a,\psi_{a,t}}$ is defined in Proposition 6.4.54.

If the support of ϕ is contained in the interval $(-R, R)$, then the support of $\psi_{a,t}$ is contained in the interval $(a - tR, a + tR)$. Since ϕ is equal to one on $[-1,1]$, then $\psi_{a,t}$ is equal to one on $[a - t, a + t]$. From the definition of $\mathcal{X}_{a,t}$ we see that for an $\ell \in H^p(\mathbb{R})$, the support of the distribution

$$\mathcal{X}_{a,t}\ell = \psi_{a,t}\ell - \sum_{j=0}^{[1/p]-1} c_j \delta_a^{(j)}$$

is contained in $(a - tR, a + tR)$. So, at least formally,

$$\mathcal{X}_{a,t}\ell \to \ell \ (t \to \infty), \quad \mathcal{X}_{a,t} \to 0 \ (t \to 0).$$

We make all this precise with the following result.

LEMMA 6.4.62. *For fixed* $p \in (0,1)$ $(1/p \notin \mathbb{N})$ *and* $a \in \mathbb{R}$, *the following are true.*

1. *The operators* $\{\mathcal{X}_{a,t}\}_{t>0}$ *are uniformly bounded in operator norm.*

2. *For each fixed $\ell \in H^p(\mathbb{R})$ and $a \in \mathbb{R}$,*
$$\lim_{t \to 0} \mathcal{X}_{a,t} \ell = 0$$
$$\lim_{t \to \infty} \mathcal{X}_{a,t} \ell = \ell$$
where the limits are in $H^p(\mathbb{R})$.

PROOF. For fixed $a \in \mathbb{R}$ and $t > 0$, the operator
$$(D_{a,t} f)(z) = t^{-1/p} f\left(\frac{z-a}{t}\right)$$
is continuous on $\mathcal{H}^p(\mathbb{C}\backslash\mathbb{R})$ with uniformly bounded operator norm (in t). Moreover
$$(D_{a,t}^{-1} f)(z) = t^{1/p} f(tz + a).$$
A straightforward integral computation [36] shows that for any atom b we have
$$D_{a,t} \mathcal{X}_{0,1} D_{a,t}^{-1}(b) = \mathcal{X}_{a,t}(b).$$
But since finite linear combinations of atoms are dense in $H^p(\mathbb{R})$ (from the atomic decomposition), we have
$$D_{a,t} \mathcal{X}_{0,1} D_{a,t}^{-1} = \mathcal{X}_{a,t}$$
and so $\mathcal{X}_{a,t}$ has uniform operator norm (in t). This proves statement (1).

To prove statement (2), observe that for an atom b,
$$\|\mathcal{X}_{a,t} b\| = \|Y_a(\psi_{a,t} b)\| \leq C_a \|\psi_{a,t} b\|_{L^1}\ [37]$$
which goes to zero as $t \to 0$ by the dominated convergence theorem. Thus, for finite linear combinations of atoms, statement (2) holds. Now use the atomic decomposition of an $\ell \in H^p(\mathbb{R})$ as well as the uniform boundedness of $\|\mathcal{X}_{a,t}\|$ to finish the proof. □

6.4.4. The main approximation for $1/p \notin \mathbb{N}$. Let us review the situation thus far. We started with a closed (but not necessarily compact) set $F \subset \mathbb{R}$ and a function $k : F \to [1, n_p] \cap \mathbb{N}$ which is permissible; that is, $k|F\backslash F_0 = n_p$, where F_0 is the set of isolated points of F. We also considered the space $E^p(F, k)$ of functions $f \in \mathcal{H}^p(\mathbb{C}\backslash\mathbb{R})$ with

$$f \in \mathcal{H}^p(\mathbb{C}\backslash\mathbb{R})_* \text{ i.e., } \lim_{y \to 0^+} \{ f(x + iy) - f(x - iy) \} = 0 \text{ a.e.}$$

f has an analytic continuation across $\mathbb{R}\backslash F$

At each point $t \in F_0$, f has a pole of order no more than $k(t)$.

The main object is to prove that the rational functions in $E^p(F, k)$ form a dense set in $E^p(F, k)$.

In order to utilize our results on distributions, we must first be able to view $E^p(F, k)$ as a subspace of $H^p(\mathbb{R})$. From Chapter 4 and our earlier work, we will be identifying a distribution $\ell \in H^p(\mathbb{R})$ with the corresponding function $F_\ell = (2\pi i)^{-1} \ell((\cdot - z)^{-1}) \in \mathcal{H}^p(\mathbb{C}\backslash\mathbb{R})$ and a function $F \in \mathcal{H}^p(\mathbb{C}\backslash\mathbb{R})$ with the boundary distribution $\ell = \lim_{y \to 0^+} (F(\cdot + iy) - F(\cdot - iy)) \in H^p(\mathbb{R})$. We leave it to the reader

---

[36] Here we identify the atom b (as a distribution in $H^p(\mathbb{R})$) with the analytic function $\frac{1}{2\pi i} \int b(t)(t-z)^{-1}\, dt$ in $\mathcal{H}^p(\mathbb{C}\backslash\mathbb{R})$.

[37] This estimate is justified since we can assume (since $t \to 0$) that the support of $\psi_{a,t}$ is contained in some fixed interval J. Now use the continuity of $Y_a : L^1(J) \to H^p(\mathbb{R})$.

to verify, using Corollary 4.11.5 and Example 4.11.3, that a distribution ℓ is the boundary distribution of a distribution $F_\ell \in E^p(F,k)$ if and only if

$$F_\ell \in \mathcal{H}^p(\mathbb{C}\setminus\mathbb{R})_*,$$

the support of ℓ is contained in F,

and for each $t \in F_0$, $\ell|U_t = \sum_{j=0}^{k(t)-1} c_j \delta_t^{(j)}$,

where U_t is any open neighborhood of t such that $U_t \cap F = \{t\}$. Notice also that if the support of ℓ is a finite set, then $\ell = \sum_{j,k} c_{j,k} \delta_{a_k}^{(j)}$ and so

$$F_\ell(z) = \frac{1}{2\pi i} \ell\left(\frac{1}{\cdot - z}\right) = \frac{1}{2\pi i} \sum_{j,k} \frac{j!\,c_{j,k}}{(a_k - z)^{j+1}}$$

is a rational function belonging to $\mathcal{H}^p(\mathbb{C}\setminus\mathbb{R})$. As we have done in the past for this section, we will equate the boundary distribution $\ell \in H^p(\mathbb{R})$ with its associated analytic function $F_\ell \in \mathcal{H}^p(\mathbb{C}\setminus\mathbb{R})$. We will also call a distribution corresponding to a rational function (i.e., of the form $\ell = \sum_{j,k} c_{j,k} \delta_{a_k}^{(j)}$) a "rational" distribution.

From the previous subsection, recall the operators $\mathcal{X}_{a,t}$ on $H^p(\mathbb{R})$ defined by

$$\mathcal{X}_{a,t}\ell = \psi_{a,t}\ell - \sum_{j=0}^{n_p-1} \ell\left(\frac{1}{j!}\psi_{a,t}(\cdot)(a-\cdot)^j\right)\delta_a^{(j)},$$

where $\phi \in C_0^\infty$ with $0 \leq \phi \leq 1$ and $\phi|[-1,1] = 1$ and $\psi_{a,t}(x) = \phi((x-a)/t)$. These operators have support in $B(a,tR) = (a-tR, a+tR)$, where the support of ϕ is contained in $B(0,R)$, and have the property that $\mathcal{X}_{a,t}\ell \to \ell$ as $t \to \infty$ and $\mathcal{X}_{a,t}\ell \to 0$ as $t \to 0$ in the norm of $H^p(\mathbb{R})$. With these reminders and preliminary remarks, we are now ready to prove the final piece of the argument for the case when $1/p \notin \mathbb{N}$.

PROOF OF THEOREM 6.4.11 FOR $1/p \notin \mathbb{N}$. Since $\mathcal{X}_{0,t}\ell \to \ell$ as $t \to \infty$ and $\mathcal{X}_{0,t}\ell$ has support in $B(0,tR)$, it suffices to approximate $\mathcal{X}_{0,t}\ell$ with rational functions from $E^p(F,k)$. That is to say, we can assume that F is compact.

We will first deal with the case where F is a countable set. Assume $\ell \in E^p(F,k)$ and $\varepsilon > 0$ be given. Let

$$F' = \{a_n : n \in \mathbb{N}\}$$

denote the set of limit points of F and define, inductively, the following sequence

$$\ell_0 := \ell$$

(6.4.63) $$\ell_{n+1} := \ell_n - \mathcal{X}_{a_{n+1},t_n}\ell_n, \quad n \in \mathbb{N},$$

where the t_n are chosen so small that

$$\|\ell_{n+1} - \ell_n\| = \|\mathcal{X}_{a_{n+1},t_n}\ell_n\| \leq \varepsilon\, 2^{-(n+1)/p}.$$

This is possible by Lemma 6.4.62. It is straightforward to derive from the above inequality that the sequence $\{\ell_n : n \in \mathbb{N}\}$ is a Cauchy sequence in $H^p(\mathbb{R})$ and so has a limit $r \in H^p(\mathbb{R})$. Moreover, from the same inequality, one can also show that

$$\|r - \ell\| \leq \varepsilon.$$

We are using the suggestive letter r to represent the limiting distribution since we will now show that r is a rational distribution in $E^p(F,k)$.

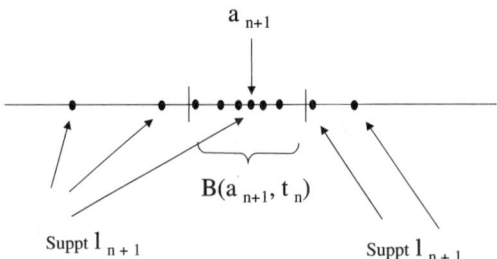

FIGURE 3. The support of ℓ_{n+1} omits the set $B(a_{n+1}, t_n)\setminus\{a_{n+1}\}$.

To do this, we first observe that if $\eta \in C_0^\infty(B(a_{n+1}, t_n)\setminus\{a_{n+1}\})$, then

$$\begin{aligned}
\ell_{n+1}(\eta) &= \ell_n(\eta) - (\mathcal{X}_{a_{n+1}, t_n} \ell_n)(\eta) \\
&= \ell_n(\eta) - \Big\{ \psi_{a_{n+1}, t_n} \ell_n(\eta) - \sum_{j=0}^{n_p - 1} c_j \delta_{a_{n+1}}^{(j)}(\eta) \Big\} \\
&= \ell_n(\eta) - \ell_n(\psi_{a_{n+1}, t_n} \eta) - 0 \\
&= \ell_n\big(\eta(1 - \psi_{a_{n+1}, t_n})\big) \\
&= 0
\end{aligned}$$

since ψ_{a_{n+1}, t_n} is equal to one on $B(a_{n+1}, t_n)$. Thus, a_{n+1} is a isolated point of $\mathrm{suppt}(\ell_{n+1}) \cup \{a_{n+1}\}$ which means that the support of ℓ_{n+1} omits the set $B(a_{n+1}, t_n)\setminus\{a_{n+1}\}$ (see Figure 3).

The set F' is a compact set (since F is compact) and the intervals $B(a_{n+1}, t_n)$ form an open cover for F'. Thus

$$F' \subset \bigcup_{n=0}^{N} B(a_{n+1}, t_n)$$

for some $N \in \mathbb{N}$. But since F' are the accumulation points of F, then ℓ_{N+1} will have its support in the set

$$\{a_{n+1} : n = 0, \cdots, N\} \bigcup \Big\{ F \setminus \bigcup_{n=0}^{N} B(a_{n+1}, t_n) \Big\}$$

which is a finite set since F was assumed to be countable and $F' \subset \cup_{n=0}^N B(a_{n+1}, t_n)$. Hence ℓ_n, $n \geq N+1$, and the limit distribution r (since, from eq.(6.4.63), the support of ℓ_{n+1} is contained in the support of ℓ_n for all n) will be a "rational" distribution in $H^p(\mathbb{R})$. Note that here we are using eq.(4.8.12) which says that if a distribution has support on a finite set, then it has finite order and is a finite linear combination of derivatives of point masses.

It remains to show that $r \in E^p(F, k)$. We will do this by showing that $\ell_n \in E^p(F, k)$ for all sufficiently large n and then note that $E^p(F, k)$ is closed and r is the limit of ℓ_n as $n \to \infty$. Since $\ell_0 = \ell$, suppt $\ell_{n+1} \subset$ suppt ℓ_n, and suppt $\ell \subset F$, then suppt $\ell_n \subset F$ for all n. For sufficiently large n, the support of ℓ_n is a finite set and so F_{ℓ_n} is a rational function which certainly belongs to $\mathcal{H}^p(\mathbb{C}\backslash\mathbb{R})_*$. To finish, we will show that for every $t \in F_0$, there are constants c_j such that

$$\ell_n|U_t = \sum_{j=0}^{k(t)-1} c_j \delta_t^{(j)} \tag{6.4.64}$$

for every neighborhood U_t of t which does not intersect any other point of F. We proceed by induction. Since $\ell_0 = \ell \in E^p(F,k)$, we have $\ell|U_t = \sum_{j=0}^{k(t)-1} c_j \delta_t^{(j)}$. Now suppose eq.(6.4.64) is true for ℓ_n. Then

$$\begin{aligned}\ell_{n+1}|U_t &= \ell_n|U_t - \psi_{a_{n+1},t_n}\ell_n|U_t - \sum_{j=0}^{n_p-1} c_j \delta_{a_{n+1}}^{(j)}|U_t \\ &= \ell_n|U_t - \psi_{a_{n+1},t_n}\ell_n|U_t - 0\end{aligned}$$

since a_{n+1} is never an isolated point of F (and hence $a_{n+1} \notin U_t$). By our inductive hypothesis,

$$\ell_n|U_t = \sum_{j=0}^{k(t)-1} c_j \delta_t^{(j)}$$

and it is an easy exercise to show that the same is true for the distribution

$$\psi_{a_{n+1},t_n}\ell_n|U_t.$$

Thus we have shown Theorem 6.4.11 (for $1/p \notin \mathbb{N}$) in the case where F is a countable set.

We will now prove Theorem 6.4.11 (for $1/p \notin \mathbb{N}$) for an arbitrary compact set F. Let F_1 denote the set of points $a \in F$ with the property that $F \cap B(a,r)$ is an uncountable set for every $r > 0$. The set F_1 is a perfect set and the set $F\backslash F_1$ is countable. Let $\{J_\gamma, \gamma \in \mathbb{N}\}$ be the (disjoint) complimentary intervals to F_1. By applying Proposition 6.4.36 to the set F_1, we observe that any $f \in E^p(F,k) \subset \mathcal{H}^p(\mathbb{C}\backslash\mathbb{R})$ can be decomposed as

$$f = \sum_\gamma f_\gamma + f_{F_1},$$

where $f_\gamma \in \mathcal{H}^p_{J_\gamma^-}$, $f_{F_1} \in \mathcal{H}^p_{F_1}$, and the convergence is in $H^p(\mathbb{R})$. For a particular $\gamma_0 \in \mathbb{N}$, we have

$$f_{\gamma_0} = f - \sum_{\gamma \neq \gamma_0} f_\gamma - f_{F_1}$$

and using that fact that $f \in E^p(F,k)$ and that each f_γ, $\gamma \neq \gamma_0$, as well as f_{F_1} has an analytic continuation across J_{γ_0}, we can conclude slightly more, namely

$$f_{\gamma_0} \in E^p(F \cap J_{\gamma_0}^-, k|F \cap J_{\gamma_0}^-).$$

Furthermore, a similar argument shows

$$f_{F_1} \in E^p(F_1, k|F_1).$$

For each γ, the set $J_\gamma^- \cap F$ can be decomposed as

$$J_\gamma^- \cap F = \{J_\gamma^- \cap (F\backslash F_1)\} \cup \{J_\gamma^- \cap F_1\}.$$

The first set on the right hand side of the above equality is countable (since $F\backslash F_1$ is countable) and the second set in the union is finite (since J_γ are the complimentary intervals to F_1). Thus $J_\gamma^- \cap F$ is countable and by the argument used before, f_γ can be approximated by rational functions from $E^p(F \cap J_\gamma^-, k|F \cap J_\gamma^-) \subset E^p(F, k)$. To show that f_{F_1} can be approximated by rational functions from $E^p(F, k)$ we proceed as follows. Since F_1 is a perfect set, every point is an accumulation point. Since k is permissible, then $k|F_1 = n_p$. By Theorem 6.4.35, any $f \in E^p(F_1, k|F_1)$ can be approximated by rational functions from $\mathcal{H}^p(\mathbb{C}\backslash\mathbb{R})$ whose poles are contained in F_1. Such a rational function cannot have a pole of order more than n_p at any point of F_1 (or else it would not be in $L^p(\mathbb{R})$) and hence belongs to $E^p(F_1, k|F_1) \subset E^p(F, k)$.

In summary, since

$$f = \sum_\gamma f_\gamma + f_{F_1},$$

and each f_γ as well as f_{F_1} can be approximated by rational functions from $E^p(F, k)$, the same is true for f. $\square$

6.4.5. Preliminaries for $1/p \in \mathbb{N}$. Throughout this section, $n \in \mathbb{N}$ and $p = 1/(n+1)$.

LEMMA 6.4.65. *Let I be a closed interval in the real line and $a_1, a_2 \in I$ with $|a_1 - a_2| \geq |I|/A$, where $1 \leq A < \infty$. Then, for any $f \in L^1(I)$, there are unique complex numbers $d_0, \cdots, d_n$ so that the distribution $f - \sum_{s=0}^{n-1} d_s \delta_{a_1}^{(s)} - d_n \delta_{a_2}$ belongs to $H^p(\mathbb{R})$. Moreover*

$$\Big\| f - \sum_{s=0}^{n-1} d_s \delta_{a_1}^{(s)} - d_n \delta_{a_2} \Big\|_{H^p(\mathbb{R})} \leq C(A, n)|I|^n \|f\|_{L^1}.$$

PROOF. This lemma is quite technical and will be shown over the course of several claims. First, we make the observation (from Example 4.11.3) that the distribution

$$\ell := f - \sum_{s=0}^{n-1} d_s \delta_{a_1}^{(s)} - d_n \delta_{a_2}$$

belongs to $H^p(\mathbb{R})$ if and only if

$$\ell(x^k) = 0, \ \forall\, k = 0, \cdots, n.$$

A computation using the definition of the derivative of a distribution shows that

(6.4.66) $$\ell(x^k) = \int t^k f(t)\, dt - \sum_{s=0}^{n-1} d_s (-1)^s \frac{k!}{(k-s)!} a_1^{k-s} - d_n a_2^k.$$

Thus, to find $d_0, \cdots, d_n$ such that $\ell(x^k) = 0$ for all $k = 0, \cdots, n$, we must solve the matrix equation

(6.4.67) $$B(a_1, a_2, n) \begin{pmatrix} d_0 \\ \vdots \\ d_n \end{pmatrix} = \begin{pmatrix} \int f(t)\, dt \\ \int t f(t)\, dt \\ \vdots \\ \int t^n f(t)\, dt \end{pmatrix},$$

where
$$B(a_1, a_2, n) := \begin{pmatrix} 1 & 0 & 0 & \cdots & 0 & 1 \\ a_1 & -1 & 0 & \cdots & 0 & a_2 \\ a_1^2 & -2a_1 & 2 & 0 & \cdots & a_2^2 \\ \vdots & \vdots & \vdots & \vdots & \vdots & \vdots \\ a_1^{n-1} & -(n-1)a_1^{n-2} & \cdots & \cdots & (-1)^n(n-1)! & a_2^{n-1} \\ a_1^n & -na_1^{n-1} & \cdots & \cdots & (-1)^n n! a_1 & a_2^n \end{pmatrix}$$

Claim 1: The $(n+1) \times (n+1)$ matrix $B(a_1, a_2, n)$ is invertible.

Proof of Claim 1: We will show that the rows $R_0, \cdots, R_n$ of $B(a_1, a_2, n)$ are linearly independent. It is routine to check that the rows $R_0, \cdots, R_{n-1}$ are linearly independent and so it suffices to prove there are no constants $c_0, \cdots, c_{n-1}$ such that

(6.4.68) $$c_0 R_0 + \cdots + c_{n-1} R_{n-1} + R_n = 0.$$

Looking at the $n-1$-th entry of eq.(6.4.68), it is easy to check that
$$c_{n-1} = (-1) \binom{n}{n-1} a_1$$

and an induction argument shows
$$c_s = (-1)^{n-s} \binom{n}{s} a_1^{n-s}, \quad s = 0, \cdots, n-1.$$

Looking at the n-th entry of eq.(6.4.68) and using the above identity, we get
$$0 = \sum_{s=0}^{n} c_s a_2^s = \sum_{s=0}^{n} (-1)^{n-s} \binom{n}{s} a_1^{n-s} a_2^s = (a_2 - a_1)^n$$

which is impossible since we are assuming $a_1 \neq a_2$. This proves Claim 1.

From the matrix equation eq.(6.4.67), observe that for each $s = 0, \cdots, n$,

(6.4.69) $$d_s = \sum_{j=0}^{n} C_{s,j} \int t^j f(t)\, dt$$

for some constants $C_{j,s}$ which depend only on a_1 and a_2.[38] Thus the mapping

(6.4.70) $$f \to f - \sum_{s=0}^{n-1} d_s \delta_{a_1}^{(s)} - d_n \delta_{a_2}$$

is a linear mapping from $L^1(I)$ to $H^p(\mathbb{R})$ which we will denote by ℓ_{I,a_1,a_2}.

To get started on estimating the $H^p(\mathbb{R})$ norm of the distribution $\ell_{I,a_1,a_2} f$, we will first look at a special case of Lemma 6.4.65 when $a_1 = -1, a_2 = 1$, and $I \subset [-A, A]$. In this case, observe that
$$|a_1 - a_2| = 2 \geq |I|/A.$$

Claim 2: Given $f \in L^1(I)$, there are unique complex constants $d_0, \cdots, d_n$ so that
$$\left\| f - \sum_{s=0}^{n-1} d_s \delta_{-1}^{(s)} - d_n \delta_1 \right\| \leq C(A, n) \|f\|_{L^1}.$$

---

[38] $(C_{s,j})_{j=0}^n$ is the s-th row of $B(a_1, a_2, n)^{-1}$.

Proof of Claim 2: To prove this claim, we will show that the corresponding analytic function F_ℓ on $\mathbb{C}\backslash\mathbb{R}$

$$F_\ell(z) = \frac{1}{2\pi i}\ell\left(\frac{1}{\cdot - z}\right)$$

$$= \frac{1}{2\pi i}\int \frac{f(t)}{t-z}\,dt - \frac{1}{2\pi i}\left\{\sum_{s=0}^{n-1}\frac{d_s s!}{(-1-z)^{s+1}} - \frac{d_n}{1-z}\right\}$$

belongs to $\mathcal{H}^p(\mathbb{C}\backslash\mathbb{R})$ and $\|F_\ell\| \leq C(A,n)\|f\|_{L^1}$.

To this end, let $y \neq 0$ and write

$$\int_\mathbb{R}|F_\ell(x+iy)|^p\,dx = \int_{|x|\leq 4A} + \int_{|x|>4A}.$$

To estimate the first integral, observe (recall $p=1/(n+1)$)

$$\int_{|x|\leq 4A}|F_\ell(x+iy)|^p\,dx$$

$$\leq C\left\{\int_{|x|\leq 4A}\left|\int\frac{f(t)}{t-(x+iy)}\,dt\right|^p\,dx\right.$$

$$+\sum_{s=0}^{n-1}|d_s|^p\int_{|x|\leq 4A}\frac{dx}{|-1-(x+iy)|^{(s+1)/(n+1)}}$$

$$\left.+|d_n|^p\int_{|x|\leq 4A}\frac{dx}{|1-(x+iy)|^{1/(n+1)}}\right\}$$

$$\leq C\left\{\int_{|x|\leq 4A}\left|\int\frac{f(t)}{t-(x+iy)}\,dt\right|^p\,dx\right.$$

$$+\sum_{s=0}^{n-1}|d_s|^p\int_{|x|\leq 4A}\frac{dx}{|(-1-x)^2+y^2|^{(s+1)/2(n+1)}}$$

$$\left.+|d_n|^p\int_{|x|\leq 4A}\frac{dx}{|(1-x)^2+y^2|^{1/2(n+1)}}\right\}$$

$$\leq C\left\{\int_{|x|\leq 4A}\left|\int\frac{f(t)}{t-(x+iy)}\,dt\right|^p\,dx\right.$$

$$+\sum_{s=0}^{n-1}|d_s|^p\int_{|x|\leq 4A}\frac{dx}{|-1-x|^{(s+1)/(n+1)}}$$

$$\left.+|d_n|^p\int_{|x|\leq 4A}\frac{dx}{|1-x|^{1/(n+1)}}\right\}$$

$$\leq C\int_{|x|\leq 4A}\left|\int\frac{f(t)}{t-(x+iy)}\,dt\right|^p\,dx$$

$$+C(A,n)\left\{\sum_{s=0}^{n}|d_s|^p\right\}.$$

Since $d_s = \sum_{j=0}^{n}C_{s,j}\int t^j f(t)\,dt$ (as in eq.(6.4.69)) and $|I|\leq 2A$, we have

(6.4.71) $$|d_s|^p \leq C(A,n)\|f\|_{L^1}^p$$

6.4. RATIONAL APPROXIMATION

and from above,

$$\int_{|x|\leq 4A} |F_\ell(x+iy)|^p dx \leq C \int_{|x|\leq 4A} \Big| \int \frac{f(t)}{t-(x+iy)} dt \Big|^p dx + C(A,n)\|f\|_{L^1}^p$$

for all $y \neq 0$. Recalling the definitions of the Poisson and conjugate Poisson integrals U_f and V_f as well as their non-tangential maximal functions U_f^* and V_f^* from Chapter 4, we observe

$$\int_{|x|\leq 4A} \Big| \int \frac{f(t)}{t-(x+iy)} dt \Big|^p dx$$
$$= C \left\{ \int_{|x|\leq 4A} |U_f(x+iy)+iV_f(x+iy)|^p dx \right\}$$
$$\leq C \left\{ \int_{|x|\leq 4A} |U_f^*(x)|^p dx + \int_{|x|\leq 4A} |V_f^*(x)|^p dx \right\}.$$

To estimate the first integral above, note that

$$\int_{|x|\leq 4A} |U_f^*(x)|^p dx = C_n \int_0^\infty \lambda^{p-1} \text{meas}\left\{ x \in [-4A, 4A] : U_f^*(x) > \lambda \right\} d\lambda$$
$$\leq C_n \int_0^{\|f\|} + C_n \int_{\|f\|}^\infty$$
$$\leq C_n 8A \int_0^{\|f\|} \lambda^{p-1} d\lambda + C_n \int_{\|f\|}^\infty$$
$$\leq C(A,n)\|f\|_{L^1}^p + C_n \int_{\|f\|}^\infty$$

Using the estimate

$$\text{meas}\left\{ x : U_f^*(x) > \lambda \right\} \leq C\|f\|_{L^1}/\lambda,$$

from Proposition 4.4.5, on the remaining integral, it is not too difficult to show that the sum above is bounded by $C(A,n)\|f\|_{L^1}^p$. The estimate of

$$\int_{|x|\leq 4A} |V_f^*(x)|^p dx$$

is similar. Thus

$$\int_{|x|\leq 4A} |F_\ell(x+iy)|^p dx \leq C(A,n)\|f\|_{L^1}^p \quad \forall\, y \neq 0.$$

Now we estimate

$$\int_{|x|>4A} |F_\ell(x+iy)|^p dx.$$

For $|z| > 4A$, $2\pi i F_\ell(z)$ has a power series about infinity equal to

$$-\sum_{k=0}^{\infty} \frac{1}{z^{k+1}} \int t^k f(t) dt - \sum_{s=0}^{n-1} d_s \sum_{j=0}^{\infty} (-1)^{j+s+1} \frac{(j+s)!}{j!} \frac{1}{z^{j+s+1}} + d_n \sum_{k=0}^{\infty} \frac{1}{z^{k+1}}$$

$$= -\sum_{k=0}^{\infty} \frac{1}{z^{k+1}} \int t^k f(t) \, dt - \sum_{k=0}^{\infty} \frac{1}{z^{k+1}} \sum_{s=0}^{n-1} d_s \frac{k!}{(k-s)!} (-1)^{k+1} + d_n \sum_{k=0}^{\infty} \frac{1}{z^{k+1}}$$

$$= \sum_{k=0}^{\infty} \frac{1}{z^{k+1}} \left\{ -\int t^k f(t) \, dt - \sum_{s=0}^{n-1} d_s \frac{k!}{(k-s)!} (-1)^{k+1} + d_n \right\}.$$

Using the condition eq.(6.4.66), we see the above series becomes

$$-\sum_{k=n+1}^{\infty} \frac{1}{z^{k+1}} \left\{ \int t^k f(t) \, dt - \sum_{s=0}^{n-1} d_s \frac{k!}{(k-s)!} (-1)^k - d_n \right\}$$

which is equal to

$$-\frac{1}{z^{n+2}} \left\{ \sum_{j=0}^{\infty} \frac{1}{z^j} \left(\int t^{n+j+1} f(t) \, dt - \sum_{s=0}^{n-1} d_s (-1)^{n+j+1} \frac{(n+j+1)!}{(n+j+1-s)!} - d_n \right) \right\}$$

Let's look at the terms of the sum separately. First observe that for $|z| > 4A$

$$\left| \sum_{j=0}^{\infty} \frac{1}{z^j} \int t^{n+j+1} f(t) \, dt \right| \leq \sum_{j=0}^{\infty} \frac{1}{(4A)^j} (2A)^{n+j+1} \|f\|_{L^1}$$

$$\leq C(A,n) \|f\|_{L^1}$$

Also, for $|z| > 4A$,

$$\left| \sum_{j=0}^{\infty} \frac{1}{z^j} \sum_{s=0}^{n-1} d_s (-1)^{n+j+1} \frac{(n+j+1)!}{(n+j+1-s)!} \right| \leq \sum_{j=0}^{\infty} \frac{1}{(4A)^j} \sum_{s=0}^{n-1} |d_s| \frac{(n+j+1)!}{(n+j+1-s)!}$$

$$\leq C(A,n) \|f\|_{L^1}$$

Finally observe that for $|z| > 4A$,

$$|d_n| \left| \sum_{j=0}^{\infty} \frac{1}{z^j} \right| \leq C(A,n) \|f\|_{L^1}.$$

Note that in the above two estimates, we are using the estimate

$$|d_s| \leq C(A,n) \|f\|_{L^1}, \quad s = 0, \cdots, n$$

from eq.(6.4.71). Putting this altogether, we get

$$|F_\ell(z)| \leq \frac{1}{|z|^{n+2}} C(A,n) \|f\|_{L^1}, \quad |z| > 4A.$$

This means that for all $y \neq 0$,

$$\int_{|x|>4A} |F_\ell(x+iy)|^p dx \leq C(A,n) \|f\|_{L^1}^p \int_{|x|>4A} \frac{1}{|x|^{(n+2)p}} \leq C(A,n) \|f\|_{L^1}^p.$$

Combining this with our previous estimate gives us the desired result,

$$\|F_\ell\| \leq C(A,n) \|f\|_{L^1}.$$

This proves Claim 2.

To prove the result in general, let $I = [\alpha, \beta]$ with $a_1, a_2 \in I$ and $|a_1 - a_2| \geq |I|/A$. Map the interval $[a_1, a_2]$ onto $[-1, 1]$ with the affine transformation

$$\phi(x) = 1 + \frac{2}{a_2 - a_1}(x - a_2)$$

and notice that ϕ will map $[\alpha, \beta]$ onto an interval $[\phi(\alpha), \phi(\beta)]$ with

$$2 \geq |\phi(\alpha) - \phi(\beta)|/A \quad \text{and} \quad [\phi(\alpha), \phi(\beta)] \subset [-2A, 2A].$$

If the operator C_ϕ is defined by $C_\phi g = g \circ \phi$, then it is easy to check that for an $\mathcal{H}^p(\mathbb{C}\backslash\mathbb{R})$ function g we have (since trivially $[a_1, a_2] \subset I$)

$$\|C_\phi g\| = |a_2 - a_1|^{1/p}\|g\| \leq |I|^{1/p}\|g\| = |I|^{n+1}\|g\|.$$

A computation shows that

$$C_\phi \ell_{[\phi(\alpha),\phi(\beta)],-1,1} C_\phi^{-1} = \ell_{I,a_1,a_2} \quad ^{39}$$

and so from Claim 2 (use $2A$ for the 'A' in the claim) and the above,

$$\begin{aligned}\|\ell_{I,a_1,a_2}f\| &\leq |I|^{n+1}C(A,n)\|f \circ \phi^{-1}\|_{L^1} \\ &\leq C(A,n)|I|^n\|f\|_{L^1}.\end{aligned}$$

$\square$

COROLLARY 6.4.72. *Let the conditions of Lemma 6.4.65 hold. If $f \in L^\infty(I)$, then*

$$\|\ell_{I,a_1,a_2}f\| \leq C(A,n)|I|^{n+1}\|f\|_{L^\infty}.$$

REMARK 6.4.73. Let the conditions of Lemma 6.4.65 hold and $d\mu = \sum_{j=1}^{J} c_j \delta_{x_j}$, where $x_j \in I$. Nearly the same proof as Lemma 6.4.65 shows that there are numbers $d_0, \cdots, d_n$ so that $\mu - \sum_{s=0}^{n-1} d_s \delta_{a_1}^{(s)} - d_n \delta_{a_2} \in H^p(\mathbb{R})$ and moreover

$$\|\mu - \sum_{s=0}^{n-1} d_s \delta_{a_1}^{(s)} - d_n \delta_{a_2}\| \leq C(A,n)|I|^{n+1} \sum_{j=1}^{J} |c_j|.$$

This next result is the analog of Proposition 6.4.54.

LEMMA 6.4.74. *Let $\phi \in C_0^\infty$; a_1, a_2 real numbers; I a closed interval such that $\mathrm{suppt}(\phi) \cup \{a_1, a_2\} \subset I$; and $|a_1 - a_2| \geq |I|/A$ for some $1 \leq A < \infty$. Then for each $\ell \in H^p(\mathbb{R})$, there are constants $d_0, \cdots, d_n$ such that the distribution*

$$\phi\ell - \sum_{s=0}^{n-1} d_s \delta_{a_1}^{(s)} - d_n \delta_{a_2}$$

belongs to $H^p(\mathbb{R})$. Moreover the operator

$$\Phi_{a_1,a_2}\ell := \phi\ell - \sum_{s=0}^{n-1} d_s \delta_{a_1}^{(s)} - d_n \delta_{a_2}$$

is a bounded linear operator on $H^p(\mathbb{R})$ with

$$\|\Phi_{a_1,a_2}\| \leq C(A,n)|I|^n\|\phi^{(n)}\|_{L^\infty}.$$

---

[39]Recall the definitions of $\ell_{[\phi(\alpha),\phi(\beta)],-1,1}$ and ℓ_{I,a_1,a_2} from eq.(6.4.70).

PROOF. Recall from Lemma 6.4.65 and our previous discussion that given $f \in L^1(I)$, there are constants $d_0, \cdots, d_n$ such that $f - \sum_{s=0}^{n-1} d_s \delta_{a_1}^{(s)} - d_n \delta_{a_2}$ belongs to $H^p(\mathbb{R})$. Moreover the constants d_s are the solution to the system in eq.(6.4.67) and so

$$(6.4.75) \qquad d_s(f) := \sum_{j=0}^{n} c_j(a_1, a_2) \int t^j f(t)\, dt,$$

where $c_j(a_1, a_2)$ are constants which depend only on a_1 and a_2.

The previous paragraph allows us to first define our operator Φ_{a_1,a_2} on atoms b by

$$\Phi_{a_1,a_2} b := \phi b - \sum_{s=0}^{n-1} d_s(\phi b) \delta_{a_1}^{(s)} - d_n(\phi b) \delta_{a_2} = \ell_{I,a_1,a_2} \phi b.$$

If $\ell \in H^p(\mathbb{R})$ has atomic decomposition

$$\ell = \sum_{k=1}^{\infty} \lambda_k b_k,$$

the above suggests that $\Phi_{a_1,a_2} \ell$ should be defined by

$$\Phi_{a_1,a_2} \ell := \phi\ell - \sum_{s=0}^{n-1} d_s \delta_{a_1}^{(s)} - d_n \delta_{a_2},$$

where

$$d_s(\phi\ell) = \sum_{j=0}^{n} c_j(a_1, a_2)(\phi\ell)(t^j).$$

Observe that

$$\lim_{K \to \infty} \Phi_{a_1,a_2} \Big(\sum_{k=1}^{K} \lambda_k b_k \Big) = \Phi_{a_1,a_2} \ell,$$

in the sense of tempered distributions, and so it remains to show the above limit converges in $H^p(\mathbb{R})$ and that the desired norm estimate holds.

Since the $H^p(\mathbb{R})$ norm of an atom is uniformly bounded, we will meet our two goals once we show that

$$(6.4.76) \qquad \|\Phi_{a_1,a_2} b\| \leq C(A,n)|I|^n \|\phi^{(n)}\|_{L^\infty}$$

for all atoms b.[40]

To this end, let b be an atom supported in the interval Δ_b. By Remark 4.9.5, we can assume that the atoms chosen in the atomic decomposition satisfy the additional property

$$(6.4.77) \qquad \int Qb\, dt = 0$$

for every polynomial Q with $\deg(Q) \leq 2n$. We divide our argument into two cases:

$$|\Delta_b| \geq |I| \quad \text{and} \quad |\Delta_b| \leq |I|.$$

---

[40]Indeed, if eq.(6.4.76) holds, then given $\ell \in H^p(\mathbb{R})$ with atomic decomposition $\ell = \sum_k \lambda_k b_k$,
$$\|\phi_{a_1,a_2}\ell\|^p \leq C(A,n)|I|^{np}\|\phi^{(n)}\|_{L^\infty}^p \sum_k |\lambda_k|^p \leq C(A,n)|I|^{np}\|\phi^{(n)}\|_{L^\infty}^p \|\ell\|^p.$$

Case 1: $|\Delta_b| \geq |I|$. By Corollary 6.4.52, $|\phi| \leq C\|\phi^{(n)}\|_{L^\infty}|I|^n$ [41] and so

$$|b\phi| \leq C\frac{1}{|\Delta_b|^{n+1}}\|\phi^{(n)}\|_{L^\infty}|I|^n \leq \frac{C\|\phi^{(n)}\|_{L^\infty}}{|I|}.$$

Thus, by Corollary 6.4.72,

$$\|\Phi_{a_1,a_2}b\| = \|\ell_{I,a_1,a_2}\phi b\| \leq C(A,n)|I|^{n+1}\|\phi b\|_{L^\infty} \leq C(A,n)|I|^n\|\phi^{(n)}\|_{L^\infty}.$$

Case 2: $|\Delta_b| \leq |I|$. Since

$$\Phi_{a_1,a_2}b = 0 \quad \text{whenever} \quad \Delta_b \cap I = \emptyset,$$

we can assume that not only is $|\Delta_b| \leq |I|$, but $\Delta_b \subset J$, where J is the interval with the same center as I but twice the length. With this notation, note that

$$\Phi_{a_1,a_2}b = \ell_{J,a_1,a_2}\phi b$$

and that a_1, a_2, J satisfy the hypothesis of Lemma 6.4.65 and Corollary 6.4.72 (by doubling the 'A' in the statement of those results).

By Lemma 6.4.50, there is a polynomial P with $\deg(P) \leq n$ so that

$$|P - \phi| \leq C|\Delta_b|^n\|\phi^{(n)}\|_{L^\infty}$$

on Δ_b. Also observe that

(6.4.78) $$\Phi_{a_1,a_2}b = \ell_{J,a_1,a_2}(\phi - P)b + \ell_{J,a_1,a_2}Pb.$$

Apply Lemma 6.4.65 to get

$$\|\ell_{J,a_1,a_2}(\phi - P)b\|$$
$$\leq C(2A,n)|J|^n \int_{\Delta_b} |\phi - P||b|\, dt$$
$$\leq C(A,n)2^n|I|^n\|\phi^{(n)}\|_{L^\infty}|\Delta_b|^n \frac{1}{|\Delta_b|^{n+1}}|\Delta_b|$$
$$\leq C(A,n)|I|^n\|\phi^{(n)}\|_{L^\infty}.$$

Since

$$\int Qb\,dt = 0,$$

for every polynomial Q with $\deg Q \leq 2n$, it follows from eq.(6.4.75) that

$$\ell_{J,a_1,a_2}(Pb) = Pb.$$

To estimate the $H^p(\mathbb{R})$-norm of Pb, we note, from the moment conditions in eq.(6.4.77), that

$$\frac{Pb}{\|P|\Delta_b|\|_{L^\infty}}$$

is an atom and hence has a universal upper bound on its $H^p(\mathbb{R})$-norm. Moreover, using the fact that $|\Delta_b| \leq |I|$, we have

$$|P|\Delta_b| \leq |(P - \phi)|\Delta_b| + |\phi|\Delta_b|$$
$$\leq C_n\|\phi^{(n)}\|_{L^\infty}|\Delta_b|^n + C_n|I|^n\|\phi^{(n)}\|_{L^\infty}$$
$$\leq C_n|I|^n\|\phi^{(n)}\|_{L^\infty}.$$

---

[41] In the proof of Lemma 6.4.50 and Corollary 6.4.52, one can check (using the fact that $\alpha \in \mathbb{N}$) that the constant C_ϕ in the statement of those results is bounded above by $2\|\phi^{(n)}\|_{L^\infty}$.

Hence
$$\|Pb\| = \left\| \|P|\Delta_b\|_\infty \frac{Pb}{\|P|\Delta_b\|_\infty} \right\| \leq C_n |I|^n \|\phi^{(n)}\|_{L^\infty}.$$

From our previous estimates along with eq.(6.4.78) we obtain
$$\|\Phi_{a_1,a_2} b\| \leq C(A,n) |I|^n \|\phi^{(n)}\|_{L^\infty}.$$

□

LEMMA 6.4.79. *Assume the following hold.*
1. $\{\phi_\gamma : \gamma \in \Gamma\}$ *is a finite family of functions in* C_0^∞
2. $\{a_\gamma : \gamma \in \Gamma\}$ *and* $\{a'_\gamma : \gamma \in \Gamma\}$ *are two finite families of real numbers.*
3. $\{I_\gamma : \gamma \in \Gamma\}$ *are closed intervals such that*

$$\{a_\gamma, a'_\gamma\} \cup \operatorname{suppt} \phi_\gamma \subset I_\gamma \quad \forall \, \gamma \in \Gamma.$$

$$|a_\gamma - a'_\gamma| \geq |I_\gamma|/A \quad \forall \, \gamma \in \Gamma,$$

for some $1 \leq A < \infty$.

Then
$$\left\| \sum_{\gamma \in \Gamma} (\Phi_\gamma)_{a_\gamma, a'_\gamma} \right\| \leq C(A,n) s^{n+1} \max_{\gamma \in \Gamma} \left\{ \|\phi_\gamma^{(n)}\|_{L^\infty} |I_\gamma|^n \right\},$$

where
$$s = \left\| \sum_{\gamma \in \Gamma} \chi_{I_\gamma} \right\|_{L^\infty}.$$

REMARK 6.4.80. The operators $(\Phi_\gamma)_{a_\gamma, a'_\gamma}$ are created from Lemma 6.4.74.

PROOF. From the atomic decomposition of a distribution in $H^p(\mathbb{R})$ (and the universal upper bound on the $H^p(\mathbb{R})$ norm of atoms), it suffices to show that

(6.4.81) $$\left\| \sum_{\gamma \in \Gamma} (\Phi_\gamma)_{a_\gamma, a'_\gamma} b \right\| \leq C(A,n) s^{n+1} \max_{\gamma \in \Gamma} \left\{ \|\phi_\gamma^{(n)}\|_{L^\infty} |I_\gamma|^n \right\}$$

for all atoms b.

To this end, let b be an atom supported on the interval Δ_b. Define
$$\Gamma_1 := \{\gamma \in \Gamma : I_\gamma^\circ \subset \Delta_b\}$$
$$\Gamma_2 := \{\gamma \in \Gamma : I_\gamma^\circ \cap \partial \Delta_b \neq \emptyset\}.$$

With this notation,

(6.4.82) $$\left\| \sum_{\gamma \in \Gamma} (\Phi_\gamma)_{a_\gamma, a'_\gamma} b \right\|^p \leq \left\| \sum_{\gamma \in \Gamma_1} (\Phi_\gamma)_{a_\gamma, a'_\gamma} b \right\|^p + \left\| \sum_{\gamma \in \Gamma_2} (\Phi_\gamma)_{a_\gamma, a'_\gamma} b \right\|^p.$$

Observe that the terms where $I_\gamma \cap \Delta_b^- = \emptyset$ need not be considered since $(\Phi_\gamma)_{a_\gamma, a'_\gamma} b = 0$. To estimate the first term, we use Corollary 6.4.72, the fact that $p = 1/(n+1)$, and the estimate
$$\|\phi_\gamma\|_{L^\infty} \leq C |I_\gamma|^n \|\phi_\gamma^{(n)}\|_{L^\infty},$$

6.4. RATIONAL APPROXIMATION

to get

$$\Big\| \sum_{\gamma \in \Gamma_1} (\Phi_\gamma)_{a_\gamma, a'_\gamma} b \Big\|^p \leq \sum_{\gamma \in \Gamma_1} \big\| (\Phi_\gamma)_{a_\gamma, a'_\gamma} b \big\|^p$$

$$\leq \sum_{\gamma \in \Gamma_1} C(A,n) |I_\gamma|^{(n+1)p} \|\phi_\gamma b\|_{L^\infty}^p$$

$$\leq \sum_{\gamma \in \Gamma_1} C(A,n) |I_\gamma|^{(n+1)p} |I_\gamma|^{np} \frac{\|\phi_\gamma^{(n)}\|_{L^\infty}^p}{|\Delta_b|^{(n+1)p}}$$

$$\leq C(A,n) \sum_{\gamma \in \Gamma_1} \Big(\frac{|I_\gamma|}{|\Delta_b|} \Big) |I_\gamma|^{np} \|\phi_\gamma^{(n)}\|_{L^\infty}^p$$

To show this sum is bounded above by

$$C(A,n) \cdot s \cdot \max_{\gamma \in \Gamma} \big\{ |I_\gamma|^{np} \|\phi_\gamma^{(n)}\|_{L^\infty}^p \big\},$$

we use the following convexity argument: First argue that since $I_\gamma^\circ \subset \Delta_b$ for all $\gamma \in \Gamma_1$, and

$$\Big\| \sum_{\gamma \in \Gamma_1} \chi_{I_\gamma} \Big\|_{L^\infty} \leq s,$$

then

$$\sum_{\gamma \in \Gamma_1} \frac{|I_\gamma|}{|\Delta_b|} \leq s. \quad ^{42}$$

Now choose $r \geq 1$ so that

$$\sum_{\gamma \in \Gamma_1} \frac{r}{s} \frac{|I_\gamma|}{|\Delta_b|} = 1.$$

From this, it follows that the sum

$$\sum_{\gamma \in \Gamma_1} \frac{r}{s} \frac{|I_\gamma|}{|\Delta_b|} |I_\gamma|^{np} \|\phi_\gamma^{(n)}\|_{L^\infty}^p$$

is a convex combination of the numbers

$$|I_\gamma|^{np} \|\phi_\gamma^{(n)}\|_{L^\infty}^p$$

and so the sum is bounded above by

$$\max_{\gamma \in \Gamma_1} \big\{ |I_\gamma|^{np} \|\phi_\gamma^{(n)}\|_{L^\infty}^p \big\}.$$

Hence

$$C(A,n) \sum_{\gamma \in \Gamma_1} \Big(\frac{|I_\gamma|}{|\Delta_b|} \Big) |I_\gamma|^{np} \|\phi_\gamma^{(n)}\|_{L^\infty}^p$$

$$\leq C(A,n) \frac{s}{r} \max_{\gamma \in \Gamma_1} \big\{ |I_\gamma|^{np} \|\phi_\gamma^{(n)}\|_{L^\infty}^p \big\}$$

$$\leq C(A,n) \cdot s \cdot \max_{\gamma \in \Gamma} \big\{ |I_\gamma|^{np} \|\phi_\gamma^{(n)}\|_{L^\infty}^p \big\}.$$

---

[42]To see this, let $g(x) := \sum_{\gamma \in \Gamma_1} \chi_{I_\gamma}(x)$ and define for each $k = 1, \cdots, s$, $A_k := \{x \in \Delta_b : g(x) = k\}$. The sets A_k are disjoint and each A_k is a finite union of sub-intervals of the intervals I_γ. Let $I_{\gamma,k}$ be the union of those sub-intervals of A_k which are sub-intervals of I_γ. The sub-intervals used in this union for $I_{\gamma,k}$ are disjoint when k is fixed. Also note that $|I_\gamma| = \sum_{k=1}^s |I_{\gamma,k}|$. So $\sum_{\gamma \in \Gamma_1} |I_\gamma| = \sum_{\gamma \in \Gamma_1} (\sum_{k=1}^s |I_{\gamma,k}|) = \sum_{k=1}^s (\sum_{\gamma \in \Gamma_1} |I_{\gamma,k}|) \leq \sum_{k=1}^s k|A_k| \leq \sum_{k=1}^s s|A_k| \leq s|\Delta_b|$.

To estimate the second sum in eq.(6.4.82), we employ Lemma 6.4.74 along with the fact that $|\Gamma_2| \leq 2s$ to obtain

$$\left\| \sum_{\gamma \in \Gamma_2} (\Phi_\gamma)_{a_1,a_2} b \right\|^p \leq C \sum_{\gamma \in \Gamma_2} \|(\Phi_\gamma)_{a_1,a_2}\|^p$$

$$\leq \sum_{\gamma \in \Gamma_2} C(A,n) \|\phi_\gamma^{(n)}\|_{L^\infty}^p |I_\gamma|^{np}$$

$$\leq 2s \cdot C(A,n) \max_{\gamma \in \Gamma_2} \left\{ \|\phi_\gamma^{(n)}\|_{L^\infty}^p |I_\gamma|^{np} \right\}$$

$$\leq C(A,n) \cdot s \cdot \max_{\gamma \in \Gamma} \left\{ \|\phi_\gamma^{(n)}\|_{L^\infty}^p |I_\gamma|^{np} \right\}.$$

Combining the two cases (with the reminder that $p = 1/(n+1)$), we have the desired estimate in eq.(6.4.81). □

Our final result of this section is a construction.

DEFINITION 6.4.83. Let Δ be a bounded interval and let Δ^* denote the bounded interval with the same center as Δ but with $|\Delta^*| = 2|\Delta|$.

LEMMA 6.4.84. *Let $\{\Delta_j : j = 1, \cdots, m\}$ be a family of closed intervals which satisfy*

$$\sum_{j=1}^m \chi_{\Delta_j^*} \leq 2.$$

Then there is a family of functions $\{\phi_j : j = 1, \cdots, m\} \subset C_0^\infty$ such that the following three conditions hold:

(6.4.85) $$\operatorname{suppt} \phi_j \subset \Delta_j^*.$$

(6.4.86) $$|\phi_j^{(n)}| \leq C(n)/|\Delta_j^*|^n \ \forall\, j = 1, \cdots, m.$$

(6.4.87) $$\sum_{j=1}^m \phi_j = 1 \ \text{in a neighborhood of} \ \bigcup_{j=1}^m \Delta_j.$$

PROOF. Without loss of generality, we can assume that

$$|\Delta_1| \geq |\Delta_2| \geq \cdots \geq |\Delta_m|.$$

Choose a $\phi \in C_0^\infty$ such that $\phi = 1$ in a neighborhood of $[-1,1]$ and $\operatorname{suppt} \phi \subset [-2,2]$. Let h_{Δ_j}, $j = 1, \cdots, m$, be the affine transformation which maps Δ_j onto $[-1,1]$ and define

$$\phi_{\Delta_j} := \phi \circ h_{\Delta_j}.$$

A computation using the fact that

$$h'_{\Delta_j} = 1/|\Delta_j| \ \text{and} \ h_{\Delta_j}^{(r)} = 0, \ r = 2, 3, \cdots,$$

yields the estimate

(6.4.88) $$|\phi_{\Delta_j}^{(r)}| \leq C(n)/|\Delta_j^*|^r, \ \forall\, r = 0, \cdots, n.$$

Define

$$\phi_1 := \phi_{\Delta_1}$$
$$\phi_k := \phi_{\Delta_k} - \sum_{j<k} \phi_{\Delta_j} \phi_{\Delta_k}, \ k = 2, \cdots, m.$$

6.4. RATIONAL APPROXIMATION

We will now show that these functions ϕ_k satisfy the three conditions eq.(6.4.85), eq.(6.4.86), eq.(6.4.87).

Before proceeding though, we make the remark that one can use the facts that

(6.4.89) $$\text{suppt } \phi_{\Delta_j} \subset \Delta_j^* \text{ and } \sum_{j=1}^m \chi_{\Delta_j^*} \leq 2$$

to argue that

(6.4.90) $$\phi_{\Delta_j} \phi_{\Delta_k} \phi_{\Delta_p} = 0 \ \forall \ 1 \leq j < k < p \leq m.$$

Note that eq.(6.4.85) follows immediately from the definition of ϕ_k. To prove eq.(6.4.86), observe that

$$\phi_k^{(n)} = \phi_{\Delta_k}^{(n)} - \sum_{j<k} \sum_{s=0}^n \binom{n}{s} \phi_{\Delta_j}^{(s)} \phi_{\Delta_k}^{(n-s)}$$

and using eq.(6.4.89) we note that for each $s = 0, \cdots, n$ and fixed $k = 1, \cdots, m$, the supports of

$$\phi_{\Delta_j}^{(s)} \phi_{\Delta_k}^{(n-s)} \text{ and } \phi_{\Delta_{j'}}^{(s)} \phi_{\Delta_k}^{(n-s)}, \ j \neq j',$$

are disjoint and so

$$|\phi_k^{(n)}| \leq |\phi_{\Delta_k}^{(n)}| + \max_{j<k} |\sum_{s=0}^n \binom{n}{s} \phi_{\Delta_j}^{(s)} \phi_{\Delta_k}^{(n-s)}|$$
$$\leq |\phi_{\Delta_k}^{(n)}| + C(n) \max_{1 \leq j < k, 0 \leq s \leq n} |\phi_{\Delta_j}^{(s)}||\phi_{\Delta_k}^{(n-s)}|.$$

By eq.(6.4.88) and the fact that $\{|\Delta_j| : j = 1, \cdots, m\}$ is a decreasing sequence of numbers, we conclude that

$$\|\phi_k^{(n)}\|_{L^\infty} \leq \frac{C(n)}{|\Delta_k^*|^n} + C(n) \max_{1 \leq j < k, 0 \leq s \leq n} \frac{1}{|\Delta_j^*|^s} \frac{1}{|\Delta_k^*|^{n-s}}$$
$$\leq \frac{C(n)}{|\Delta_k^*|^n} + C(n) \frac{1}{|\Delta_k^*|^s} \frac{1}{|\Delta_k^*|^{n-s}}$$
$$\leq \frac{C(n)}{|\Delta_k^*|^n}.$$

This proves eq.(6.4.86).

To prove eq.(6.4.87), use eq.(6.4.90) to show that

$$\prod_{j=1}^m (1 - \phi_{\Delta_j}) = 1 - \sum_{j=1}^m \phi_{\Delta_j} + \sum_{k=1}^m \sum_{j=k+1}^m \phi_{\Delta_k} \phi_{\Delta_j}$$

and so

$$\begin{aligned}
1 - \prod_{j=1}^{m}(1 - \phi_{\Delta_j}) &= \sum_{j=1}^{m}\phi_{\Delta_j} - \sum_{k=1}^{m}\sum_{j=k+1}^{m}\phi_{\Delta_k}\phi_{\Delta_j} \\
&= \sum_{j=1}^{m}\phi_{\Delta_j} - \sum_{j=2}^{m}\sum_{k=1}^{j-1}\phi_{\Delta_k}\phi_{\Delta_j} \\
&= \phi_{\Delta_1} + \sum_{j=2}^{m}\left\{\phi_{\Delta_j} - \sum_{k=1}^{j-1}\phi_{\Delta_k}\phi_{\Delta_j}\right\} \\
&= \sum_{j=1}^{m}\phi_j.
\end{aligned}$$

Finally, by the above calculation and the fact each ϕ_{Δ_j} is equal to one in a neighborhood of Δ_j, one can show eq.(6.4.87). $\square$

6.4.6. The main approximation for $1/p \in \mathbb{N}$. We are given a closed set $F \subset \mathbb{R}$ which we can assume is compact [43] and we wish to approximate a distribution $\ell \in E^p(F,k)$ with a "rational" distribution in $E^p(F,k)$ (one which can be written as a finite sum of point masses and their derivatives). By a shift on $\mathbb{R}$, we can also assume that F is contained in a closed interval Δ which is contained in $(-\infty, 0]$. This last assumption is merely for notational convenience as we shall see shortly.

The general idea here is to construct (from the closed set F and the surrounding interval Δ) a sequence of operators

$$X_m : H^p(\mathbb{R}) \to H^p(\mathbb{R}), \quad m = 1, 2, 3, \cdots$$

for which

1. X_m are uniformly bounded in operator norm.
2. $\ell - X_m\ell$ is a distribution with finite support (and hence a "rational" distribution).
3. If $\ell \in E^p(F,k)$, then $\ell - X_m\ell$ also belongs to $E^p(F,k)$.
4. $X_m\ell \to 0$ for all $\ell \in \mathcal{H}^p \cap \overline{\mathcal{H}^p}$.

Condition 1 will be needed to prove condition 4 since we will prove that $X_m(d\mu) \to 0$ for all measures $d\mu \in H^p(\mathbb{R})$ which are linear combinations of point masses. By Theorem 6.4.16, this set is dense in $\mathcal{H}^p \cap \overline{\mathcal{H}^p}$. [44] These four facts imply that $E^p(F,k)$ is the closure of the rational functions from $E^p(F,k)$.

We now proceed with the Aleksandrov construction.

Step 1: Fix $m \in \mathbb{N}$ and divide Δ into m equal closed intervals

$$\Delta_1^{(m)}, \Delta_2^{(m)}, \cdots, \Delta_m^{(m)}.$$

Step 2: (See Figure 4) Let J be the set of $j = 1, \cdots, m$ for which

$$F' \cap \Delta_j^{(m)} \neq \emptyset, \text{ [45]}$$

---

[43] Indeed the set F is the image of a closed set A in the circle by means of a conformal map $\phi = i(1+z)/(1-z)$ which sends the point $z = 1$ to $z = \infty$. We can assume that A is not all of the circle and so by a rotation we can assume that A does not contain the point $z = 1$. Under these conditions the image, F, will be a compact subset of the real line.

[44] Of course here, we are thinking of $\mathcal{H}^p \cap \overline{\mathcal{H}^p}$ in terms of distributions.

[45] F' are the accumulation points of F.

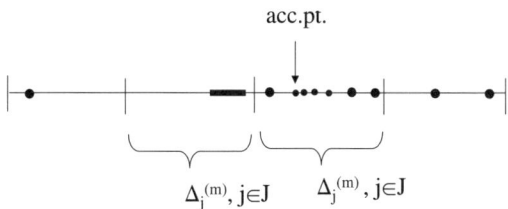

FIGURE 4. The set F along with the intervals $\Delta_j^{(m)}$ for $j \in J$

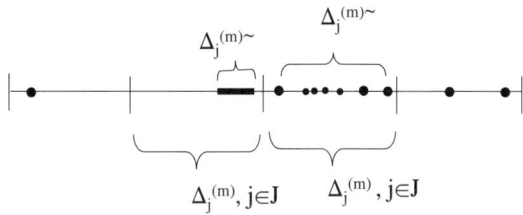

FIGURE 5. The set F along with the intervals $\Delta_j^{(m)}$ for $j \in J$ and the sub-intervals $\widetilde{\Delta_j^{(m)}}$.

$$\operatorname{card}(F \cap \Delta_j^{(m)}) \geq 2.$$

Step 3: (See Figure 5) For each $j \in J$, form the interval

$$\widetilde{\Delta_j^{(m)}} = [\min(F \cap \Delta_j^{(m)}), \max(F \cap \Delta_j^{(m)})].$$

Step 4: (See Figure 6) For each $j \in J$, choose the following two points from $\widetilde{\Delta_j^{(m)}}$:

$$a_j^{(m)} := \min(F' \cap \Delta_j^{(m)}),$$

and $a_j^{(m)'}$ which is the endpoint of $\widetilde{\Delta_j^{(m)}}$ *farthest* from $a_j^{(m)}$ so that

(6.4.91) $$|a_j^{(m)} - a_j^{(m)'}| \geq \frac{1}{2} |\widetilde{\Delta_j^{(m)}}|.$$

Observe that

$$A_m := F \setminus \bigcup_{j \in J} \widetilde{\Delta_j^{(m)}}$$

is a finite set of isolated points of F.

Step 5: For each $\alpha \in A_m$, form the interval $[\alpha, \alpha + \varepsilon]$, where ε is chosen so small that the following three conditions are satisfied:

$$[\alpha, \alpha + \varepsilon] \cap [\alpha', \alpha' + \varepsilon'] = \emptyset, \quad \alpha, \alpha' \in A_m,$$

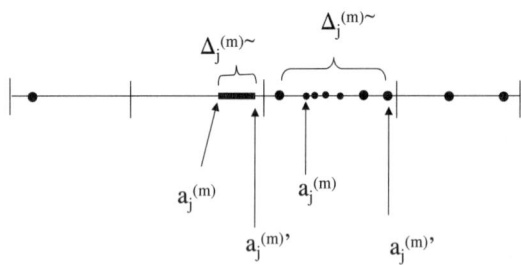

FIGURE 6. The points $a_j^{(m)}$ and $a_j^{(m)'}$ chosen from the sub-intervals $\widetilde{\Delta_j^{(m)}}$.

$$[\alpha, \alpha + \varepsilon]^* \cap (F \setminus \{\alpha\}) = \emptyset,$$

$$\sum_{j \in J \cup A_m} \chi_{(\widetilde{\Delta_j^{(m)}})^*} \leq 2,\,^{46}$$

where for $\alpha \in A_m$, we let

$$\widetilde{\Delta_\alpha^{(m)}} = [\alpha, \alpha + \varepsilon].$$

For each interval $\widetilde{\Delta_\alpha^{(m)}}$, $\alpha \in A_m$, pick the following two points $a_\alpha^{(m)} := \alpha$, $a_\alpha^{(m)'} := \alpha + \varepsilon$. [47]

Step 6: Apply Lemma 6.4.84 to the family of intervals

$$\{ \widetilde{\Delta_j^{(m)}} : j \in J \cup A_m \}$$

and the points $a_j^{(m)}, a_j^{(m)'} \in \widetilde{\Delta_j^{(m)}}$ to produce C_0^∞ functions

$$\{ \phi_j^{[m]} : j \in J \cup A_m \}$$

which satisfy the following conditions:

$$\operatorname{suppt} \phi_j^{[m]} \subset (\widetilde{\Delta_j^{(m)}})^*$$

$$|(\phi_j^{[m]})^{(n)}| \leq \frac{C(n)}{\left| (\widetilde{\Delta_j^{(m)}})^* \right|^n}$$

$$\sum_{j \in J \cup A_m} \phi_j^{[m]} = 1 \text{ in a neighborhood of } \bigcup_{j \in J \cup A_m} \widetilde{\Delta_j^{(m)}}.$$

---

[46] Just a reminder that for a bounded interval $Q \subset \mathbb{R}$, the interval Q^* is an interval with the same center as Q but with twice the length.

[47] A few words about this notation. First note that the interval Δ is contained in $(-\infty, 0]$ and so there is no confusion in using the index j as an integer (when $j \in J$) or as a non-negative real number (when $j \in A_m$). This convention helps us establish a uniform index for the intervals since, as we shall see shortly, we will be working with all of these intervals at once. Secondly, the rather artificial definition of $\widetilde{\Delta_\alpha^{(m)}}$, $\alpha \in A_m$ is also used to establish a uniform notation.

6.4. RATIONAL APPROXIMATION

Step 7: For each $j \in J \cup A_m$ use eq.(6.4.91) and Lemma 6.4.74 to create the operator
$$(\Phi_j^{[m]})_{a_j^{(m)}, a_j^{(m)'}}$$
and the operator
$$X_m := \sum_{j \in J \cup A_m} (\Phi_j^{[m]})_{a_j^{(m)}, a_j^{(m)'}}.$$
Now use Lemma 6.4.79 and Lemma 6.4.84 to show that X_m is uniformly bounded in operator norm.

REMARK 6.4.92. Before proceeding further, note that for $j \in A_m$ and $\ell \in E^p(F, k)$,
$$(\Phi_j^{[m]})_{a_j^{(m)}, a_j^{(m)'}} \ell = \phi_j^{[m]} \ell - \sum_{s=0}^{k(j)-1} d_s \delta_{a_j^{(m)}}^{(s)} - d_n \delta_{a_j^{(m)'}}.$$
Moreover, since $a_j^{(m)}$ is an isolated point of F, $\ell|U_{a_j^{(m)}}$ is a finite linear combination of point masses and their derivatives at $a_j^{(m)}$ of order less than or equal to $k(j) - 1$ (note $j = a_j^{(m)}$). Thus, the distribution $\phi_j^{[m]} \ell$ has its support at the single point $a_j^{(m)}$ [48] and so
$$\phi_j^{[m]} \ell = \sum_{s=0}^{k(j)-1} h_s \delta_{a_j^{(m)}}^{(s)}.$$
The unique constants $d_0, \cdots, d_n$ above for which $(\Phi_j^{[m]})_{a_j^{(m)}, a_j^{(m)'}} \ell \in H^p(\mathbb{R})$ are when
$$d_s = h_s \text{ for } s = 0, \cdots, k(j) - 1 \text{ and } d_n = 0.$$
All of this says that for $j \in A_m$ and $\ell \in E^p(F, k)$, the distribution $(\Phi_j^{[m]})_{a_j^{(m)}, a_j^{(m)'}} \ell$ is the zero distribution.

With that being said, note that if $\eta \in C_0^\infty$ and is zero in a neighborhood of the finite set
$$\mathcal{B}_m := \bigcup_{j \in J} \{a_j^{(m)}, a_j^{(m)'}\} \bigcup A_m,$$
and $\ell \in E^p(F, k)$, then
$$(\ell - X_m \ell)(\eta)$$
$$= \ell(\eta) - \left(\sum_{j \in J \cup A_m} (\Phi_j^{[m]})_{a_j^{(m)}, a_j^{(m)'}} \ell \right)(\eta)$$
$$= \ell(\eta) - \sum_{j \in J \cup A_m} \left(\phi_j^{[m]} \ell - \sum_{s=0}^{n-1} d_s \delta_{a_j^{(m)}}^{(s)} - d_n \delta_{a_j^{(m)'}} \right)(\eta)$$
$$= \ell\left(\eta \left\{1 - \sum_{j \in J \cup A_m} \phi_j^{[m]}\right\}\right) + \left(\sum_{j \in J \cup A_m} \left\{ \sum_{s=0}^{n-1} d_s \delta_{a_j^{(m)}}^{(s)} - d_n \delta_{a_j^{(m)'}} \right\} \right)(\eta)$$
$$= 0$$

---

[48] Recall that the intervals were chosen so small so that the support of $\phi_j^{[m]}$ does not intersect any other points of F.

since
$$\sum_{j \in J \cup A_m} \phi_j^{[m]} = 1 \text{ in a neighborhood of } F,$$

$$\eta^{(k)}(a_j^{(m)}) = \eta^{(k)}(a_j^{(m)'}) = 0 \ \forall k \in \mathbb{N}, \ j \in J$$

$$\eta^{(k)}(a_j^{(m)}) = 0 \ \forall k \in \mathbb{N}, \ j \in A_m$$

$$d_n = d_n(\phi_j^{[m]}\ell) = 0 \ \forall j \in A_m.$$

Hence the support of $\ell - X_m\ell$ is contained in the finite set $\mathcal{B}_m$.

Also notice that by our remarks concerning the case when $j \in A_m$,

$$\ell - X_m\ell = \sum_{j \in A_m} \sum_{s=0}^{k(j)-1} d_s \delta_{a_j^{(m)}}^{(s)} + \sum_{j \in J} \left(\sum_{s=0}^{n-1} d_s \delta_{a_j^{(m)}}^{(s)} - d_n \delta_{a_j^{(m)'}} \right) \quad {}^{49}$$

and so

$$\ell - X_m\ell \in E^p(\mathcal{B}_m, k|\mathcal{B}_m)$$

as desired. Also observe that $E^p(\mathcal{B}_m, k|\mathcal{B}_m)$ are rational functions from $E^p(F, k)$.

To finish the proof, we will show that for any $\ell \in \mathcal{H}^p \cap \overline{\mathcal{H}^p}$,

$$X_m\ell \to 0, \ m \to \infty.$$

Recall from Theorem 6.4.16 that distributions (measures) in $H^p(\mathbb{R})$ of the form

$$d\mu = \sum_{k=1}^{K} c_k \delta_{x_k}, \ x_k \in \mathbb{R},$$

are dense in $\mathcal{H}^p \cap \overline{\mathcal{H}^p}$. Furthermore, since the operator norms of X_m are uniformly bounded, to show that $X_m\ell \to 0$ for a general $\ell \in \mathcal{H}^p \cap \overline{\mathcal{H}^p}$ it suffices to show that for a fixed $d\mu$ as above, $X_m d\mu \to 0$ as $m \to \infty$.

To this end, note that for large enough m,

$$\text{suppt } \phi_j^{[m]} \cap \{x_1, \cdots, x_K\}$$

will be at most a single point and so by Remark 6.4.92

$$(\Phi_j^{[m]})_{a_j^{(m)}, a_j^{(m)'}} \left(\sum_{k=1}^{K} c_k \delta_{x_k} \right) = 0$$

and so

$$X_m \left(\sum_{k=1}^{K} c_k \delta_{x_k} \right) = 0.$$

---

[49] Note that for $j \in J$, $a_j^{(m)}$ is an accumulation point for F.

6.5. Spectral properties

Recall from Chapter 5 that for $p \in [1, \infty)$

$$\sigma(B|H^p) = \mathbb{D}^-, \quad \sigma_p(B|H^p) = \mathbb{D},^{50} \quad \sigma(B|H^p \cap I\overline{H_0^p}) = \overline{\sigma(I)}.$$

In this section, we will use our main theorem from the last section to examine the spectrum of $B|\mathcal{E}^p(I, F, k)$.

First notice that for $\lambda \in \mathbb{D}^-$, $(1 - \lambda z)^{-1} \in H^p$ $(0 < p < 1)$ and

$$B\left(\frac{1}{1-\lambda z}\right) = \lambda \frac{1}{1-\lambda z}.$$

Thus $\mathbb{D}^- \subset \sigma_p(B|H^p) \subset \sigma(B|H^p)$. Observe that for $|\lambda| > 1$

$$\frac{zf - f(1/\lambda)/\lambda}{\lambda z - 1} \in H^p$$

whenever $f \in H^p$. Furthermore

$$\left\| \frac{zf - f(1/\lambda)/\lambda}{\lambda z - 1} \right\| \leq \frac{1}{|\lambda| - 1} \|zf - f(1/\lambda)/\lambda\| \leq C_\lambda \|f\|$$

and so the operator

$$R_\lambda f := \frac{zf - f(1/\lambda)/\lambda}{\lambda z - 1}$$

is continuous on H^p with $R_\lambda = (\lambda I_d - B)^{-1}$. Hence

$$\sigma(B|H^p) = \sigma_p(B|H^p) = \mathbb{D}^-.$$

From the definition of $\mathcal{E}^p(I, F, k)$, one can see that $R_\lambda \mathcal{E}^p(I, F, k) \subset \mathcal{E}^p(I, F, k)$ for $|\lambda| > 1$ and so

$$\sigma(B|\mathcal{E}^p(I, F, k)) \subset \mathbb{D}^-.$$

THEOREM 6.5.1. *For $\lambda \in \mathbb{D}^-$, the following are equivalent.*
1. $(1 - \lambda z)^{-1} \in \mathcal{E}^p(I, F, k)$.
2. $\lambda \in \sigma(B|\mathcal{E}^p(I, F, k))$.
3. $\lambda \in \sigma_p(B|\mathcal{E}^p(I, F, k))$.
4. $\overline{\lambda} \in \sigma(I) \cup F$.

PROOF. The implications $(1) \Rightarrow (2)$, $(1) \Rightarrow (3)$ and $(3) \Rightarrow (2)$ are trivial.

We will now prove $(1) \Rightarrow (4)$. If $\lambda \in \mathbb{D}$ and $(1 - \lambda z)^{-1} \in \mathcal{E}^p(I, F, k)$, then $(1 - \lambda z)^{-1} \in H^p \cap I\overline{H_0^p}$ and so

$$\frac{1}{1 - \lambda \zeta} = I(\zeta)\overline{g}(\zeta)$$

almost everywhere for some $g \in H_0^p$. It follows that

$$g(z) = \frac{zI(z)}{z - \overline{\lambda}}, \quad z \in \mathbb{D}$$

from which $I(\overline{\lambda}) = 0$. Hence $\overline{\lambda} \in \sigma(I)$. If $\lambda \in \mathbb{T}$ with $(1 - \lambda z)^{-1} \in \mathcal{E}^p(I, F, k)$, then by the definition of the set F, $\overline{\lambda} \in F$.

To prove $(4) \Rightarrow (1)$, let $\overline{\lambda} \in \sigma(I) \cup F$. If $\overline{\lambda} \in F$, then

$$(1 - \lambda z)^{-1} \in \mathcal{E}^p(I, F, k).$$

---

[50]$\sigma_p(A)$ is the "point spectrum" or eigenvalues for the operator A.

If $\overline{\lambda} \in \sigma(I) \cap \mathbb{D}$, then $I(\overline{\lambda}) = 0$, hence

$$\frac{zI(z)}{z - \overline{\lambda}} \in H_0^p.$$

Moreover

$$\frac{1}{1 - \lambda \zeta} = \overline{\left\{\frac{\zeta I(\zeta)}{\zeta - \overline{\lambda}}\right\}} I(\zeta)$$

almost everywhere on $\mathbb{T}$ and so $(1 - \lambda z)^{-1} \in H^p \cap I\overline{H_0^p}$. From here it follows that $(1 - \lambda z)^{-1} \in \mathcal{E}^p(I, F, k)$.

To prove (3) $\Rightarrow$ (1), note that if $\lambda \in \sigma_p(B|\mathcal{E}^p(I, F, k))$, there is a non-zero $g \in \mathcal{E}^p(I, F, k)$ with $Bg = \lambda g$. A calculation using the definition of B yields

$$g = \frac{g(0)}{1 - \lambda z}$$

and so $(1 - \lambda z)^{-1} \in \mathcal{E}^p(I, F, k)$.

The proof of (2) $\Rightarrow$ (4) is similar to part of the proof of Theorem 5.5.2 but for the sake of completeness, we will outline it here anyway. If $\lambda \notin \overline{\sigma(I) \cup F}$, then by definition, every $f \in \mathcal{E}^p(I, F, k)$ has a pseudocontinuation Tf which is analytic in a neighborhood of $1/\lambda$. Moreover (from the proof of Proposition 6.2.1),

$$|T_f(1/\lambda)| \leq C_\lambda \|f\|$$

and an application of the closed graph theorem shows that the operator

$$R_\lambda f = \frac{zf - T_f(1/\lambda)/\lambda}{\lambda z - 1}$$

is a continuous operator from $\mathcal{E}^p(I, F, k)$ to H^p. Argue (as in the proof of Theorem 5.5.2) that $R_\lambda \mathcal{E}^p(I, F, k) \subset \mathcal{E}^p(I, F, k)$ and that

$$R_\lambda = (\lambda I_d - B|\mathcal{E}^p(I, F, k))^{-1}.$$

Hence $\lambda \notin \sigma(B|\mathcal{E}^p(I, F, k))$. □

6.6. Cyclic vectors

Though the invariant subspace structure for B on H^p $(0 < p < 1)$ is more complicated than B on H^p $(1 \leq p < \infty)$, the condition for cyclicity[51] is exactly the same (with nearly the same proof). We leave it to the reader to use the description of the B-invariant subspaces as $\mathcal{E}^p(I, F, k)$ and make the very minor adjustments to the proof of Theorem 5.9.5 to prove the following.

THEOREM 6.6.1. *If $f \in H^p$ $(0 < p < 1)$, then f is not cyclic for B if and only if f has a pseudocontinuation of bounded type.*

As in the $p \in [1, \infty)$ case (see Remark 5.9.6), a more "testable" condition for cyclicity is still very much an open problem.

---

[51]Recall from Definition 5.9.1 that a vector $f \in H^p$ is "cyclic" for B if $\bigvee \{B^n f : n \in \mathbb{N} \cup \{0\}\} = H^p$.

6.7. Duality

In Chapter 5, recall that for $p \in (1, \infty)$, the dual of $H^p \cap I\overline{H_0^p}$ can be equated with $H^q \cap I\overline{H_0^q}$ (where $1/q + 1/p = 1$) by means of the Cauchy duality. For $p \in (0, 1)$, the dual of $\mathcal{E}^p(I, F, k)$, a typical B-invariant subspace of H^p, seems difficult to identify except in some special cases. One such case is $\mathcal{E}^p(1, \mathbb{T}, n_p) = H^p \cap \overline{H_0^p}$.

THEOREM 6.7.1 (Aleksandrov [**2**]). *For $p \in (0, 1)$, every $\ell \in (H^p \cap \overline{H_0^p})^*$ can be written in the following form*

$$\ell = \Phi_g(f) := \lim_{r \to 1^-} \int_{\mathbb{T}} \{ f(r\zeta) - f(\zeta/r) \} g(\zeta) \, dm(\zeta),\,{}^{52}$$

where $g \in \Lambda_{1/p-1}$ and

$$g(t) = \ell \left(\frac{1}{1 - \bar{t}\zeta} \right) \quad \forall t \in \mathbb{T}.$$

Conversely, if $g \in \Lambda_{1/p-1}$, then Φ_g defines a continuous linear functional on $H^p \cap \overline{H_0^p}$ such that

$$\Phi_g \left(\frac{1}{1 - \bar{t}\zeta} \right) = g(t) \quad \forall t \in \mathbb{T}.$$

PROOF. If $g \in \Lambda_{1/p-1}$, then

(6.7.2)
$$\lim_{r \to 1^-} \int \left\{ \frac{1}{1 - \bar{t}r\zeta} - \frac{1}{1 - \bar{t}r/\zeta} \right\} g(\zeta) \, dm(\zeta) = \lim_{r \to 1^-} \int P_{rt}(\zeta) g(\zeta) \, dm(\zeta) = g(t).$$

Here $P_{rt}(\zeta)$ is a Poisson kernel on the disk.

Next observe that any $g \in \Lambda_{1/p-1}$ can be decomposed, via the Riesz projection operator (which acts continuously on $\Lambda_{1/p-1}$, see Definition 3.5.29) as

$$g = g_1 + \overline{g_2}, \quad g_1, g_2 \in P\Lambda_{1/p-1}, g_2(0) = 0.$$

Thus for $f \in H^p \cap \overline{H_0^p}$ we define $F \in H_0^p(\mathbb{D})$ by $F(z) := \overline{f}(1/\bar{z})$ (the complex conjugate of the pseudocontinuation of f evaluated at $1/\bar{z}$) and observe

$$\int \{ f(r\zeta) - f(\zeta/r) \} g(\zeta) \, dm(\zeta)$$
$$= \int f(r\zeta)(g_1(\zeta) + \overline{g_2}(\zeta)) \, dm(\zeta) - \int f(\zeta/r)(g_1(\zeta) + \overline{g_2}(\zeta)) \, dm(\zeta)$$
$$= f(0)g_1(0) + \int f(r\zeta)\overline{g_2}(\zeta) \, dm(\zeta) - \int \overline{F}(r\zeta)(g_1(\zeta) + \overline{g_2}(\zeta)) \, dm(\zeta)$$
$$= f(0)g_1(0) + \int f(r\zeta)\overline{g_2}(\zeta) \, dm(\zeta) - \int \overline{F}(r\zeta)g_1(\zeta) \, dm(\zeta).$$

Thus, by the Duren-Romberg-Shields result on duality (Theorem 3.5.31), Φ_g defines a continuous linear functional on $H^p \cap \overline{H_0^p}$. Moreover, by eq.(6.7.2),

$$\Phi_g \left(\frac{1}{1 - \bar{t}\zeta} \right) = g(t), \quad \forall t \in \mathbb{T}.$$

Conversely, suppose that ℓ is any continuous linear functional on $H^p \cap \overline{H_0^p}$. Then the function

$$g(t) := \ell \left(\frac{1}{1 - \bar{t}\zeta} \right)$$

---

[52] By $f(\zeta/r)$ we mean the pseudocontinuation of f (to an $H^p(\mathbb{D}_e)$ function) evaluated at ζ/r.

is a well defined function on $\mathbb{T}$. Using the fact that the function
$$t \to \frac{1}{1 - \bar{t}\zeta}$$
is an $H^p \cap \overline{H_0^p}$-valued function satisfying the $\Lambda_{1/p-1}$ smoothness condition (and the continuity of the linear functional), we can conclude more; namely $g \in \Lambda_{1/p-1}$. For this g, the above paragraph shows that Φ_g belongs to $(H^p \cap \overline{H_0^p})^*$ and by eq.(6.7.2)
$$\Phi_g \left(\frac{1}{1 - \bar{t}\zeta} \right) = \ell \left(\frac{1}{1 - \bar{t}\zeta} \right).$$
From this follows that Φ_g and ℓ agree on
$$\bigvee \left\{ \frac{1}{1 - \bar{t}z} : t \in \mathbb{T} \right\} = H^p \cap \overline{H_0^p}$$
(see Proposition 6.1.6) and so $\ell = \Phi_g$. $\square$

6.8. The commutant

Recall from Chapter 5 (Section 11) that if $p \in (0, \infty)$, then $A \in \{B\}'$ (the commutant of B) if and only if there is a unique $\psi \in H^\infty(\mathbb{D})$ such that
$$A \left(\frac{1}{1 - \lambda z} \right) = \frac{\psi(\lambda)}{1 - \lambda z} \quad \forall \lambda \in \mathbb{D}.$$
For $p \in (0, 1)$, $(1 - \lambda z)^{-1} \in H^p$ for all $\lambda \in \mathbb{T}$ and so in this case we can say that $A \in \{B\}'$ if and only if there is a $\psi \in H^\infty(\mathbb{D})$ which is also defined on $\mathbb{T}$ such that the above identity holds for $\lambda \in \mathbb{D}^-$. Again, as with the $p \in [1, \infty)$ case, $A \in \{B\}'$ if and only if $Af = T_{\overline{\psi}}f$ for all $f \in H^\infty$. Here, one needs to recall the Toeplitz operator
$$T_{\overline{\psi}} f = P(\overline{\psi} f),$$
where P is the Riesz projection and $f \in H^\infty$. Our main theorem here is the following.

THEOREM 6.8.1 (Aleksandrov [4]). *For $\psi \in H^\infty(\mathbb{D})$ and $p \in (0, 1)$, the Toeplitz operator $T_{\overline{\psi}}$ extends to be continuous on H^p if and only if $\psi \in P\Lambda_{1/p-1}$. Hence*
$$\{B\}' = \{ T_{\overline{\psi}} : \psi \in P\Lambda_{1/p-1} \}.$$

PROOF. Suppose that $T_{\overline{\psi}}$ has a continuous extension to H^p. Then the linear functional
$$\ell(f) := (T_{\overline{\psi}} f)(0)$$
is continuous on H^p. By Theorem 3.5.31, there is a $g \in P\Lambda_{1/p-1}$ such that
$$\ell(f) = \lim_{r \to 1^-} \int_{\mathbb{T}} f(r\zeta) \overline{g}(\zeta) \, dm(\zeta).$$
A power series computation shows that
$$(6.8.2) \qquad \ell\left(\frac{1}{1 - \lambda z}\right) = \overline{g}(\overline{\lambda}), \quad \lambda \in \mathbb{D},$$
and moreover,
$$\ell\left(\frac{1}{1 - \lambda z}\right) = (T_{\overline{\psi}} \frac{1}{1 - \lambda z})(0) = \frac{\psi(\lambda)}{1 - \lambda z}\bigg|_{z=0} = \psi(\lambda)$$
and so $\psi(\lambda) = \overline{g}(\overline{\lambda}) \in P\Lambda_{1/p-1}$.

6.8. THE COMMUTANT

Conversely, suppose that $\psi \in P\Lambda_{1/p-1}$. We will show that $T_{\overline{\psi}}$ has a continuous extension to H^p. Setting $g(\lambda) = \overline{\psi(\lambda)}$, we see from eq.(6.8.2) that $\ell = \ell_g \in (H^p)^*$ and

$$(6.8.3) \qquad \ell\left(\frac{1}{1-\lambda z}\right) = \psi(\lambda) = T_{\overline{\psi}}\left(\frac{1}{1-\lambda z}\right)\Big|_{z=0} \quad \forall \lambda \in \mathbb{D}.$$

For $f \in H^p$, define the analytic function Uf on $\mathbb{D}$ by

$$(Uf)(w) := \ell\left(\frac{zf - wf(w)}{z-w}\right).$$

To show $Uf \in H^p$, note that

$$\int |(Uf)(r\zeta)|^p \, dm(\zeta)$$
$$\leq \|\ell\|^p \int \left\{\int \left|\frac{zf(z) - r\zeta f(r\zeta)}{z - r\zeta}\right|^p dm(z)\right\} dm(\zeta)$$
$$\leq 2\|\ell\|^p \int \left\{\int \frac{|zf(z)|^p}{|z-r\zeta|^p} dm(z) + \int \frac{|r\zeta f(r\zeta)|^p}{|z-r\zeta|^p} dm(z)\right\} dm(\zeta)$$
$$\leq 2\|\ell\|^p \left\|\frac{1}{1-z}\right\|^p \|f\|^p.$$

Finally, observe from eq.(6.8.3) that

$$U\left(\frac{1}{1-\lambda z}\right) = T_{\overline{\psi}}\left(\frac{1}{1-\lambda z}\right), \quad \forall \lambda \in \mathbb{D}.$$

But since

$$\bigvee\left\{\frac{1}{1-\lambda z} : \lambda \in \mathbb{D}\right\} = H^p$$

(see eq.(5.11.2)), $T_{\overline{\psi}}$ has a continuous extension to H^p. $\square$

In light of Sarason's theorem which characterizes the commutant of the operator $B|H^2 \cap \overline{IH_0^2}$, it is natural examine the commutant of $B|\mathcal{E}^p(I, F, k)$. This seems to be a complicated question which can be handled only in certain special cases. One such case is $H^p \cap \overline{H_0^p} = \mathcal{E}^p(1, \mathbb{T}, n_p)$. From Proposition 6.1.6,

$$H^p \cap \overline{H_0^p} = \bigvee\left\{\frac{1}{1-\zeta z} : \zeta \in \mathbb{T}\right\}.$$

By our analysis above, if $A \in \{B|H^p \cap \overline{H_0^p}\}'$, there is a (unique) function ψ on $\mathbb{T}$ such that

$$A\left(\frac{1}{1-\zeta z}\right) = \frac{\psi(\zeta)}{1-\zeta z} \quad \forall \zeta \in \mathbb{T}.$$

Define the operator U_ψ (densely) on $H^p \cap \overline{H_0^p}$ by

$$U_\psi\left(\frac{1}{1-\zeta z}\right) = \frac{\psi(\zeta)}{1-\zeta z}, \quad \zeta \in \mathbb{T}.$$

THEOREM 6.8.4 (Aleksandrov [4]). *For a function ψ on $\mathbb{T}$, the operator U_ψ extends continuously to $H^p \cap \overline{H_0^p}$ if and only if $\psi \in \Lambda_{1/p-1}$. Hence*

$$\{B|H^p \cap \overline{H_0^p}\}' = \{U_\psi : \psi \in \Lambda_{1/p-1}\}.$$

PROOF. Suppose $\psi \in \Lambda_{1/p-1}$. By using the Riesz projection, the function ψ can be decomposed as
$$\psi = \psi_1 + \overline{\psi_2}, \quad \psi_1, \psi_2 \in P\Lambda_{1/p-1}, \psi_2(0) = 0.$$
By Theorem 6.8.1, the operators
$$T_{\overline{\psi_1}}\left(\frac{1}{1-\lambda z}\right) = \frac{\psi_1(\lambda)}{1-\lambda z}, \quad \lambda \in \mathbb{D}^-$$
$$V_{\psi_2}\left(\frac{\overline{\lambda}z}{\overline{\lambda}\overline{z}-1}\right) := \frac{\overline{\lambda}z\overline{\psi_2(\lambda)}}{\overline{\lambda}\overline{z}-1}, \quad \lambda \in \mathbb{D}^-$$
extend continuously to H^p (respectively $\overline{H_0^p}$) and
$$U_\psi = T_{\overline{\psi_1}} + V_{\psi_2}$$
on finite linear combinations of Cauchy kernels $(1-\overline{\zeta}z)^{-1}, \zeta \in \mathbb{T}$. Thus U_ψ extends continuously to $H^p \cap \overline{H_0^p}$.

Conversely, suppose that U_ψ extends continuously to $H^p \cap \overline{H_0^p}$. Define the continuous linear functional on $H^p \cap \overline{H_0^p}$ by
$$\ell(f) := (U_\psi)(0).$$
From Theorem 6.7.1, $\ell = \Phi_g$ for some $g \in \Lambda_{1/p-1}$ with
$$\Phi_g\left(\frac{1}{1-\overline{t}\zeta}\right) = g(t) \quad \forall\, t \in \mathbb{T}.$$
Finally,
$$g(t) = \ell\left(\frac{1}{1-\overline{t}\zeta}\right) = \left(U_\psi \frac{1}{1-\overline{t}\zeta}\right)(0) = \psi(\overline{t}).$$
Thus $\psi(t) = g(\overline{t}) \in \Lambda_{1/p-1}$. □

Bibliography

[1] E. Abakumov, 'Cyclicity and approximation by lacunary power series', Michigan Math. J. **42** (1995), 277 - 299.

[2] A. B. Aleksandrov, 'Approximation by rational functions and an analog of the M. Riesz theorem on conjugate functions for L^p spaces with $p \in (0,1)'$, Math. USSR Sbornik, **35** (1979), 301 - 316.

[3] A. B. Aleksandrov, 'Invariant subspaces of the backward shift operator in the space H^p $(p \in (0,1))$' (Russian), Investigations on linear operators and the theory of functions, IX, Zap. Nauchn. Sem. Leningrad Otdel. Mat. Inst. Steklov (LOMI), **92** (1979), 7 - 29.

[4] A. B. Aleksandrov, 'Essays on non locally convex Hardy spaces', Lecture Notes in Math., 864, Springer-Verlag, Berlin-New York, 1981.

[5] A. B. Aleksandrov, 'Invariant subspaces of shift operators. An axiomatic approach', Investigations on linear operators and the theory of functions, XI, Zap. Nauchn. Sem. Leningrad Otdel. Mat. Inst. Steklov (LOMI), **113** (1981), 7 - 26.

[6] A. B. Aleksandrov, 'Lacunary series and pseudocontinuations', J. Math. Sci., **92** (1998), 3550 - 3559.

[7] A. B. Aleksandrov, 'Lacunary series and pseudocontinuations, an arithemetic approach', St. Petersburg Math. J. **9** (1998), 1 - 20.

[8] A. Aleman and W. T. Ross, 'The backward shift on weighted Bergman spaces', Michigan Math. J. **43** (1996), 291 - 319.

[9] A. Aleman, S. Richter, and W. T. Ross, 'Pseudocontinuations and the backward shift', Indiana Univ. Math. J. **47** (1998), 223 - 276.

[10] T. Ando, 'On the predual of H^∞', Special issue dedicated to Władysław Orlicz on the occasion of his seventy-fifth birthday. Comment. Math. Special Issue 1 (1978), 33 - 40.

[11] S. Axler, 'Bergman spaces and their operators', *Surveys of some recent results in operator theory*, Vol. I, 1 - 50, Pitman Res. Notes Math. Ser., 171, Longman Sci. Tech., Harlow, 1988.

[12] F. Bagemihl and W. Seidel, 'Some boundary properties of analytic functions', Math. Z. **61** (1954), 186 - 199.

[13] C. Bessaga and A. Pełczyński, 'Spaces of continuous functions (IV). On isomorphical classification of spaces of continuous functions', Studia Math. **19** (1960), 53 - 62 .

[14] A. Beurling, 'On two problems concerning linear transformations in Hilbert space', Acta Math. **81** (1949), 239 - 255.

[15] M. Brelot, *Éléments de la théorie classique du potentiel*. Les Cours de Sorbonne. 3e cycle, Centre de Documentation Universitaire, Paris, 1959.

[16] L. Brown, A. Shields, K. Zeller, 'On absolutely convergent exponential sums', Trans. Amer. Math. Soc. **96** (1960), 162 - 183.

[17] D. L. Burkholder, R. F. Gundy, and M. L. Silverstein, 'A maximal function characterization of the class H^p', Trans. Amer. Math. Soc. **157** (1971), 137 - 153.

[18] S. Campanato 'Proprietà di hölderianità di alcune classi di funzioni', Ann. Scuola Norm. Sup. Pisa (3) **17** (1963), 175 - 188.

[19] L. Carleson, 'On the zeros of functions with bounded Dirichlet integrals', Math. Z. **56** (1952), 289 - 295.

[20] W. S. Cohn, 'Radial limits and star-invariant subspaces of bounded mean oscillation', Amer. J. Math. **108** (1986), 719 - 749.

[21] R. R. Coifman, 'A real variable characterization of H^p', Studia Math. **51** (1974), 269 - 274.

[22] R. R. Coifman and G. Weiss, 'Extension of Hardy spaces and their use in analysis', Bull. Amer. Math. Soc. **83** (1977), 569 - 645.

[23] E. F. Collingwood and A. J. Lohwater, *The theory of cluster sets*, Cambridge University Press, London, 1966.
[24] C. S. Davis, 'Iterated limits in $N^*(U^n)$', Trans. Amer. Math. Soc. **178** (1973), 139 - 146.
[25] M. M. Day, 'The spaces L^p with $0 < p < 1$', Bull. Amer. Math. Soc. **46** (1940), 816 - 823.
[26] L. de Branges and J. Rovnyak, 'Cannonical models in quantum scattering theory', *Perturbation Theory and Its Applications in Quantum Mechanics* (Proc. Adv. Sem. Math. Res. Center, U.S. Army, Theoret. Cham. Inst., Univ. of Wisconsin, Madison, Wis. 1965), Wiley, New York.
[27] K. de Leeuw, 'The failure of spectral analysis in L^p for $0 < p < 1$', Bull. Amer. Math. Soc. **82** (1976), 111 - 114.
[28] R.G. Douglas, H.S. Shapiro, and A.L. Shields, 'On cyclic vectors of the backward shift', Bull. Amer. Math. Soc. **73** (1967), 156 - 159.
[29] R.G. Douglas, H.S. Shapiro, and A.L. Shields, 'Cyclic vectors and invariant subspaces for the backward shift operator', Ann. Inst. Fourier (Grenoble), **20** (1970), 37 - 76.
[30] N. Dunford and J. T. Schwartz, *Linear Operators*, Part I, Wiley (Interscience), New York, 1958.
[31] P. L. Duren, *Theory of H^p spaces*, Pure and Appl. Math., Vol. 38, Academic Press, New York-London, 1970.
[32] P. L. Duren, B.W. Romberg, A.L. Shields, 'Linear functionals on H^p spaces with $0 < p < 1$', J. Reine Angew. Math. **238** (1969), 32 - 60.
[33] P. Fatou, 'Séries trigonométriques et séries de Taylor', Acta Math. **30** (1906), 335 - 400.
[34] C. Fefferman, 'Characterizations of bounded mean oscillation', Bull. Amer. Math. Soc. **77** (1971), 587 - 588.
[35] C. Fefferman, E. M. Stein, 'H^p spaces of several variables', Acta Math. **129** (1972), 137 - 193.
[36] C. Foiaş, 'A remark on the universal model for contractions of G.-C Rota', Com. Acad. R. P. Romîne, **13** (1963) 349 - 352.
[37] T. Gamelin, *Uniform Algebras*, Chelsea, New York, 1984.
[38] J. García-Cuerva and J. L. R. de Francia, *Weighted Norm Inequalities and Related Topics*, North-Holland Mathematics Studies, 116, Notas de Matemática (104), North-Holland Publishing Co., Amsterdam-New York, 1985.
[39] J. Garnett, *Bounded Analytic Functions*, Pure and Appl. Math., Vol. 96, Academic Press, New York-London, 1981.
[40] A. Gonchar, 'On quasi-analytic continuation of analytic functions across a Jordan arc', (Russian) Dokl. Akad. Nauk. SSSR **166** (1966), 1028 - 1031.
[41] G. H. Hardy, 'The mean value of the modulus of an analytic function', Proc. London Math. Soc. **14** (1914), 269 - 277.
[42] G. H. Hardy and J. E. Littlewood, 'A maximal theorem with function-theoretic applications', Acta Math. **54** (1930), 81 - 116.
[43] E. Hewitt and K. Stromberg, *Real and Abstract Analysis*, Springer-Verlag, New York, 1965.
[44] K. Hoffman, *Banach spaces of analytic functions*, Dover Publications, Inc., New York, 1988.
[45] J. P. Kahane and Y. Katznelson, 'Sur le comportement radial des fonctions analytiques', C-R Acad. Sci. Paris, Sér. A-B, t. 272, 15 mars 1971, 718 - 719.
[46] N. J. Kalton, 'Basic sequences in F-spaces and their applications', Proc. Edinburgh Math. Soc. (2) **19** (1974/75), 151 - 167.
[47] S. V. Khrushchev, 'Every ideal of the ring C_A^n is principal', (Russian) Zap. Nauch. Sem. Leningrad Otdel. Mat. Inst. Steklov (LOMI) **65** (1976), 149 - 160; translation in J. Soviet Math. **16** (1981), 1150 - 1158.
[48] S. Khrushchev and V. Peller, 'Hankel operators, best approximation, and stationary Gaussian processes, (Russian) Uspekhi Mat. Nauk **37** (1982), 53 - 124; translation in Russian Math. Surveys **37** (1982), 61 - 144.
[49] A. N. Kolmogorov, 'Sur les fonctions harmoniques conjuguées et les séries de Fourier', Fund. Math. **7** (1925), 24 - 29.
[50] A. Kolmogorov, 'Zur Normierbarkeit eines allemeinen topologischen linearen Raumes', Studia Math. **5** (1934), 29 - 33.
[51] P. Koosis, *Introduction to H_p Spaces* (Second edition), Cambridge Tracts in Mathematics, 115. Cambridge University Press, Cambridge, 1998.

[52] T. L. Kriete, 'On the Fourier coefficients of outer functions', Indiana Univ. Math. J. **20** (1970), 147 - 155.

[53] E. Lindelöf, 'Sur un principe général de l'analyse et ses applications à la théorie de la représentation conforme', Acta Soc. Sci. Fenn. **46**, no. 4 (1915), 1 - 35.

[54] J. E. Littlewood, Mathematical Notes (4), 'On the theorem of Fatou', J. London Math. Soc. **2** (1927), 172 - 176.

[55] J. E. Littlewood, Mathematical Notes (13), 'On the mean value of power series (II)', J. London. Math. Soc. **5** (1930), 179 - 182.

[56] A. E. Livingston, 'The space H^p, $0 < p < 1$, is not normable', Pacific J. Math. **3** (1953), 613 - 616.

[57] A. J. Lohwater and G. Piranian, 'The boundary behavior of functions analytic in a disk', Ann. Acad. Sci. Fenn. Ser. A. I., no. 239 (1957) 17 pp.

[58] S. Lu, *Four lectures on real H^p spaces*, World Scientific Publishing Co., Inc., River Edge, New Jersey, 1995.

[59] N. Lusin and I. I. Privalov, Sur l'unicité et la multiplicité des fonctions analytiques, Annales de l'Ecole Normale Supérieure **42**, (1925), 143 - 191.

[60] G. R. MacLane, 'Holomorphic functions, of arbitrary slowth growth, without radial limits', Michigan Math. J. **9** (1962), 21 - 24.

[61] J. Marcinkiewicz and A. Zygmund, 'A theorem of Lusin', Duke Math. J. **4** (1938), 473 - 485.

[62] N. G. Meyers, 'Mean oscillation over cubes and Hölder continuity', Proc. Amer. Math. Soc. **15** (1964), 717 - 721.

[63] J. W. Moeller, 'On the spectra of some translation invariant subspaces', J. Math. Anal. Appl. **4** (1962), p. 276 - 296.

[64] D. J. Newman, 'The non-existence of projections from L^1 to H^1', Proc. Amer. Math. Soc. **12** (1961), 98 - 99.

[65] N. K. Nikolskiĭ, *Treatise on the shift operator*, Springer-Verlag, New York, 1986.

[66] A. I. Plessner, 'Über das Verhalten analytischer Funktionen am Range ihres Definitionsbereichs', J. Riene Angew. Math. **158** (1927), 219 - 227.

[67] I. I. Privalov, 'Sur les fonctions conjuguées', Bull. Soc. Math. France, **44** (1916), 100 - 103.

[68] I. I. Privalov, 'Intégrale de Cauchy' (in Russian), *Saratov*, 1919, p. 1 - 104.

[69] I. I. Privalov, *Randeigenschaften analytischer Funktionen*, VEB Deutscher Verlag der Wissenschaften, Berlin, 1956.

[70] S. Richter and C. Sundberg, 'Invariant subspaces of the Dirichlet shift and pseudocontinuations', Trans. Amer. Math. Soc. **341** (1994), 863 - 879.

[71] F. Riesz, 'Über die Randwerte einer analytischen Funktion', Math. Z. **18** (1923), 87 - 95.

[72] F. Riesz and M. Riesz, 'Über die Randwerte einer analytischen Funktion', Quatrième Congrès des Math. Scand. Stockholm (1916), 27 - 44.

[73] M. Riesz, 'Les fonctions conjuguées et les séries de Fourier', C. R. Acad. Sci. Paris, Sér. A - B **178** (1924), 1464 - 1467.

[74] M. Riesz, 'Sur les fonctions conjuguées', Math. Z. **27** (1927), 218 - 244.

[75] M. Rosenblum and J. Rovnyak, *Hardy classes and operator theory*, Oxford Mathematical Monographs, Oxford University Press, New York, 1985.

[76] G.-C Rota, 'On models for linear operators', Comm. Pure Appl. Math., **13** (1960) 469 - 472.

[77] W. Rudin, 'Radial cluster sets of analytic functions', Bull. Amer. Math. Soc. **60** (1954), 545.

[78] W. Rudin, *Functional Analysis* (Second Edition), McGraw Hill, Inc., New York, 1991.

[79] D. Sarason, 'Generalized interpolation in H^∞', Trans. Amer. Math. Soc. **127** (1967), 179 - 203.

[80] D. Sarason, 'Functions of vanishing mean oscillation', Trans. Amer. Math. Soc. **207** (1975), 391 - 405.

[81] D. Sarason, *Function theory on the unit circle*, Notes for lectures given at a conference at Virginia Polytechnic Institute and State University, Blasksburg, Virginia, June 19 - 23, 1978, Department of Mathematics, Virginia Polytechnic Institute and State University, 1978.

[82] F. A. Shamoyan, 'Toeplitz operators and division by an inner function in some spaces of analytic functions', Akad. Nauk Armyan. SSR Dokl. **76** (1983), 109 - 113.

[83] H. S. Shapiro, 'Weakly invertible elements in certain function spaces, and generators in ℓ_1', Michigan Math. J. **11** (1964), 161 - 165.

[84] H. S. Shapiro, 'Generalized analytic continuation', Symposia on Theor. Phys. and Math. **8** (1968), Plenum Press, New York, 151 - 163.

[85] H. S. Shapiro, 'Functions nowhere continuable in a generalized sense', Publ. Ramanujan Inst. No. **1** (1968/1969), 179 - 182.
[86] H. S. Shapiro, private communication, 1999.
[87] J. H. Shapiro and A. L. Shields, 'Unusual topological properties of the Nevanlinna class', Amer. J. Math. **97** (1975), 915 - 936.
[88] N. A. Shirokov, 'Closed ideals of algebras of type $B_{p,q}^{\alpha}$', Izv. Akad. Nauk SSSR, Ser. Mat. **46** (1982), 1316 - 1332; translation in Math. USSR - Ivz. **21** (1983), 585 - 600.
[89] V. I. Smirnov, 'Sur les valeurs limites des fonctions régulières à l'intérieur d'un cercle', Journal de la Société Phys.-Math. de Léningrade, **2** (1929), no. 2, 22 - 37.
[90] S. Spanne, 'Sur l'interpolation entre les espaces $\mathcal{L}_k^{p,\Phi}$', Ann. Scuola Norm. Sup. Pisa (3) **20** (1966), 625 - 648.
[91] D. Spencer, 'A function theoretic identity', Amer. J. Math. **65** (1943), 147 - 160.
[92] D. Stegenga, 'Bounded Toeplitz operators on H^1 and applications of the duality between H^1 and the functions of bounded mean oscillation', Amer. J. Math. **98** (1976), 573 - 589.
[93] E. Stein, Singular integrals, harmonic functions and differentiability properties of functions of several variables, Proc. Sympos. in Pure Math. **10** (1967), 316 - 335
[94] E. Stein, *Singular integrals and differentiability properties of functions*, Princeton University Press, Princeton, New Jersey, 1970.
[95] E. Stein, *Harmonic Analysis*, Princeton Universiry Press, Princeton, New Jersey, 1993.
[96] E. Stein and G. Weiss, *Fourier analysis on Euclidean spaces*, Princeton University Press, New Jersey, 1971.
[97] C. Sundberg, 'Truncations of BMO functions', Indiana Univ. Math. J., **33** (1984), 749 - 771.
[98] B. Sz-Nagy and C. Fioaş, *Harmonic analysis of operators on Hilbert space*, North-Holland, Amsterdam-London, 1970.
[99] B. A. Taylor and D. L. Williams, 'Zeros of Lipschitz functions in the unit disk', Michigan. Math. J. **18** (1971), 129 - 139.
[100] G. C. Tumarkin, 'Description of a class of functions admitting an approximation by fractions with preassigned poles' (Russian) Izv. Akad. Nauk Armjan. SSR Ser. Mat. **1** (1966), 89 - 105.
[101] I. N. Vekua, *Generalized Analytic Functions*, Addison-Wesley, Reading, Mass., 1962.
[102] J. Wolff, 'Sur les séries $\sum A_k/(z - z_k)$, Comptes Rendus, **173** (1921), 1057 - 1058, 1327 - 1328.
[103] N. Yanagihara, 'Multipliers and linear functionals for the class N^+', Trans. Amer. Math. Soc. **180** (1973), 449 - 461.
[104] A. Zygmund, 'Smooth functions', Duke Math. J. **12** (1945), 47 - 76.
[105] A. Zygmund, 'On a theorem of Littlewood', Summa Brasil. Math. **2** (1949), 51 - 57.
[106] A. Zygmund, *Trigonometric Series*, Cambridge University Press, Cambridge, 1988.

Index

A (disk algebra), 35
Abakumov, E., 14, 108
Aleksandrov, A. B., 4–6, 14, 15, 43, 97, 99, 101, 108, 111, 116, 119, 180, 187–189
Aleman, A., 8, 87, 89, 91, 97, 108
Ando, T., 37
annihilator ($\mathcal{M}^\perp$), 28
atom, 72
atomic decomposition, 47, 72, 153, 161, 164, 174, 176
    Coifman's theorem, 72
Axler, S., 87

B (backward shift operator), 1
backward shift operator
    commutant, 109, 188, *see also* commutant
    cyclic vectors, 105, 186, *see also* cyclic vectors
    definition, 81
    invariant subspaces
        $BMOA$, 100
        H^1, 101
        H^p, $0 < p < 1$, 119
        H^p, $1 < p < \infty$, 82
        $L^1/\overline{H^1_0}$, 99
        L^p_a, 88, 89
        $VMOA$, 99
    spectrum, 86, 88, 185, *see also* $\sigma(I)$
Bagemihl, F., 10, 12
balayage (sweep), 90
Bergman space
    backward shift invariant subspaces, 88
    cyclic vectors for B, 108
    definition, 88
    dual, 88
Bessaga, C., 37
Beurling, A., 3, 81, 82, 85, 97, 98
Beurling's theorem
    for H^p, 81
    for Smirnov class, 97
Blaschke product, 18
BMO (functions of bounded mean oscillation), *see also* bounded mean oscillation
$BMOA$ (analytic functions of bounded mean oscillation), *see also* bounded mean oscillation

Borel series, 15
bounded mean oscillation
    backward shift invariant subspaces, 100
    definition, 30
    Fefferman duality theorem, 32
    John-Nirenberg theorem, 31
    Riesz projection operator, 31
    truncation, 32
bounded type, 41
Brelot, M., 90
Brown, L., 15
Burkholder, D., 54, 61

Calderon, A., 54
Campanato, S., 66, 160
Carleson, L., 108
$\mathcal{C}$ (the cyclic vectors for B), 107
Cauchy integral formula
    for Hardy space of the disk, 22
    for Hardy space of the upper half plane, 51, 75
Cauchy transform
    Aleksandrov's characterization, 26
    definition, 23
    Fatou's jump theorem, 91
    Kolmogorov's theorem, 23
    norm, 27
$C_0^\infty(\mathbb{R})$, 62
$\mathbb{C}_\infty$ (extended complex plane), 9
$C(\mathbb{T})/\overline{zA}$, 35
$C\mu$ (Cauchy transform on a measure), 23
Cohn, W., 109
Coifman, R., 6, 47, 72, 75
Coifman's theorem (on atomic decompositions), 72
Collingwood, E., 9, 41
commutant
    of B on $H^2 \cap \overline{IH^2_0}$, 111
    of B on $H^p \cap \overline{H^p_0}$ ($0 < p < 1$), 189
    of B on H^p ($0 < p < 1$), 188
    of B on H^p ($1 \leq p < \infty$), 111
conjugate Poisson kernel, 52
$\mathbb{C}_+$ (upper half plane), 9
$C(\mathbb{T})$ (continuous functions on $\mathbb{T}$), 35
cyclic vectors
    for B on D_q, 108

for B on H^p $(0 < p < 1)$, 186
for B on H^p $(1 \leq p < \infty)$, 105
for B on L^p_a, 108

Davis, C., 40
Day, M., 37, 144
de Branges, L., 1
de Francia, J., 49, 66, 73, 149
de Leeuw, K., 115
δ_a (evaluation functional at a), 63
Dirac distribution, 63
Dirichlet space
  cyclic vectors for B, 108
  definition, 88
$\mathbb{D}$ (unit disk), 9
disk algebra, 35
$\mathbb{D}_e$ (extended exterior disk), 9
$\mathcal{D}'$ (Schwartz distributions), 63
distributions, 62
dm (normalized Lebesgue measure on the circle), 2
dm_2 (normalized Lebesgue area measure on the disk), 88
dominating sequence, 15
Douglas, R., ix, 3, 14, 105, 107
D_q (Dirichlet space), 88
duality
  BMO, 32, 73
  $BMOA$, 31
  $C(\mathbb{T})/\overline{zA}$, 36
  D_q, 88
  $\mathcal{H}^p_h$, 73
  H^1, 30
  $H^p \cap I\overline{H^p_0}$, 109
  $H^p \cap \overline{H^p_0}$ $(0 < p < 1)$, 187
  $H^p(\mathbb{R})$, 73
  $H^p, 0 < p < 1$, 37
  $H^p, 1 < p < \infty$, 29
  $L^1/\overline{H^1_0}$, 37
  $L^p, 0 < p < 1$, 37
  L^p_a, 88
  $\Lambda_{1/p-1}$, 38
  $\Re H^1$, 32, 36
Ando's theorem , 37
annihilator, 28
Bessaga, C. and Pełczyński, A., 37
Day's theorem, 37
Duren, P., Romberg, B., Shields, A., 37
Fefferman duality theorem, 32
Hahn-Banach separation property, 39
Hahn-Banach theorem, 28
Kalton's theorem, 39
Lipschitz class, 73
locally convex, 39
pre-annihilator, 28
Sarason's theorem, 35
VMOA, 36
Zygmund class, 73

Dunford, N., 38, 86
Duren, P., ix, 7, 10, 37, 38, 40, 41, 47, 48, 51, 73, 116, 123

$\mathcal{E}^p(I, F, k)$, 5, 119
$e^p(I, F, k)$, 120
$E^p(F, k)$, 138

F. and M. Riesz theorem (on measures), 21
F. and M. Riesz theorem (uniqueness), 12
factorization theorem, 18
Fatou, P., 7, 11–14, 18, 25, 40, 49, 54, 67, 85, 87, 91, 102, 116, 124, 150
Fatou's jump theorem, 91, see also Cauchy transform
Fatou's theorem, 11, 49
Fefferman, C., 31, 60, 61, 65, 71
Fefferman duality theorem, 32
F_g (Cauchy integral of a function), 75
F_ℓ (Cauchy "integral" of a distribution ℓ), 75
Foiaş, C., 1
F-space, 21

Gamelin, T., 41, 42, 82
$\Gamma_\alpha(\zeta)$ (non-tangential approach region), 9
García-Cuerva, J., 49, 66, 73, 149
Garnett, J., ix, 7, 19, 24, 31, 47, 48, 53, 82, 84
Gonchar, A., 13
greatest common divisor (of inner functions), 82
Gundy, R., 54, 61

$\text{Har}(\Omega)$ (the harmonic functions on Ω), 49
Hardy, G. H., 17, 52, 54, 61
Hardy spaces of the disk
  boundary values, 18
  Cauchy integral formula, 22
  definition, 17
  dual, 30, 38
  Fourier series, 21
  growth estimates, 20
  of exterior disk, 21
  Riesz factorization theorem, 18
  topology, 20, 38
Hardy spaces of the upper half plane
  and H^p of the disk, 48, 138
  atomic decomposition, 72
  boundary values, 48
  Cauchy integral formula, 51, 75
  definition, 48
  Fatou's theorem, 49
  growth estimates, 48
  Poisson integral, 49
harmonic conjugation operator, 23, 52
harmonic Hardy space
  atomic decomposition, 72

Burkholder, D., Gundy, R., and Silverstein, M., 61
definition, 60
distributions, 65
dual, 73
Fefferman, C., Stein, E., 61, 65, 71
growth condition, 60
Hardy, G. H. and Littlewood, J. E., 61
Hewitt, E., 25
Hf (Hilbert transform of f), 54
Hilbert transform, 145, 148
and harmonic conjugation, 55
definition, 54
H^∞ (bounded analytic functions), 17
Hoffman, K., 7, 19, 23, 82
$\mathfrak{H}(\Omega)$ (holomorphic functions on Ω), 9
H^p (Hardy space of the disk), 17
$\mathcal{H}^p(\mathbb{C}\backslash\mathbb{R})$ (Hardy space of $\mathbb{C}\backslash\mathbb{R}$), 76
$\mathcal{H}^p(\mathbb{C}\backslash\mathbb{R})_*$, 137
$H^p(\mathbb{D}_e)$ (Hardy space of the exterior disk), 21
$\mathcal{H}^p_E$, 151
$\mathcal{H}^p_h$ (harmonic Hardy space - upper half plane), 60
H^p_0 (H^p functions vanishing at the origin), 3
$H^p \cap I\overline{H^p_0}$, 3
$\mathcal{H}^p \cap \overline{\mathcal{H}^p}$, 136
$H^p(\mathbb{R})$ (distributional Hardy space), 67
$\mathcal{H}^p(\mathbb{R})$ (boundary functions for $\mathcal{H}^p$ functions), 48
$H^p(\mathbb{T})$ (boundary values of H^p functions), 21
$\mathcal{H}^p$ (Hardy space of the upper half plane), 47

I_f (inner factor of f), 19
inner function
analytic continuation, 20
Blaschke product, 18
definition, 19
singular inner function, 18
spectrum, 19
(I,p)-permissible, 119

John, F., 31

K (space of Cauchy transforms), 26
$K_a (\simeq L^1/\overline{H^1_0})$ (Cauchy transforms of L^1 functions), 27
K_s (Cauchy transforms of singular measures), 27
Kahane, J., 10, 89
Kalton, N., 5, 39
Katznelson, Y., 10, 89
Khrushchev, S., 87, 89
Kolmogorov, A. N., 14, 23, 38, 53, 90, 105
Kolmogorov's theorem, 23, see also Cauchy transform
Koosis, P., 7, 38, 52, 82

Korenblum, B., 12
Kriete, T., 107

$L^1/\overline{H^1_0}$
backward shift invariant subspaces, 99
definition, 36
duality, 37
lacunary series, 14
$\Lambda^n_\alpha(\mathbb{T})$ (Lipschitz class of circle), 37
$\Lambda^n_\alpha(\mathbb{R})$ (Lipschitz class of line), 66
$\Lambda^n_*(\mathbb{T})$ (Zygmund classes of circle), 37
$\Lambda^n_*(\mathbb{R})$ (Zygmund class of the line), 66
$\Lambda_{1/p-1}(\mathbb{T})$, 38
$\Lambda_{1/p-1}(\mathbb{R})$, 66
L^p_a (Bergman space), 88
Lebesgue point, 50
lim inf zero set (for an inner function), 19
Lindelöf, E., 11
Lipschitz class, 37, 65, 66, 73
Littlewood, J. E., 10, 11, 52, 54, 61, 91, 116
Littlewood subordination principle, 116
Livingston, A., 39
local Dirichlet integral, 89
$\log L$, 42
Lohwater, A., 9, 10, 12, 41
Lu, S., 75
Lusin, N., 7, 11–14, 18, 40, 61
Lusin area function, 11, 61
Lusin-Privalov uniqueness theorem, 12

M_a (absolutely continuous measures), 27
MacLane, G., 10
Marcinkiewicz, J., 11
maximal function
and conjugate Poisson integral, 53
and Poisson integral, 53
definition (Hardy-Littlewood), 52
definition (non-tangential), 53
Hardy Littlewood theorem, 52
Kolmogorov, A. N., 53
Riesz, M., 54
$\mathfrak{M}(\Omega)$ (meromorphic functions on Ω), 9
Meyers, N., 66, 160
Mf (Hardy-Littlewood maximal function), 52
$M_\infty(r,f)$ (maximum modulus function), 17
model theory, 1
Moeller, J., 4, 94
Morera's theorem, 79, 84, 130
$M_p(r,f)$ (p-th integral means - disk), 17
m_p (p-th integral means - upper half plane), 48
M_s (singular measures), 27
$M(\mathbb{T})$ (finite Borel measures on $\mathbb{T}$), 23

N (Nevanlinna class), 40
N^+ (Smirnov class), 41
N^-, 43
$\mathcal{N}$ (the non-cyclic vectors for B), 107

$\mathfrak{N}(\mathbb{D})$ (functions of bounded type), 41
Nevanlinna class
  Aleksandrov, A. B., 43
  Davis, C., 40
  definition, 40
  metric, 40
  of exterior disk, 42
  Shapiro, J., 40
  Smirnov class, 41
  Yanagihara, N., 40, 42
Newman, D. J., 23
Nikolskiĭ, N., ix, 1, 111
Nirenberg, L., 31
non-tangential limits
  definition, 9
  Fatou's theorem, 11
  Hardy space functions, 18
  Lindelöf's theorem, 11
  Lusin area function, 11, 61
  Lusin-Privalov uniqueness theorem, 12
  Plessner's theorem, 11
non-tangential maximal function, 53, see also maximal function
$n_p(=[1/p])$, 5

O_f (outer factor of f), 19
outer function, 19

P (Riesz projection), 22
Pełczyński, A., 37
Peller, V., 87
permissible, 119, see also (I,p)-permissible
$\Phi_{a_1,a_2}\ell$, 173
Piranian, G., 10, 12
$P\Lambda_\alpha^n$ (analytic Lipschitz class), 38
$P\Lambda_{1/p-1}$, 38
$P\Lambda_*^n$ (analytic Zygmund class), 38
Plessner, A., 7, 11, 13
Plessner's theorem, 11
Poisson kernel function, 25, 49
pre-annihilator ($^\perp\mathcal{N}$), 28
Privalov, I. I., 7, 11–14, 18, 38, 40, see also Lusin-Privalov uniqueness theorem
pseudocontinuations, 3
  and the Lusin-Privalov uniqueness theorem, 13
  cyclic vectors, 105, 186
  definition, 13
  generalized analytic continuations, 13
  of an inner function, 19
  of bounded type, 106
  Shapiro, H. S., 3, 14
$P_z(\zeta)$ (Poisson kernel function - disk), 25
$P_z(t)$ (Poisson kernel function - upper half plane), 49

$Q_z(t)$ (conjugate Poisson kernel - upper half plane), 52

radial limit
  definition, 9
  F. and M. Riesz uniqueness theorem, 12
  Fatou's theorem, 11
  pathological behavior, 10
real H^p, 75
$\Re H^1$ (real H^1), 32
$\Re H^p$ (real H^p), 75
Richter, S., 8, 87, 89, 91, 97, 108
Riesz factorization theorem, 18
Riesz projection operator
  and $BMOA$, 31
  and Cauchy transforms, 24
  and Lipschitz and Zygmund classes, 38
  and L^p, 22
  and VMOA, 35
  definition, 22
Riesz, F., 7, 12, 18, 21, 22, 27, 29, 36, 55, 83, 85, 93, 97, 99, 100, 104
Riesz, M., 7, 12, 21, 22, 27, 29, 36, 52, 54, 55, 83, 85, 93, 97, 99, 100, 104
Romberg, B., 37, 38, 73
Ross, W. T., 8, 87, 89, 91, 97, 108
Rota, G.-C., 1
Rovnyak, J., 1
Rudin, W., ix, 10, 28, 38, 62, 63, 141

Sarason, D., 30, 35, 105, 111
Schwartz, J., 38, 86
Seidel, W., 10, 12
Shapiro, H. S., ix, 3, 7, 13, 14, 105, 107
Shapiro, J., 40
Shields, A., ix, 3, 14, 15, 37, 38, 73, 105, 107
Shirokov, N., 89
$\sigma(I)$ (spectrum of an inner function), 19
Silverstein, M., 54, 61
singular inner function, 18
Smirnov class
  Beurling's theorem, 97
  definition, 41
  of exterior disk, 42
Smirnov, V., 14, 18, 41
Spanne, S., 31
spectrum
  of an inner function, 19
  of B on $\mathcal{E}^p(I,F,k)$, 185
  of B on $H^p \cap I\overline{H_0^p}$ ($1 \leq p < \infty$), 95
  of B on H^p ($1 \leq p < \infty$), 86
Spencer, D., 11
$\mathcal{S}$ (test functions), 62
$\mathcal{S}'$ (tempered distributions), 62
Stegenga, D., 111
Stein, E., ix, 7, 11, 31, 49, 53–55, 60–62, 65, 67, 69–73, 88, 162
Stromberg, K., 25
Sundberg, C., 32, 89
sweep of a measure (balayage), 90
Sz-Nagy, B., 1

Taylor, B., 108
T_f (the pseudocontinuation of a function f), 13
$\mathbb{T}$ (unit circle), 9
Toeplitz operator, 110, 188, *see also* commutant
truncation of BMO functions, 32
Tumarkin, G., 107

U_f (Poisson integral of a function f), 49
U_ℓ (Poisson "integral" of a distribution ℓ), 69
U_μ (Poisson integral of a measure μ), 50
$u*$ (non-tangential maximal function), 53

vanishing mean oscillation
    backward shift invariant subspaces, 99
    definition, 34
    dual, 35, 36
    Riesz projection, 35
    Sarason's theorem, 35
Vekua, I., 91
V_f (conjugate Poisson integral of f), 52
V_ℓ (conjugate Poisson "integral" of a distribution ℓ), 75
VMO (functions of vanishing mean oscillation), *see also* vanishing mean oscillation
$VMOA$ (analytic functions of vanishing mean oscillation), *see also* vanishing mean oscillation

weak-type $(1,1)$, 53
Weiss, G., 55, 62, 75
Williams, D., 108
Wolff, J., 15

$\mathcal{X}_{a,t}$, 163

Yanagihara, N., 40, 42

Zeller, K., 15
Zygmund, A., 7, 11, 38, 54
Zygmund class, 37, 66, 73

Selected Titles in This Series

(*Continued from the front of this publication*)

47 **A. D. Elmendorf, I. Kriz, M. A. Mandell, and J. P. May** (with an appendix by **M. Cole**), Rings, modules, and algebras in stable homotopy theory, 1997
46 **Stephen Lipscomb,** Symmetric inverse semigroups, 1996
45 **George M. Bergman and Adam O. Hausknecht,** Cogroups and co-rings in categories of associative rings, 1996
44 **J. Amorós, M. Burger, K. Corlette, D. Kotschick, and D. Toledo,** Fundamental groups of compact Kähler manifolds, 1996
43 **James E. Humphreys,** Conjugacy classes in semisimple algebraic groups, 1995
42 **Ralph Freese, Jaroslav Ježek, and J. B. Nation,** Free lattices, 1995
41 **Hal L. Smith,** Monotone dynamical systems: an introduction to the theory of competitive and cooperative systems, 1995
40.4 **Daniel Gorenstein, Richard Lyons, and Ronald Solomon,** The classification of the finite simple groups, number 4, 1999
40.3 **Daniel Gorenstein, Richard Lyons, and Ronald Solomon,** The classification of the finite simple groups, number 3, 1998
40.2 **Daniel Gorenstein, Richard Lyons, and Ronald Solomon,** The classification of the finite simple groups, number 2, 1995
40.1 **Daniel Gorenstein, Richard Lyons, and Ronald Solomon,** The classification of the finite simple groups, number 1, 1994
39 **Sigurdur Helgason,** Geometric analysis on symmetric spaces, 1994
38 **Guy David and Stephen Semmes,** Analysis of and on uniformly rectifiable sets, 1993
37 **Leonard Lewin, Editor,** Structural properties of polylogarithms, 1991
36 **John B. Conway,** The theory of subnormal operators, 1991
35 **Shreeram S. Abhyankar,** Algebraic geometry for scientists and engineers, 1990
34 **Victor Isakov,** Inverse source problems, 1990
33 **Vladimir G. Berkovich,** Spectral theory and analytic geometry over non-Archimedean fields, 1990
32 **Howard Jacobowitz,** An introduction to CR structures, 1990
31 **Paul J. Sally, Jr. and David A. Vogan, Jr., Editors,** Representation theory and harmonic analysis on semisimple Lie groups, 1989
30 **Thomas W. Cusick and Mary E. Flahive,** The Markoff and Lagrange spectra, 1989
29 **Alan L. T. Paterson,** Amenability, 1988
28 **Richard Beals, Percy Deift, and Carlos Tomei,** Direct and inverse scattering on the line, 1988
27 **Nathan J. Fine,** Basic hypergeometric series and applications, 1988
26 **Hari Bercovici,** Operator theory and arithmetic in H^∞, 1988
25 **Jack K. Hale,** Asymptotic behavior of dissipative systems, 1988
24 **Lance W. Small, Editor,** Noetherian rings and their applications, 1987
23 **E. H. Rothe,** Introduction to various aspects of degree theory in Banach spaces, 1986
22 **Michael E. Taylor,** Noncommutative harmonic analysis, 1986
21 **Albert Baernstein, David Drasin, Peter Duren, and Albert Marden, Editors,** The Bieberbach conjecture: Proceedings of the symposium on the occasion of the proof, 1986
20 **Kenneth R. Goodearl,** Partially ordered abelian groups with interpolation, 1986
19 **Gregory V. Chudnovsky,** Contributions to the theory of transcendental numbers, 1984

For a complete list of titles in this series, visit the
AMS Bookstore at **www.ams.org/bookstore/**.